Prestressed Concrete Bridges

Prestressed Concrete Bridges

Prestressed Concrete Bridges

N. Krishna Raju BE, MSc, PhD

Emeritus Professor of Civil Engineering
MS Ramaiah Institute of Technology
Bangalore

CBS Publishers & Distributors Pvt. Ltd.
New Delhi • Bengaluru • Chennai • Kochi • Kolkata • Mumbai
Hyderabad • Nagpur • Patna • Pune • Vijayawada

ISBN: 978-81-239-1700-9

First Edition: 2009
Reprint: 2010, 2012, 2014, 2016

Copyright © Author & Publisher

All rights reserved. No part of this book may be reproduced or transmitted in any form or by any means, electronic or mechanical, including photocopying, recording, or any information storage and retrieval system without permission, in writing, from the publisher.

Published by:
Satish Kumar Jain for CBS Publishers & Distributors Pvt. Ltd.,
4819/XI Prahlad Street, 24 Ansari Road, Daryaganj, New Delhi - 110002
delhi@cbspd.com, cbspubs@airtelmail.in • www.cbspd.com
Ph.: 23289259, 23266861, 23266867 • Fax: 011-23243014

Corporate Office: 204 FIE, Industrial Area, Patparganj, Delhi - 110 092
Ph: 49344934 • Fax: 011-49344935
E-mail: publishing@cbspd.com • publicity@cbspd.com

Branches:
- *Bengaluru:* 2975, 17th Cross, K.R. Road, Bansankari 2nd Stage, Bengaluru - 70 • Ph: +91-80-26771678/79 • Fax: +91-80-26771680
  E-mail: cbsbng@gmail.com, bangalore@cbspd.com
- *Chennai:* No. 7, Subbaraya Street, Shenoy Nagar, Chennai - 600030
  Ph: +91-44-26681266, 26680620 • Fax: +91-44-42032115
  E-mail: chennai@cbspd.com
- *Kochi:* Ashana House, 39/1904, A.M. Thomas Road, Valanjambalam, Ernakulum, Kochi • Ph: +91-484-4059061-65
  Fax: +91-484-4059065 • E-mail: cochin@cbspd.com
- *Kolkata:* 6-B, Ground Floor, Rameshwar Shaw Road, Kolkata - 700014
  Ph: +91-33-22891126/7/8 • E-mail: kolkata@cbspd.com
- *Mumbai:* 83-C, Dr. E. Moses Road, Worli, Mumbai - 400018
  Ph: +91-9833017933, 022-24902340/41 • E-mail: mumbai@cbspd.com

Representatives:
- Hyderabad: 0-9885175004
- Patna: 0-9334159340
- Vijayawada: 0-9000660880
- Nagpur: 0-9021734563
- Pune: 0-9623451994

Printed at:
India Binding House, Noida, UP (India)

dedicated

to

the pioneers and research workers
Eugene Freyssinet, Yves Guyon, Gustav Magnel
Ulrich Finisterwalder, M. Birkenmaier, A. R. Cusens
R. E. Rowe, Paul W. Abeles, T. Y. Lin, Fritz Leonhardt
W. Podolony, J. Muller, M. S. Troitsky, N. Burns
Ben C. Gerwick Jr, T. N. Subba Rao, V. K. Raina
and a host of others who toiled incessantly
for the development and widespread use of
prestressed concrete in bridge construction

dedicated to

the pioneers and researchers...

PREFACE

During the last fifty years, rapid advances in the development and widespread use of new materials like high strength concrete, high tensile steel, epoxy resins, polymeric materials in conjunction with phenomenal developments in computer and construction technology have facilitated creative bridge engineers to design and build structurally sound, aesthetically superior bridge structures which are economical and also durable using the revolutionary building material *prestressed concrete*.

The prophetic words of Yves Guyon *there is probably no structural problem to which prestressed concrete cannot provide a solution and often a revolutionary one* have been amply justified if one scans through the tremendous progress achieved in the field of analysis, design and construction of prestressed concrete bridges throughout the world.

At present, India has embarked on a gigantic highway project involving the golden quadrilateral connecting the north-south and east-west corridors connecting the capital cities of various states. Naturally this massive highway project necessarily involves the design and construction of innumerable number of bridges to cross the east and west flowing rivers.

The purpose of this book is to present under one cover the whole gamut of theory and design of prestressed concrete bridges of different types to the extent required by undergraduate and postgraduate students of civil, structural and highway engineering streams. The subject matter in this book should also serve as useful reference material for practising highway and structural engineers dealing with the design, construction and maintenance of prestressed concrete bridges.

The book presents a concise and lucid exposition of the various types of prestressed concrete bridges, illustrated by examples relevant to design practice. The designs presented conform to various National Codes such as IS: 1343-1980 (2000), IS: 456-2000 and the Indian Roads Congress Codes IRC: 6-2000, IRC: 18-2000 and IRC: 21-2000.

The book traces the historical evolution of bridges together with the advantages of pretensioned and post-tensioned prestressed concrete bridges in Chapter 1. Chapter 2 deals with the highway Bridge Loading Standards followed in various countries together with their comparative analysis. A concise description of the materials used in prestressed concrete bridges is presented in Chapter 3, while Chapter 4 deals with the limit state design concepts of reinforced and prestressed concrete bridge decks along with numerous design examples. Chapter 5 presents the analysis and design of prestressed concrete slab bridge decks followed by the load distribution methods of tee beam and slab bridge decks in Chapter 6. Chapter 7 is devoted to the design of prestressed concrete continuous girder bridge decks while Chapter 8 presents the analysis and design of rigid frame bridges generally used in urban flyover crossings. The design of prestressed concrete cellular box girder bridges is presented in Chapter 9 while prestressed cable stayed bridges are treated exhaustively in Chapter 10. The planning, economical aspects and cost considerations of different types of prestressed concrete bridges are detailed in Chapter 11. Chapter 12 presents the constructional aspects dealing with concrete, assembly of prestressing steel, expansion joints together with the various techniques developed for the construction of long span post-tensioned bridges. The maintenance, repairs and rehabilitation of prestressed concrete bridge components together with practical examples of restoration and case studies are included in Chapter 13. Illustrations of prominent prestressed concrete bridges built in India and various other countries are presented along with their structural details in Chapter 14.

Most of the chapters contain design examples together with exercises for practice by students and list of references are included at the end of each chapter. Appendices 1, 2, 3 and 4 contain the properties of commonly used prestressing steels, constants for beam sections and data regarding the prominent proprietary post-tensioning anchorage systems and grouting of post-tensioned ducts. The data compiled is useful in the design of various types of prestressed concrete bridge decks.

Numerous figures have been included in keeping with the spirit of *drawing is the language of the engineer*. Finally the author welcomes constructive criticisms and useful suggestions which will immensely help in updating the text material.

N. Krishna Raju

ACKNOWLEDGEMENTS

The author gratefully acknowledges various institutions, associations, societies, journals and authors of technical publications and monographs for the reproduction of certain design data, tables, figures and reference material mentioned throughout the text.

Indian Roads Congress, Bureau of Indian Standards, American Concrete Institute, American Association of State Highway Organization, Prestressed Concrete Institute, Federation International Precontrainte, European Concrete Committee, Institution of Engineers (India), International Association for Bridge and Structural Engineering, Structural Engineering Research Centre, Roorkee, Japan Prestressed Concrete Industry, Bridge Loading Standards of UK, Germany, France and various other European countries.

The complete details of the sources are listed in the form of references at the end of each chapter.

The author is deeply indebted to Dr. E.W. Bennett, Prof. R.H. Evans and Prof. A.M. Neville for inspiring and initiating the author to the field of prestressed concrete during his doctoral research work under the Commonwealth Fellowship Programme at the University of Leeds (UK) during 1965-68. The author also records his gratitude to Prof. Fritz Leonhardt, Prof. Ben. C. Gerwick Jr, Dr. V.K. Raina and various other authors whose innumerable publications served as an invaluable source material and inspired the author in the preparation of this book.

The text material in the book has been compiled from the lecture notes of the author, prepared for teaching the masters degree courses in structural engineering at Karnataka Regional Engineering College, Surathkal, University of Basrah, Iraq, M.S. Ramaiah Institute of Technology, Bangalore, and the material prepared to deliver expert lectures during the short-term courses at various technical institutions in India during his active career, spanning four decades. The author has benefited immensely by the feedback comments of students, research workers, fellow teachers and participants in various short-term courses.

The author is grateful to his wife Pramila for extending the fullest cooperation in the preparation of the manuscript and assisting in every possible way for its successful completion.

N. Krishna Raju

Contents

Preface .... *vii*
List of Symbols .... *xv*

1. INTRODUCTION 1
    1.1   Historical Evolution of Bridges 1
    1.2   Advantages of Prestressed Concrete Bridges 8
    1.3   Pretensioned Prestressed Concrete Bridge Decks 11
    1.4   Post Tensioned Prestressed Concrete Bridge Decks 11
    1.5   Modern Trends in Prestressed Concrete Bridges 13
           References 16

2. BRIDGE LOADING STANDARDS 18
    2.1   Evolution of Bridge Loading Standards 18
    2.2   Indian Highway Bridge Loading Standards 18
    2.3   Highway Bridge Loading Standards of Different Countries 23
    2.4   Impact Factors 34
    2.5   Comparative Analysis of Highway Bridge Loading Standards 37
    2.6   Indian Railway Bridge Loading Standards 39
           References 42

3. MATERIALS FOR PRESTRESSED CONCRETE BRIDGES 43
    3.1   Grades of Concrete 43
    3.2   High Strength Concrete Mixes 43
    3.3   High Tensile Steel 46
    3.4   Untensioned Steel or Supplementary Reinforcement 48
    3.5   Permissible Stresses in Concrete 49
    3.6   Permissible Stresses in Steel 50
    3.7   Anchorages and Sheathing Ducts 51
           References 52

4. LIMIT STATE DESIGN OF REINFORCED AND PRESTRESSED CONCRETE BRIDGE DECK SECTIONS ... 54

- 4.1 Design Philosophy of Reinforced and Prestressed Concrete Structures ... 54
- 4.2 Elastic Design Coefficients for Reinforced Concrete Sections ... 54
- 4.3 Flexural and Shear Strength of Reinforced Concrete Sections ... 55
- 4.4 Design of Prestressed Concrete Sections for Service Loads ... 56
- 4.5 Flexural and Shear Strength of Prestressed Concrete Sections ... 58
- 4.6 Torsional Resistance of Prestressed Concrete Sections ... 60
- 4.7 Forces in End Blocks ... 62
- 4.8 Design Examples ... 63
- References ... 69
- Exercises ... 69

5. PRESTRESSED CONCRETE SLAB BRIDGE DECKS ... 72

- 5.1 General Features ... 72
- 5.2 Analysis of Slab Decks ... 72
- 5.3 Design Aids and Tables for Prestressed Concrete Bridge Deck Slabs ... 87
- 5.4 Minimum Reinforcements in Slabs ... 89
- 5.5 Design Example of Prestressed Concrete Slab Deck for IRC Loads ... 89
- References ... 96
- Exercises ... 96

6. PRESTRESSED CONCRETE TEE BEAM AND SLAB BRIDGE DECK ... 98

- 6.1 General Features ... 98
- 6.2 Structural Components of Tee Beam and Slab Bridge Decks ... 98
- 6.3 Load Distribution Methods for Beam and Slab Bridge Decks ... 101
- 6.4 Comparative Analysis of Various Load Distribution Methods ... 113
- 6.5 Design of Post Tensioned Prestressed Concrete Tee Beam and Slab Bridge Deck ... 121
- References ... 139
- Exercises ... 139

7. PRESTRESSED CONCRETE CONTINUOUS SPAN BRIDGE DECKS ... 141

- 7.1 Advantages of Continuous Span Bridge Decks ... 141
- 7.2 Methods of Prestressing Continuous Bridge Decks ... 142
- 7.3 Cross-sections of Prestressed Concrete Continuous Bridge Decks ... 142
- 7.4 Design of Two Span Continuous Prestressed Concrete Bridge Deck ... 143
- References ... 163
- Exercises ... 163

8. DESIGN OF PRESTRESSED CONCRETE RIGID FRAME BRIDGES ... 165

- 8.1 General Features ... 165
- 8.2 Advantages of Rigid Frame Bridges ... 165
- 8.3 Design Principles of Prestressed Concrete Portal Frames ... 167
- 8.4 Design Example ... 170
- References ... 185
- Exercises ... 185

9. PRESTRESSED CONCRETE CELLULAR BOX GIRDER BRIDGES ... 187

- 9.1 General Features ... 187
- 9.2 Advantages of Segmental Box Girder Construction for Long Span Bridge Decks ... 187
- 9.3 Typical Cross-sections of Cellular Box Decks ... 188
- 9.4 Analysis of Box Girder Bridge Decks ... 190
- 9.5 Design Principles ... 192
- 9.6 Design Example of Cellular Box Girder Bridge Deck ... 196
  - References ... 210
  - Exercises ... 211

10. PRESTRESSED CONCRETE CABLE STAYED BRIDGES ... 212

- 10.1 Evolution of Cable Stayed Bridges ... 212
- 10.2 Advantages of Cable Stayed Bridge Decks ... 213
- 10.3 Structural Components of Cable Stayed Bridges ... 214
- 10.4 Towers or Pylons ... 215
- 10.5 Cable Stays ... 217
- 10.6 Longitudinal Cable Profiles ... 221
- 10.7 Superiority of Cable Stayed Bridges over Conventional Bridges ... 223
- 10.8 Basic Principles of Structural Analysis ... 223
- 10.9 Structural Analysis of Cable Stayed Bridges ... 225
- 10.10 Structural Anchorages ... 237
- 10.11 Dynamic Behaviour and Aerodynamic Stability ... 237
- 10.12 Construction Methods ... 240
- 10.13 Economic Studies ... 241
- 10.14 Design Examples of Prestressed Concrete Cable Stayed Bridge ... 243
  - References ... 254
  - Exercises ... 254

11. PLANNING AND ECONOMICAL ASPECTS OF PRESTRESSED CONCRETE BRIDGES ... 256

- 11.1 Introduction ... 256
- 11.2 Structural Forms for Bridges ... 256
- 11.3 Cost Considerations of Different Types of Bridge Decks ... 257
- 11.4 Economical Evaluation ... 264
  - References ... 266

12. CONSTRUCTION OF PRESTRESSED CONCRETE BRIDGES ... 267

- 12.1 Introduction ... 267
- 12.2 High Strength Concrete Mix Considerations ... 268
- 12.3 Batching and Mixing of Concrete ... 269
- 12.4 Placing of Concrete in Forms ... 270
- 12.5 Compaction of Concrete by Vibration ... 270
- 12.6 Rheodynamic Concrete ... 271

12.7	Expansion Joints for Bridge Decks	271
12.8	Assembly of Prestressing Steel and Grouting of Ducts	274
12.9	Long Span Bridge Construction Techniques	274
	References	289

13. MAINTENANCE AND REHABILITATION OF PRESTRESSED CONCRETE BRIDGES ... 290

13.1	Introduction	290
13.2	General Features of Bridge Maintenance and Rehabilitation	290
13.3	Maintenance Methodology	291
13.4	Inspection of Bridges	292
13.5	Inspection Instrumentation	293
13.6	Cracks in Prestressed Concrete Bridges	294
13.7	Repairs and Rehabilitation of Bridges	302
13.8	Repairs of Girders Damaged by Collision	306
13.9	Restoration of Damaged Prestressed Concrete Beams	306
13.10	Strengthening of Beams by Externally Bonded Steel Plates	310
13.11	Case studies of Repairs and Rehabilitation of Bridges	312
	References	317

14. WORLD'S PROMINENT PRESTRESSED CONCRETE BRIDGES ... 318

14.1	General Aspects	318
14.2	World's Long Span Prestressed Concrete Bridges	319
14.3	Notable Examples of Prestressed Concrete Bridges	319
14.4	General Remarks	334
	References	336

Appendices ... 337

Appendix 1 Properties of Prestressing Steels ... 339

Appendix 2 Constants for Beam Sections ... 341

Appendix 3 Post-tensioning Systems ... 346

Appendix 4 Grouting of Post-tensioned Ducts ... 352

Author Index ... 355

Subject Index ... 359

LIST OF SYMBOLS

A	Cross-sectional area of member
A_{br}	Bearing area
A_{ps}	Area of prestressing tendons
A_{pun}	Punching area
A_s	Area of non prestressed reinforcement
A_{sv}	Area of transverse reinforcement for shear
B_f	Effective width of flange
D_f	Thickness of flange
E_c	Modulus of elasticity of concrete
E_s	Modulus of elasticity of steel
F_{bst}	Bursting tension
F_d	Design load
F_k	Characteristic load
G	Permanent dead load
G_k	Characteristic dead load
I	Second moment of area of section
K	Constant for solid slabs
L	Effective span
M	Bending moment
M_d	Design moment
M_g	Bending moment due to dead loads
M_o	Cracking moment
M_q	Bending moment due to live loads
M_u	Ultimate moment
P	Prestressing force
P_k	Characteristic load in tendon
SG	Superimposed dead load
T	Torsional moment
T_u	Ultimate torsional moment
U	Ultimate load
V	Shear force

V_c	Ultimate shear resistance of concrete
V_g	Dead load shear force
V_q	Live load shear force
V_{cw}	Ultimate shear resistance of concrete in a section uncracked in flexure
V_{cf}	Ultimate shear resistance of concrete in a section cracked in flexure
V_u	Ultimate shear force
Z	Section modulus
Z_t	Section modulus of top fibre of member
Z_b	Section modulus of bottom fibre of member
EI	Flexural rigidity
CJ	Torsional rigidity
b	Width of rectangular section or web of flanged section
b_e	Effective width of slab
b_w	Breadth of web
d	Effective depth of tension reinforcement
e	Eccentricity of prestressing force
f_{br}	Range of stress at bottom fibre
f_c	Compressive stress
f_{cp}	Compressive stress at the centroidal axis due to prestress after all losses
f_{cr}	Flexural tensile strength of concrete
f_{ct}	Allowable compressive stress in concrete at initial transfer of prestress
f_{ck}	Characteristic compressive strength of concrete
f_{cw}	Allowable compressive stress in concrete under service loads
f_{inf}	Prestress in concrete at bottom of section (inferior)
f_{pu}	Characteristic strength of prestressing tendons
f_s	Stress in reinforcement
f_t	Tensile strength of concrete
f_{sup}	Prestress in concrete at top (superior)
f_{tt}	Allowable tensile stress in concrete at initial transfer of prestress
f_{tw}	Allowable tensile stress in concrete under service loads
f_y	Characteristic strength of reinforcement
g	Distributed dead load
h	Overall depth of member
h_f	Thickness of compression flange
h_{max}	Larger dimension of section
h_{min}	Smaller dimension of section
j	Lever arm factor
m	Modular ratio
m_1, m_2	Moment coefficients
n	Neutral axis depth factor
Q	Moment factor
q	Distributed live load
s	Spacing of stirrups or links
t	Thickness of flange
x	Linear distance

List of Symbols

x_1	The smaller centre to centre dimension of the links
y	Vertical distance of a point from centroid of concrete section
y_1	The larger centre to centre dimension of the links
y_b	Distance of lowest (inferior) point from centroid of concrete section
$2y_o$	Side of end block
$2y_{po}$	Side of loaded area
y_t	Distance of highest (superior) point from centroid of concrete section
α	Angle, ratio or dimensionless coefficient
η	Reduction factor for loss of prestress or loss ratio
τ_c	Ultimate shear stress in concrete
$\tau_{c,\,max}$	Maximum permissible shear stress in concrete
τ_u	Ultimate shear stress under design ultimate loads
τ_v	Shear stress due to transverse shear force
τ_t	Torsional shear stress
τ_{tu}	Maximum torsional shear stress
$\tau_{t,\,min}$	Minimum torsional shear stress
ϕ	Diameter of reinforcing bar or tendon
Σ	Sum
ξ	Design shear strength of concrete
μ	Coefficient of friction
σ_{cb}	Permissible bending stress in concrete
σ_{sy}	Permissible tensile stress in steel
σ_{co}	Permissible direct compressive stress in concrete

1

INTRODUCTION

1.1 HISTORICAL EVOLUTION OF BRIDGES

The glorious history of bridges is closely associated with the progress of human civilization spread over several centuries. The earliest bridge on record is that built on river Nile by king Menes of Egypt as early as 2650 BC. Wooden bridges were built over the river Eupharetes during the reign of queen of Babylon in Iraq during 783 BC. Around 320 BC, Alexander the Great built floating bridges for the passage of his army. Raina[1] cites examples of early timber beam and crude suspension bridges built over Min River in China and the Himalayas as shown in Fig. 1.1. Etruscan and Romans developed the art of building arched bridges using stones and bricks. The optimum profile of stone arch developed by early builders intuitively has seen very little changes over the years. The old stone arched bridge has been extensively used in almost all the countries for road as well as railway crossings. The series of stone masonry arched bridges across the river Seine in Paris shown in Fig. 1.2 are unique examples of human ingenuity. The proven durability of material and the long experience in intuitive proportioning made stone masonry arch bridges the most popular form of construction in the early days of Railways until iron bridges made their way in 17th century. Arched bridges continued to be the popular choice with materials like iron, steel and concrete mainly due to its superior aesthetical qualities besides structural efficiency of utilizing the entire cross sectional area to resist the compressive forces.

In mid 19th century, demand for stronger and bigger bridges over larger rivers resulted in the use of cast iron and wrought iron replacing timber and stones for bridges. The first recorded use of iron in bridges was a chain bridge built in 1734 by the German army across the Oder River in Prussia. Cast iron being brittle was not found very suitable for building large bridges. An effective combination of cast iron for compression members and wrought iron for tension members was first used in trussed bridges around 1840 especially for railway bridges.

The development of steel by Bessemer in 1856 and the open hearth process by Sieman and Martin in 1861 paved the way for extensive use of steel and caught the imagination of bridge builders. Firth of Forth Cantilever bridge of 520 m span and Roebling's Suspension bridge of 490 m span were a few of the famous bridges of the 19th century and marked the beginning of modern era of bridge engineering.

Fibre rope suspension form over Min River in China (total length 552 m)

Crude suspension form the Himalayas

Fig. 1.1: Early Timber Suspension Bridges over Min River in China and Himalayas

By the turn of the century, widespread use and availability of structural steel sections and development of methods of analysis and design paved the way for large span, continuous, cantilever and arched bridges in steel exceeding spans of 500 m. Howrah bridge built in 1943 at Calcutta is a typical example of several outstanding bridges built using steel in the early 20th century.

Rapid advances in the development of theoretical analysis of the load response of the structural system of suspension bridges in the early part of 20th century resulted in the construction of many elegant bridges. The prominent among them are Lindernathal's 450 m span Manhattan Bridge built during 1909, Steinman's Florianopolis Bridge of span 340 m and the Delaware River Bridge of 530 m span in 1926. The giant leap came with the construction of George Washington Bridge with a span of 1060 m breaking the barrier of 1000 m. According to Stussi, Washington Bridge is acclaimed as a *"Great and most important step in the evolution of the art of bridge engineering"*. The great architect Le Corbusier exclaimed that, "The Washington Bridge is the most beautiful bridge in the world". This eight lane major road way bridge without stiffening girders built by Amman, a Swiss engineer, is a significant breakthrough according to Navier and Roebling. Amman considered as the foremost bridge engineer in the world during that period also built Verrazano

Narrows Bridge spanning 1300 m in New York which is considered as a master piece and opened for traffic in 1964 just a few months before he died. Figure 1.3 shows a typical steel trussed bridge of 360 m span at Baltimore, U.S.A.

Fig. 1.2: Stone Masonny Arched Bridges across the River Siene in Paris

Postwar years saw the emergence of reinforced concrete as a suitable material for short and medium span bridges with the added advantage of durability against aggressive environmental conditions in comparison with steel. Tee beam and slab bridge decks (Fig. 1.4) were popular along with Bow string (Fig. 1.5), Balanced cantilever (Fig. 1.6), Continuous girder (Fig. 1.7) and Open spandrel arched bridge (Fig. 1.8) were also widely used throughout the world.

Fig. 1.3: Steel Trussed Bridge (360 m) at Baltimore, U.S.A.

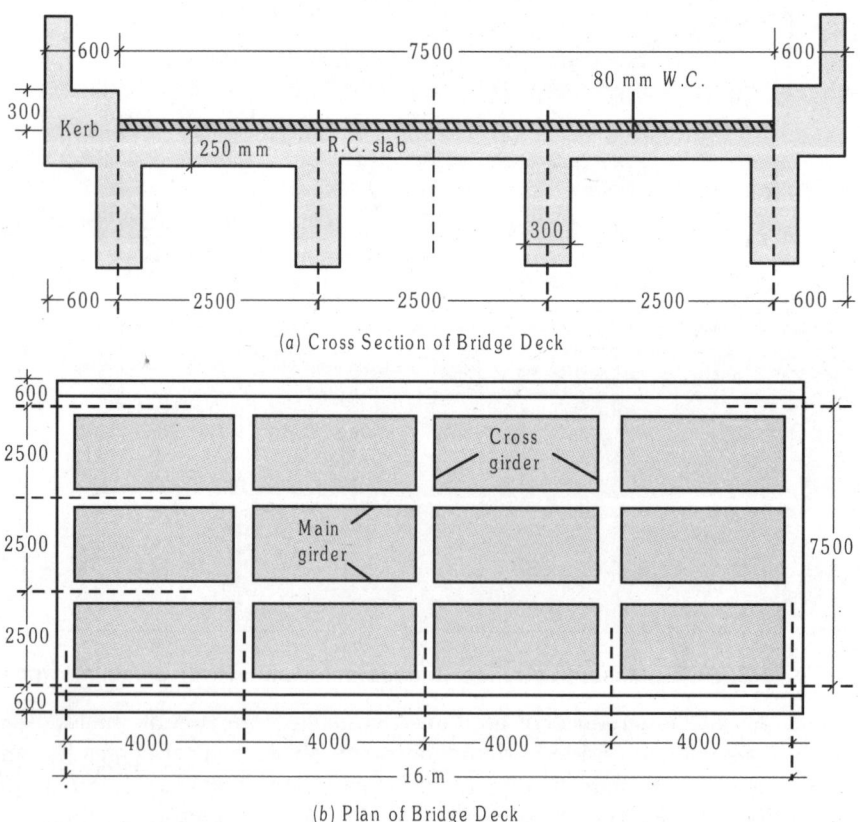

Fig. 1.4: Tee Beam and Slab Bridge Deck

Introduction

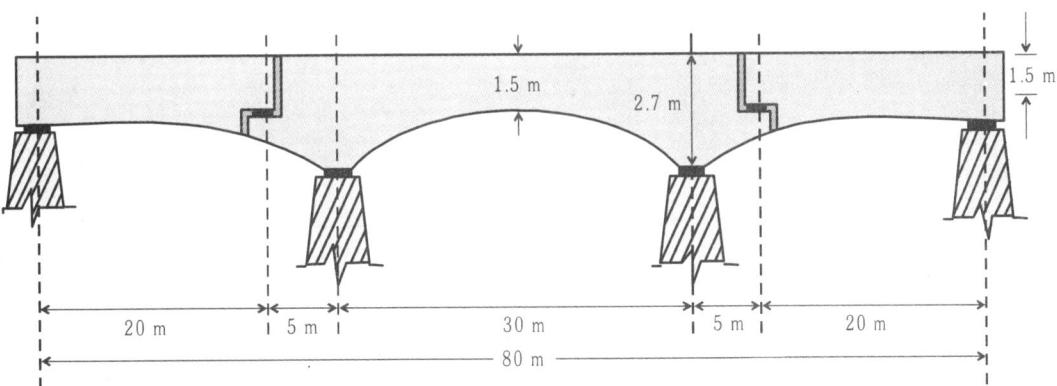

Fig. 1.5: Bow String Girder Bridge

Fig. 1.6: Balanced Cantilever Bridge

A revolutionary and a path breaking achievement in materials technology was witnessed in 1928 when Eugene Freyssinet[2], a French engineer introduced a new construction material designated as **"Prestressed Concrete"**. Freyssinet, who struggled for seven years for the cause of prestressed concrete found in the beginning there were no buyers for his most exciting construction material. The big boom in prestressed concrete was witnessed after the Second World War. In the beginning, prestressed concrete flourished in Europe and soon spread to all parts of the world.

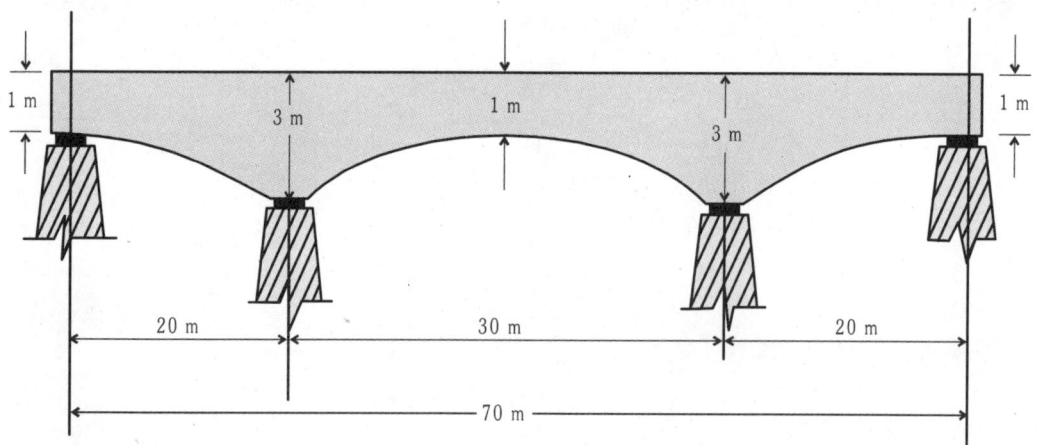

Fig. 1.7: Continuous Girder Bridge

Fig. 1.8: Open Spandrel Arched Bridge

The development of high strength concrete associated with significant improvements in the quality of cements and high strength steels of different forms, paved the way for its widespread application in bridge construction. In 1950's, prestsressed concrete came to be used mostly for bridges of ever increasing spans coupled with rapidity and ease of construction and competing in costs with other alternative types like steel and reinforced concrete.

Among the 500 bridges built in Germany during 1949-53, seventy per cent of them used prestresed concrete. In 1949 prestressed concrete was introduced in U.S.A. in the construction of Magnel's Walnut Bridge and around the same time in India for the construction of Coleron Bridge using Freyssinet system of anchorages. During the last 50 years, the achievements in building long

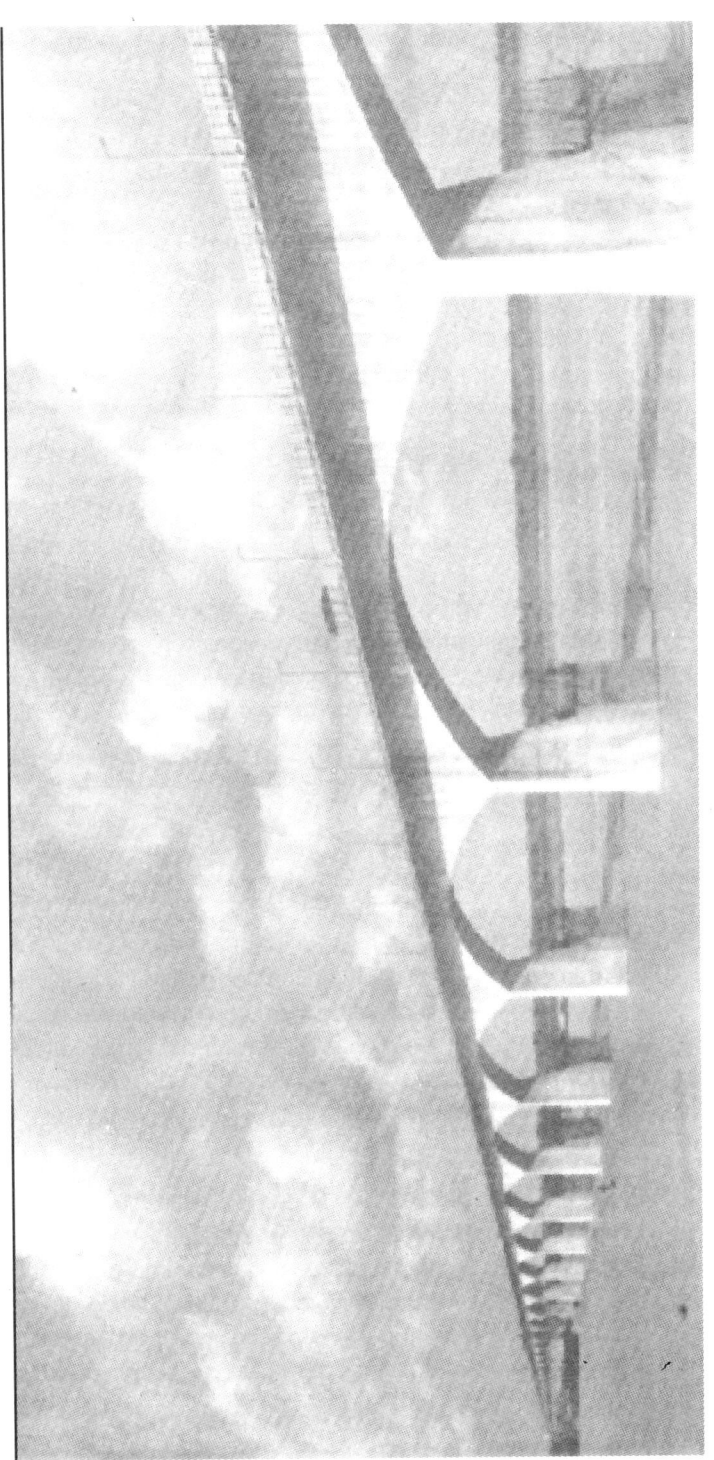

Fig. 1.9: Ganga Bridge at Patna

span prestressed concrete bridges throughout the world are too numerous to document and the developments of new forms and construction techniques are growing at a dizzy pace. Ganga Bridge at Patna having 46 spans of 122 m each shown in Fig. 1.9 is a typical example of continuous prestressed concrete bridge ideally suited for long spans.

Another revolutionary approach in bridge construction first conceived by Dischinger, in 1938 and later put into practice in the construction of first modern cable stayed bridge is the Stromsund Bridge in Sweeden around 1953 by DEMAG. This paved the way for the construction of number of famous Rhine family cable stayed bridges with spans up to and exceeding 300 m.

According to Leonhardt[3], cable stayed system is technically, economically, aerodynamically and aesthetically superior to the classical suspension bridges for spans in the range of 750 to 1500 m. In the year 1960, beam bridges had reached spans up to 160 m and by 1970, the spans of beams had touched 230 m in Japan and for cable stayed bridges, designs were in progress for spans of 300 m. From 1970 onwards, the development of long span bridges in prestressed concrete with land marks in span, form and construction technology is growing at an unprecedented pace. The panorama is so vast that it defies a simple survey.

The innovative Cantilever construction method developed by Ulrich Finisterwalder[4] has stretched the span range of prestressed concrete bridges beyond 200 m. The Hamana Bridge in Japan with a main span of 240 m is considered as a classic example of long span prestressed concrete bridges erected by cantilever construction method.

The combination of cable stays with box girder prestressed concrete decks have significantly extended the span range of high way bridges. Detailed studies by Fritz Leonhardt indicates that cable stayed bridges are structurally efficient and cost effective for low, medium and long span ranging from 40 to 1800 m. Pedestrian bridges spanning 40 m comprising a prestressed concrete deck with a depth of 250 to 300 mm supported by a few cable stays have been successfully built in Germany.

Vidyasagar Sethu (Second Hooghly Bridge) at Kolkata (Fig. 1.10) is an excellent example of a cable stayed three lane high way bridge comprising a main span of 457 m and two side spans of 183 m. The composite deck is made up of a concrete slab 230 mm thick with two outer steel I-girders 28 m apart and a central girder.

The phenomenal development of prestressed concrete during the last five decades has expanded its scope of application and has led to increasingly imaginative forms of construction. Over the years, prestressed concrete has emerged as the choicest material for all kinds of structures. But the idea of prestressing arose out of bridges and naturally its most impressive and novel forms are found in long span prestressed concrete bridges.

1.2 ADVANTAGES OF PRESTRESSED CONCRETE BRIDGES

Prestressed concrete comprising high strength concrete and high tensile steel is ideally suited for the construction of bridges due to its distinct advantages over other types of materials. Prestressed concrete offers immense technical advantages in comparison with other forms of construction such as steel and reinforced concrete. In fully prestressed members (type-1), the cross-section is efficiently utilized in resisting the stresses developed due to superimposed loads in comparison with a reinforced concrete section which is cracked under service loads. Within certain limits, a permanent dead load may be counteracted by increasing the eccentricity of the prestressing force in a prestressed concrete member, thus effecting savings in the quantity of material used in the structure.

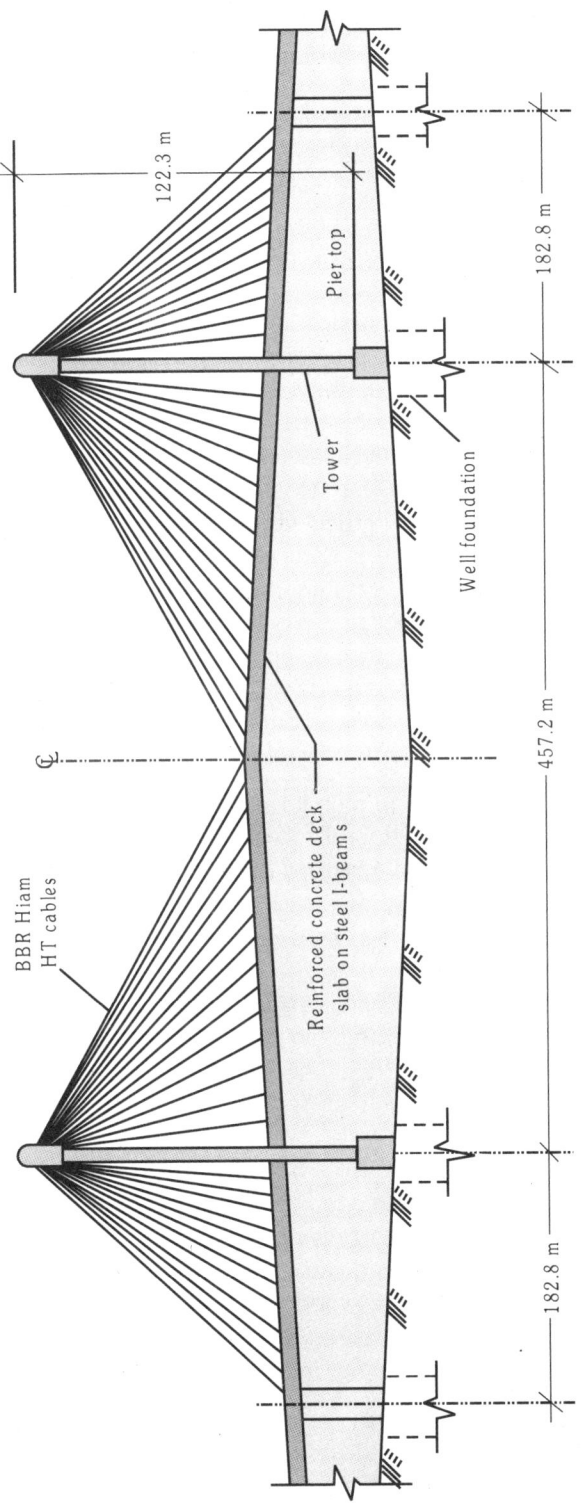

Fig. 1.10: Vidyasagar Sethu Cable Stayed Bridge

In comparison with reinforced concrete members, prestressed members possess improved resistance to shear forces due to the effect of compressive prestress which reduces the principal tensile stress, thus eliminating the development of diagonal tension cracks. Curved cables generally used in long span members, reduce the shear forces developed at the support sections. The economy of prestressed concrete is well established for long span bridge structures.

According to Dean[5], standardized precast bridge beams between 10 and 30 m long and precast prestressed piles have proved to be more economical than steel and reinforced concrete in the United States. According to Abeles[6], precast prestressed concrete is economical for floors, roofs and bridges of spans up to 30 m and for cast *in situ* work up to 100 m. In the long span range, prestressed concrete is generally more economical than reinforced concrete and steel.

The deformation characteristics of prestressed concrete girders in bridges are superior to that of reinforced concrete since prestressed members have considerable resilience due to its capacity for completely recovering from substantial effects of overloading without undergoing any serious damage. Leonhardt[7] has reported that in prestressed girders of bridges, cracks which temporarily develop under occasional overloading close up completely on removal of the loads.

Bridge girders are normally subjected to repetitive loads leading to fatigue of members. Since the fatigue strength of prestressed concrete is comparatively better than that of other materials mainly due to the small stress variation in prestressing steel, it is preferred for dynamically loaded structures such as railway and roads bridges and machine foundations.

The experience accumulated in building prestressed concrete bridges over the last three decades in different countries has clearly indicated the superiority of prestressed concrete in terms of economy, slender form, aesthetics, performance in service and durability. The author in a separate manograph[8] has summarized the advantage of prestressed concrete bridges as follows:

1. The use of high strength concrete and high tensile steel results in slender sections which are aesthetically superior coupled with durability and overall economy.
2. Prestressed concrete bridges can be designed as class I type structures without any tensile stresses under service loads resulting in a crack free structure.
3. In comparison with steel bridges, prestressed concrete bridges require very little periodical maintenance.
4. Prestressed concrete is ideally suited for cantilever method of construction, both *in situ* and precast resulting in faster rate of construction of long span bridges.
5. Post tensioned prestressed concrete finds extensive applications in long span continuous girder bridges of variable cross-section resulting in sleek structures with considerable savings in the overall cost of construction.
6. Prestressed concrete is ideally suited for composite bridge construction in which precast prestressed girders support the cast *in situ* concrete deck slab. This type of construction is very popular since it involves minimum disruption of traffic.
7. Combination of prestressed girders with cable stays are ideally suited for long span bridges with proven economy throughout the world.
8. In recent years, limited or partially prestressed concrete (type-3 structures) is preferred for bridge construction with significant savings in the quantity of costly high tensile steel used in bridge deck girders.

Twenty first century will witness the application of prestressed concrete with cable stays for increasingly longer spans resulting in the reduction of piers and the associated foundation costs.

Introduction

The future bridges will be sleeker and aesthetically superior and blend with the local natural surroundings.

1.3 PRETENSIONED PRESTRESSED CONCRETE BRIDGE DECKS

The **Long line system**[9] of pretensioning developed by Hoyer is ideally suited for casting pretensioned prestressed concrete bridge deck units of standard sizes and of various cross sectional shapes such as single *tee*, double *tee*, hollow box, solid and voided slab, I and Y-shaped girders. Pretensioned prestressed concrete bridge decks generally comprise precast pretensioned units used in conjunction with cast *in situ* concrete topping slab resulting in composite bridge decks ideally suited for small and medium spans in the range of 10 to 30 m. In general, pretensioned beams are prestressed using straight tendons. The use of seven wire strands have been found to be advantageous in comparison with plain or indented high tensile wires. Deflected strands are employed in larger girders in U.S.A.

In the United Kingdom, the precast prestressed I and inverted tee beam have been standardized by the Cement and Concrete Association, London for use in the construction of bridge decks of spans in the range of 7 to 36 m. Standard I and Tee section beams are widely employed in high way bridge decks in U.S.A. Russia and European countries. Recently in U.K. Y-beams have been developed to replace the M-beams introduced in 1960. The design and development of the Y-beams which are superior to M-beams are explicitly suited for medium spans in the range of 15 to 30 m. The salient features of widely used composite bridge decks with precast pretensioned standard beams are shown in Fig. 1.11.

Post war construction in Japan has extensively used precast pretensioned girders for both railway and high way bridges. Pretensioned girders with I and Tee sections have been used on a large scale for the Tokyo-Nagoya and Osaka-Kobe Express ways[10].

1.4 POST TENSIONED PRESTRESSED CONCRETE BRIDGE DECKS

Long span bridge decks are generally constructed using post tensioned girders, since the self weight of these girders being large, the cost of transportation and positioning of pretensioned girders is prohibitive. Post tensioning is generally adopted for girder spans exceeding 20 m. bridge decks with pre cast or cast *in situ* post tensioned girders of either Tee or Box type in conjunction with a cast *in situ* slab is commonly adopted for spans exceeding 30 m. Another advantage of post tensioned girders is the use of curved cables with varying eccentricity to suit the bending moments along the span in comparison with pretensioned girders in which the tendons are normally straight. The use of curved cables will enhance the shear resistance of post tensioned girders in the vicinity of supports.

Post tensioning is ideally suited for prestressing long span girders at the site of construction without the need for costly factory type installations like pretensioning beds. Modern long span bridges are constructed using segmental construction. In this method a number of segments of modular length can be combined by post tensioning cables resulting in an integrated structure. In India a large number of long span bridges have been constructed using the Cantilever method of construction. Some of the notable examples being the Barak bridge at Silchar built in 1960 with a main span of 130 m and the Lubha bridge[10] in Assam with a span of 172 m between the bearings. Long span continuous prestressed concrete bridges are invariably built using modular multicelled box segments of variable depth using the post tensioning system. Typical cross sections of post tensioned prestressed concrete bridge decks are shown in Fig. 1.12.

Japan has been using post tensioned prestressed concrete girders for the construction of Railway and High way bridges on a large scale during the post war period. Otogawa rail way bridge with a span of 30 m was constructed on the Shigaragi line in 1954. The Harumi Railway Bridge with six spans and three of which being continuous spans of 21.3 m was constructed in 1957.[10]

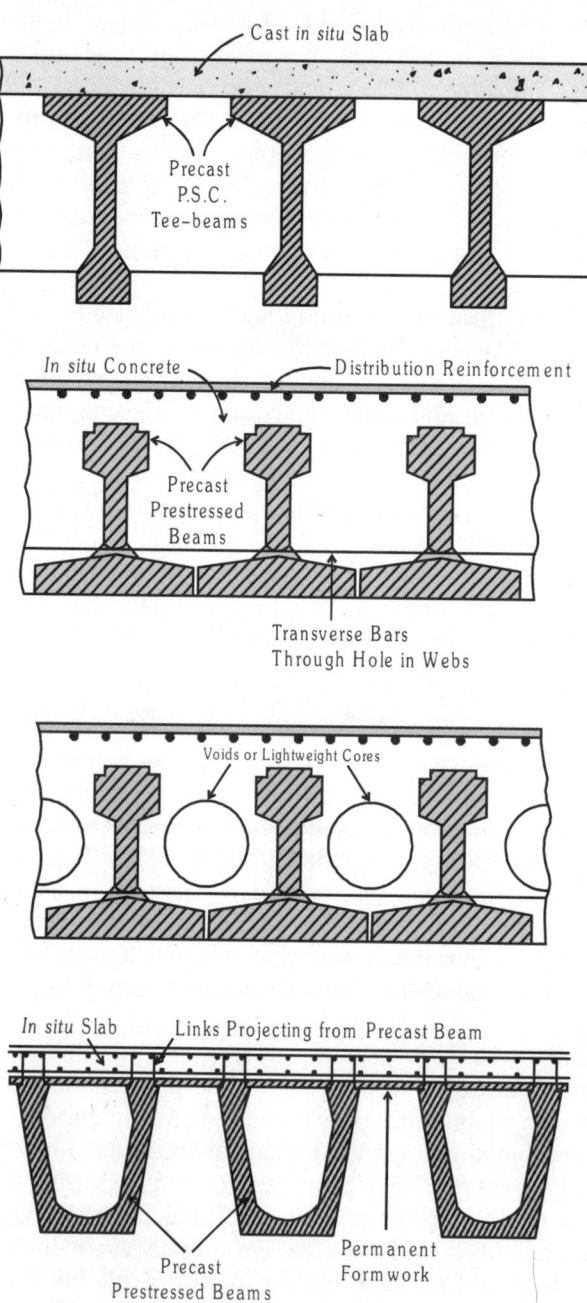

Fig. 1.11: Precast Pretensioned Bridge Decks

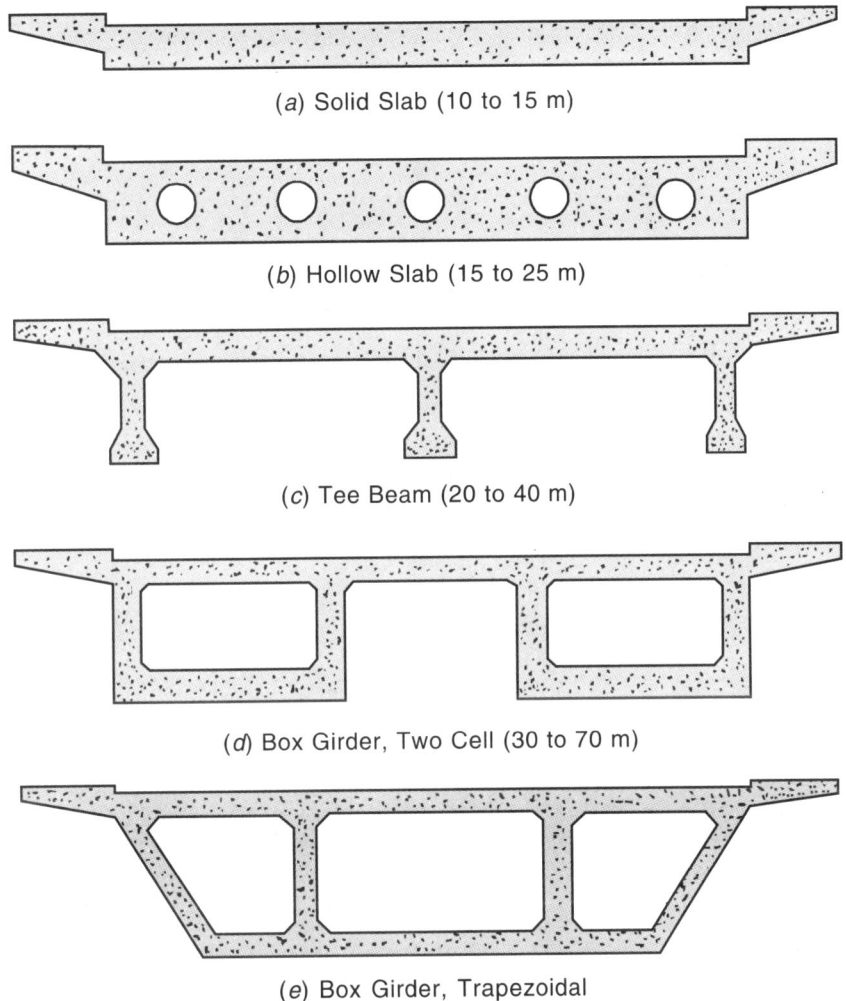

(a) Solid Slab (10 to 15 m)

(b) Hollow Slab (15 to 25 m)

(c) Tee Beam (20 to 40 m)

(d) Box Girder, Two Cell (30 to 70 m)

(e) Box Girder, Trapezoidal

Fig. 1.12: Post Tensioned Prestressed Concrete Bridge Decks

1.5 MODERN TRENDS IN PRESTRESSED CONCRETE BRIDGES

Recent developments in the field of concrete technology indicates that it is possible to produce ultra high strength concrete[11] with a characteristic compressive strength exceeding 100 N/mm^2 and high tensile steel cable of superior quality and strength required for new types of prestressed concrete bridges. Innovations in construction techniques coupled with rapid advances in the design philosophy of complex bridge forms has paved the way for the development and widespread use of new types of prestressed concrete bridge decks for long spans.

The dawn of 21st century has seen the evolution of long span cable stayed bridges with hybrid decks using steel girders, concrete slabs and high tensile steel cable stays. The cantilever method of erection is the latest and the most economical and popular method for the construction of long span precast or cast *in situ* prestressed concrete segmental bridges. This method is also ideally suited for the construction of cable stayed bridge decks.

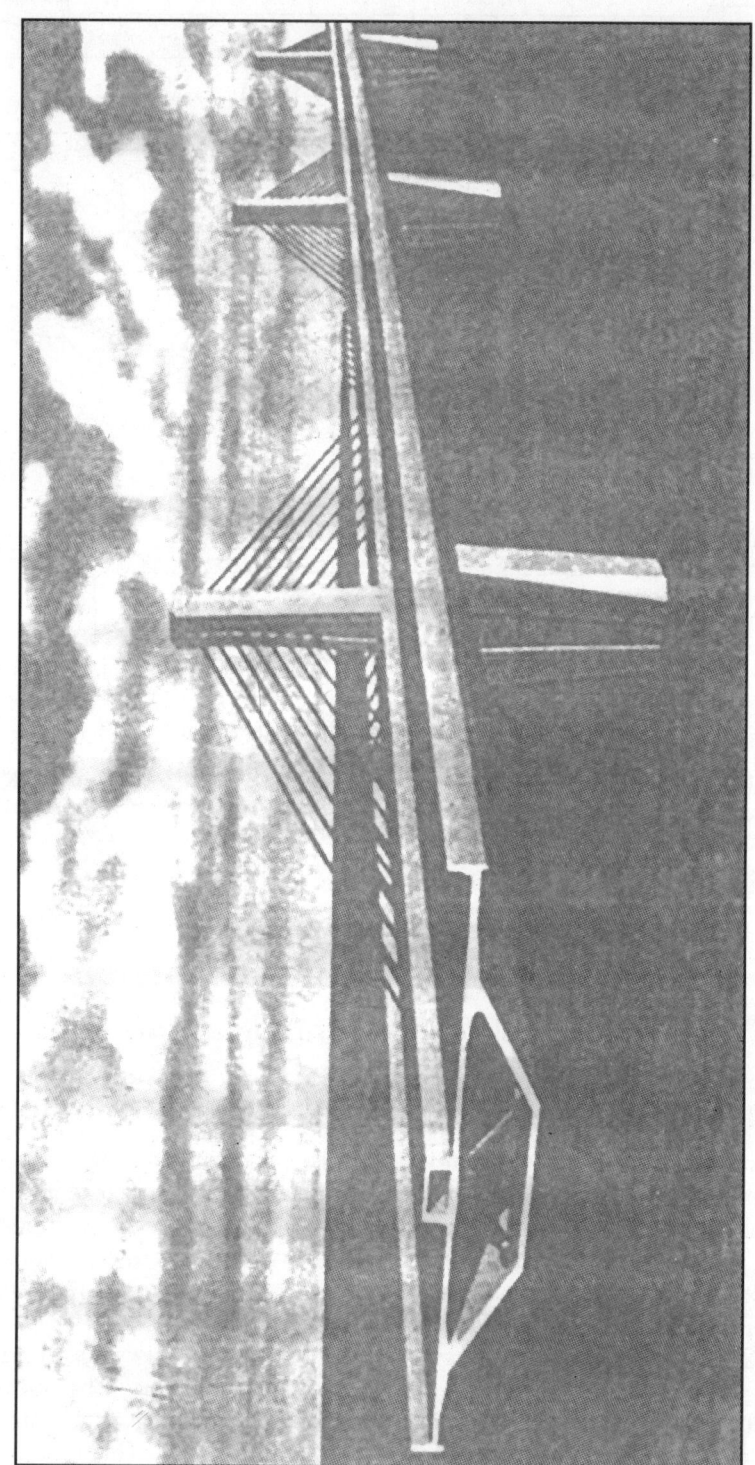

Fig. 1.13: Second Vivekananda Extradosed P.S.C. Bridge across Hooghley River Near Kolkata

Fig. 1.14: World's Tallest and Longest Cable Stayed Bridge in France

The newly planned second Vivekananda tollway bridge just north of Kolkata is an excellent example of an extradosed bridge, comprising a hybrid structure with elements of cable stayed post tensioned prestressed concrete box girders. The nine span extradosed bridge stretching across India's Hooghly River is considered as Asia's first multi span extradosed bridge and one of only three extradosed bridges in Asia outside Japan according to Egeman Ayna[12], principal engineer of the International Bridge Technologies (IBT) who are the design consultants for the bridge project. The modern bridge having a total length of 880 m comprises of seven spans of 110 m and two 55 m long spans.

The second Vivekananda Bridge with a width of 28.6 m is an unusually wide extradosed structure catering to 8 lane traffic according to IRC Standards[13]. The bridge deck (Fig. 1.13) comprising of post tensioned prestressed concrete box girders are supported by a single central suspension system instead of two planes of cable stays, eight cables composed of 63 to 73 strands each of 15 mm diameter extend from both sides of the 2 m wide pylons. The pylons of height 14 m are supported on Caisson foundations having a diameter of 11 m and sunk to a depth of 45 m below the river bed. The concrete wall thickness of the Caisson is 2 m.

According to Ayna, a typical box girder bridge would have had a depth of 6 m. However, the second Vivekananda Bridge lowers that profile by approximately 2.5 m. Also the bridges constant depth profile is a departure from the variable depth seen in other extradosed bridges. At present the bridge project is under execution by the well known construction firm M/S. Larsen and Toubro Ltd. with its head quarters at Mumbai.

France being the homeland of the innovator of pretsressed concrete, Eugene Freyssinet, it is befitting that the tallest and longest cable stayed bridge recently opened for traffic in December 2004 is located outside the French town of Millau. The bridge extending over a length of 2.46 km is considered as an engineering feat since some of the bridge pillars rise gracefully to a height of more than 300 m. The bridge designed by the famous British architect Sir Norman Foster is currently the world's tallest and longest cable stayed bridge shown in Fig. 1.14.

A critical survey of long span bridges constructed in various countries during the last two decades indicates that the modern trend is to adopt cable stayed bridges with prestressed concrete decks as it is economical, structurally efficient and aesthetically superior in comparison with other types of bridges.

REFERENCES

1. RAINA, V.K., *Concrete Bridge Practice, Analysis, Design and Economics*. Tata McGraw Hill Publishing Co, New Delhi, 1991, pp. 587-617.
2. FREYSSINET, E., The Birth of Prestressing, *Cement and Concrete Association*, London, Translation No. CJ-59, 1956, p. 44.
3. LEONHARDT FRITZ and ZELLNER, W., Cable Stayed Bridges, *IABSE Surveys S-13/80* and *IABSE Periodical*, 2/1980, May, 1980.
4. FINISTERWALDER, U., Modern Designs for Prestressed Concrete Bridges. *Concrete and Constructional Engineering*, London, Vol. 60, No. 3, 1965, pp. 99-103.
5. DEAN, E.E., Prestressed Concrete Difficulties Overcome in Florida Bridge Practice. *Civil Engineering*, Vol. 27, January 1957, pp. 404-408.
6. ABELES, P.W., *An Introduction to Prestressed Concrete*, Vol. II. Concrete Publications Ltd, London, 1965, p. 555.
7. LEONHARDT, F., *Prestressed Concrete, Design and Construction*. Wilhelm Ernst and Sohn, Berlin, 1964, pp. 13-14.

8. KRISHNA RAJU, N., *Design of Bridges* (Third Edition), Oxford and IBH Publishing Co, New Delhi, 1998, pp. 156-157
9. Lin, T.Y. and BURNS, N., *Design of Prestressed Concrete Structures* (3rd Edition) (S.I.Version), John Wiley and Sons, New York, 1982, pp. 8-40.
10. Prestressed Concrete in Japan, Published by the Japan Association of Prestressed Concrete Industry, Tokyo, Japan, 1970, pp. 1-56.
11. KRISHNA RAJU, N., *Design of Concrete Mixes* (4th Edition), C.B.S. Publishers, New Delhi, 2002, pp. 210 215.
12. JESSICA BINNS, Extradosed Bridge Distinguishes Tollway Project in India. *Civil Engineering*, Vol.75, No. 2, February 2005, p. 20.
13. IRC: 6-2000, Standard Specifications and Code of Practice for Road Bridges, Section II, Indian Roads Congress, New Delhi, 1985, pp. 1-58.

BRIDGE LOADING STANDARDS

2.1 EVOLUTION OF BRIDGE LOADING STANDARDS

The earliest bridge loading standards in some countries were first formulated to regulate heavy military vehicles and were generally specified by local authorities. The loadings often consisted of steam rollers and some form of traction engines. The earliest specifications of highway bridge loadings originated from the need to transport heavy military vehicles in U.K. and Europe. This resulted in the introduction of the Ministry of Transport's first 'Standard Loading Train' in the U.K. in 1932 and the original loading standards of many European countries.

In the United Kingdom, these standards formed the basis for the present type HA loading of BS: 153[1]. In the U.S.A, a loading standard consisting of truck trains and equivalent loads was introduced by the American Association of State Highway Officials (AASHO) in 1935. It is significant to note that in some of the developing countries like India, the first loading standards were introduced nationally in 1937. Critical studies have been done to study the significant differences in the loading standards of various countries by Seni[2] and Rajagopalan[3]. The impact allowance was observed to vary considerably in different countries. In a recent survey made for the International Federation, Galambos[4] has reported that many countries are planning revisions of their highway bridge loading standards based on the research investigations.

The first loading standard (IRC: 6) in India was published by the Indian Roads Congress in 1958 and subsequently reprinted in 1962 and 1963. The section-II of the code dealing with loads and stresses was revised in the second revision published in 1964. The metric version was introduced in the third revision of 1966. The IRC: 6 Code has been revised to include the combination of loads, forces and permissible stresses in the Fourth revision published in 2000[5].

2.2 INDIAN HIGHWAY BRIDGE LOADING STANDARDS

Highway bridge decks have to be designed to withstand the live loads specified by the Indian Roads Congress. The different categories of loadings were first formulated in 1958 and they have not changed in the subsequent revisions of 1964, 1966 and 2000.

The standard IRC loads specified in IRC: 6-2000 are grouped under four categories as detailed below:

Bridge Loading Standards

1. **IRC Class AA Loading:** Two different types of vehicles are specified under this category grouped as tracked and wheeled vehicles. The IRC Class AA tracked vehicle (simulating an army tank) of 700 kN and a wheeled vehicle (heavy duty army truck) of 400 kN are shown in Fig. 2.1.

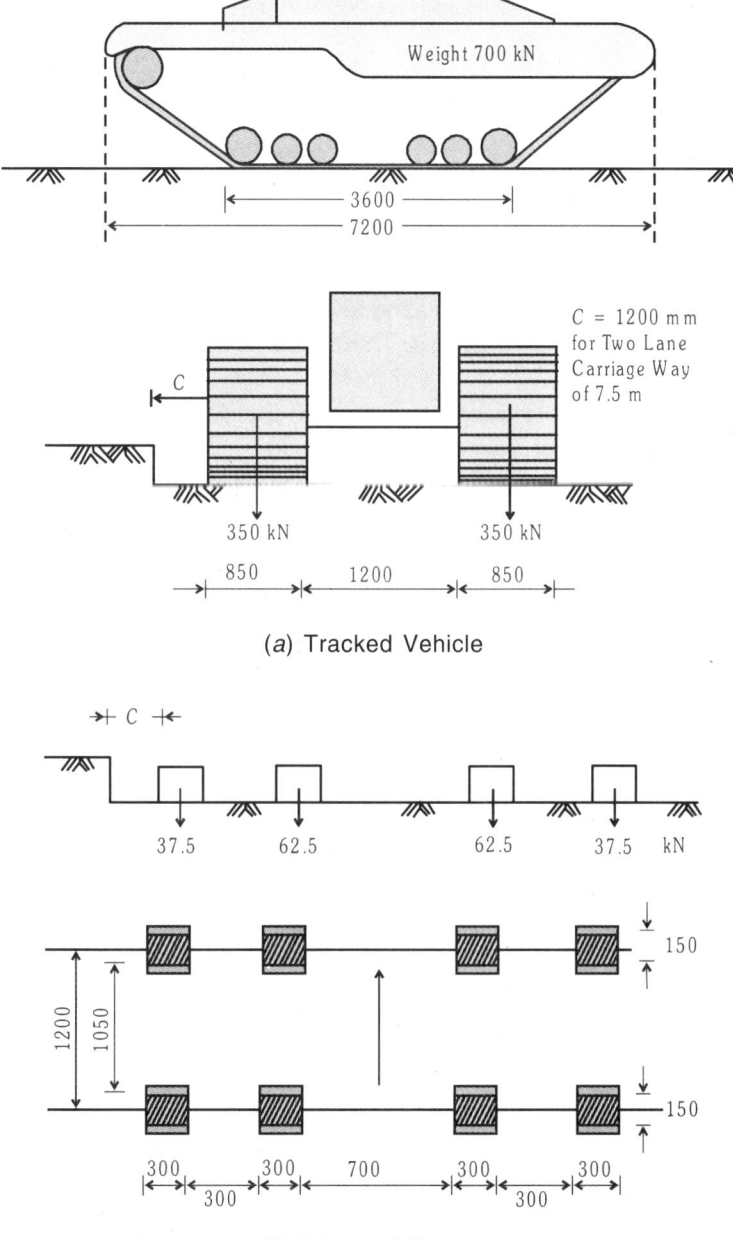

(a) Tracked Vehicle

(b) Wheeled Vehicle

Fig. 2.1: IRC Class AA Tracked and Wheeled Vehicles

All the bridges located on National Highways and State Highways have to be designed for this heavy loading. These loadings are also adopted for bridges located within certain specified municipal localities and along specified highways. Alternatively, another type of loading designated as Class 70 R is specified instead of Class AA loading.

2. **IRC Class 70 R Loading:** IRC Class 70 R loading consists of the following three types of vehicles.

   (a) Tracked vehicle of total load 700 kN with two tracks each weighing 350 kN.
   (b) Wheeled vehicle comprising 4 wheels, each with a load of 100 kN totaling 400 kN
   (c) Wheeled vehicle with a train of vehicles on seven axles with a total load of 1000 kN.

   The tracked vehicle is somewhat similar to that of Class AA, except that the contact length of the track is 4.87 m, the nose to tail length of the vehicle is 7.92 m and the specified minimum spacing between successive vehicles is 30 m. The wheeled vehicle is 15.22 m long and has seven axles with the loads totaling to 1000 kN. The bogie axle type loading with 4 wheels totaling 400 kN is also specified. The details of IRC Class 70 R loading vehicles are shown in Fig. 2.2.

   The 700 kN tracked vehicle is common to both the classes, the only difference being the loaded length which is slightly more for the Class 70 R. The second category is the wheeled type comprising 1000 kN train of vehicles on seven axles for the Class 70 R and a 400 kN bogie axle type vehicle for the Class AA.

   The Class A loading is a 554 kN train of wheeled vehicles on eight axles. Impact is to be allowed for all the loadings as per the specified formulae which is different for steel and concrete bridges.

   The various categories of loads are to be separately considered and the worst effect has to be considered in design. Only one lane of Class 70 R or Class A load is considered whereas both the lanes are assumed to be occupied by Class A loading if that gives the worst effect.

3. **IRC Class A Loading:** IRC Class A type loading consists of a wheel load train comprising a truck with trailers of specified axle spacing and loads as shown in Fig. 2.3. The heavy duty truck with two trailers transmits loads from 8 axles varying from a minimum of 27 kN to a maximum of 114 kN. The Class A loading is a 554 kN train of wheeled vehicles on eight axles. Impact has to be allowed as per the formulae recommended in the IRC: 6-2000. The impact factor is inversely proportional to the length of the span and is different for steel and concrete bridges. This type of loading is recommended for all roads on which permanent bridges and culverts are constructed.

4. **IRC Class B Loading:** Class B type of loading is similar to Class A loading except that the axle loads are comparatively of lesser magnitude. The axle loads of Class B are a 332 kN train of wheeled vehicles on eight axles as shown in Fig. 2.3. This type of loading is adopted for temporary structures and timber bridges. Combinations of different types of live loads are recommended for the design of bridges in clause 207.4 of IRC: 6-2000. The carriageway live load combination recommended for design is compiled in Table 2.1.

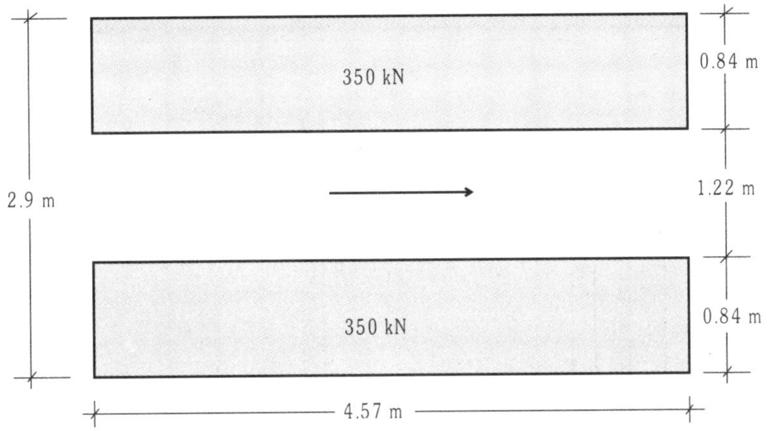

Bridge Loading Standards

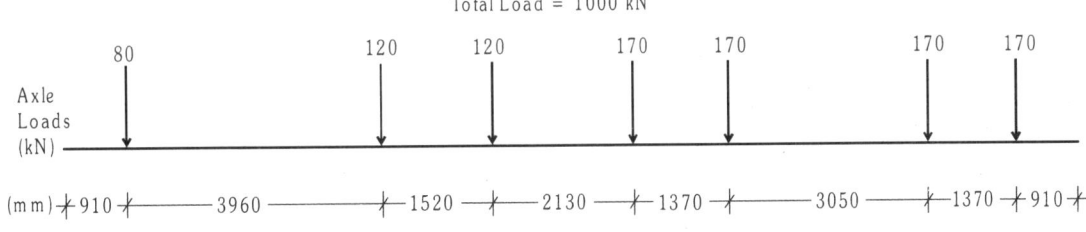

(a) Class 70 R Tracked Vehicle

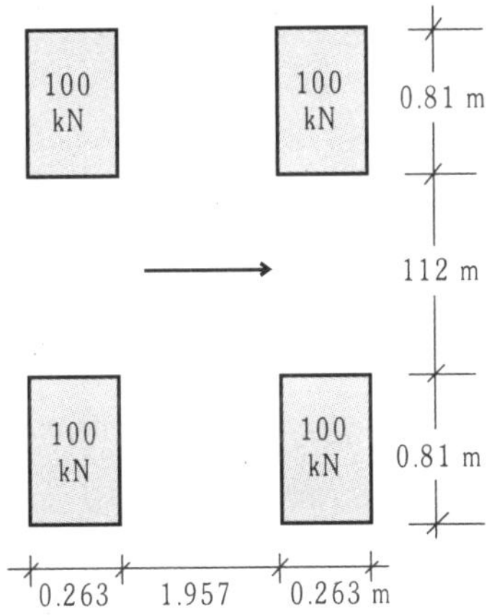

(b) Class 70 R Bogie Axle Type Vehicle

(c) Class 70 R Wheeled Vehicle Loading

Fig. 2.2: IRC Class 70 R Tracked and Wheeled Vehicles

The IRC Code also provides for the reduction of the longitudinal effects on bridges accommodating more than two traffic lanes due to the low probability of all lanes not subjected to the characteristic loads simultaneously. The reduction in longitudinal effect recommended is 10 per cent for three lanes and 20 per cent for four lanes or more. However, it should be ensured that the reduced longitudinal effects are not less severe than the longitudinal effect resulting from simultaneous load on two adjacent lanes.

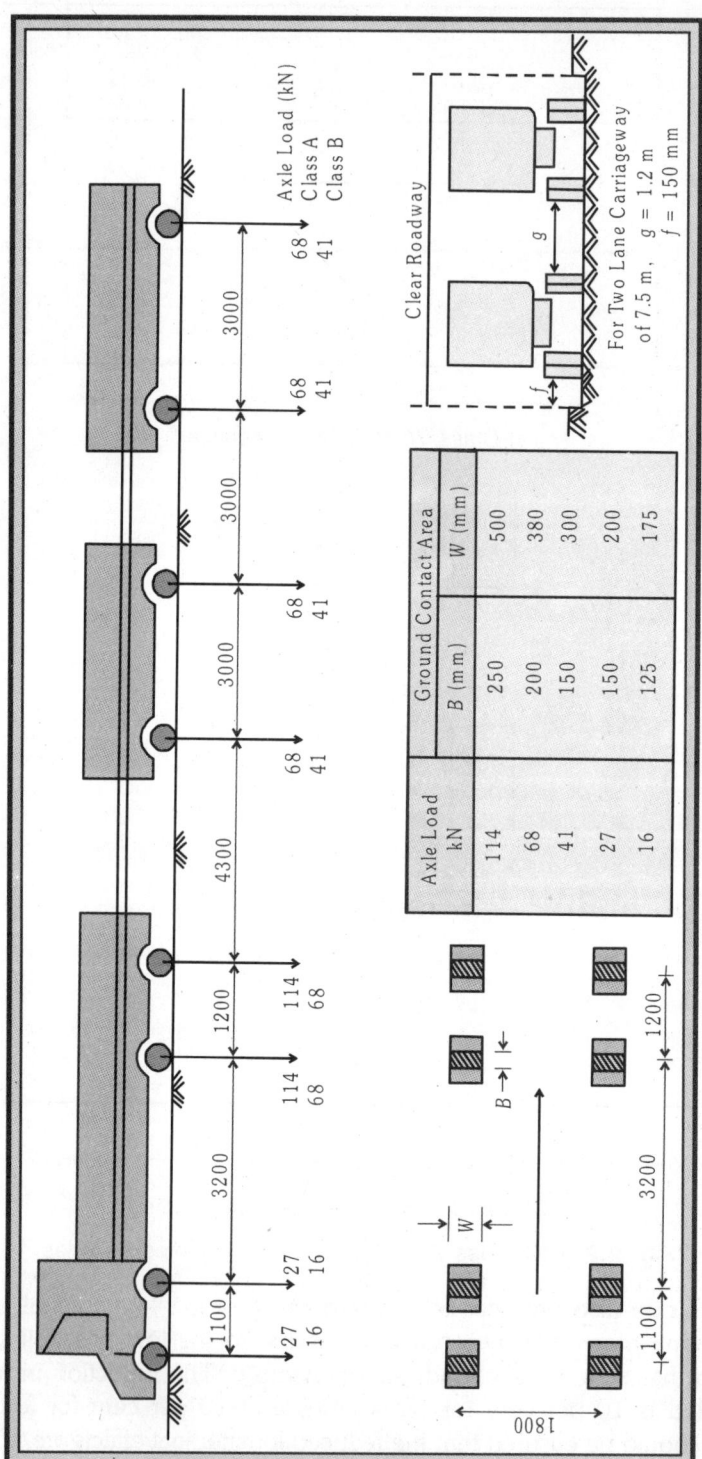

Fig. 2.3: IRC Class A and B Type Loading Vehicle

Bridge Loading Standards

Table 2.1: Live Load Combinations (Table-2 of IRC: 6-2000)

Sl.No.	Carriage Way Width	Number of Lanes for Design Purposes	Load Combination
1	Less than 5.3 m	1	One lane of Class A considered to occupy 2.3 m. The remaining width of carriage way shall be loaded with 5 kN/m^2
2	5.3 m and above but less than 9.6 m	2	One lane of Class 70 R or two lanes of Class A
3	9.6 m and above but less than 13.1 m	3	One lane of Class 70 R for every two lanes with one lane of Class A on the remaining lane or 3 lanes of Class A
4	13.1 m and above but less than 16.6 mm	4	One lane of Class 70 R for every two lanes with one lane of Class A for the remaining lanes, if any, or one lane of Class A for each lane.
5	16.6 m and above but less than 20.1 m	5	------------ do ------------
6	20.1 m and above but less than 23.6 m	6	------------ do ------------

Note: The width of the two lane carriage way shall be 7.5 m as per clause 112.1 of IRC: 5-1998.

2.3 HIGHWAY BRIDGE LOADING STANDARDS OF DIFFERENT COUNTRIES

A brief survey of the highway bridge loadings prescribed in some countries based on work of Raina[6] is compiled below:

1. **British Standard Loadings:** The British standards prescribe two main types of loadings designated as HA and HB loading. The HA type loading is designated as the normal design loading and consists of:

    (a) A uniformly distributed loading carrying from 318.6 kN/m for 1 m loaded length (span) to 5.8 kN/m for 900 m loaded length (span) (shown in Fig. 2.4) and a knife edge load of 120 kN per lane which are inclusive of impact. There is no reduction in the intensity of HA loading for up to two lanes of traffic.

    (b) And alternative axle load is also specified on which impact must be considered. This loading system consists of two loads each of 112 kN in line transversely to the direction of traffic flow speed at 0.9 m. The uniformly distributed load has a constant value of 31.5 kN/m of lane for loaded length for 6.5 m to 23 m. For span below 6.5 m, separate curves for the uniformly distributed load are specified. Two lanes are always considered as occupied by full type HA loading while all other lanes in excess of two are considered as occupied by one third of the full lane loading. The standard design lane width is 3 m. In considering the effects of the 112 kN wheel loads, an over stress 25% is permitted. An impact allowance of 25% is specified for this type of HA loading.

    (c) Type HB loading is an abnormal unit loading. The number of units per axle (4 axles in all) specified in the UK for bridges carrying the heaviest class of load is 45, amounting to a total load of 1800 kN. This is idealized on four axles which allows for the weight of tractors accompanying trailers. With this loading an overstress of 25% is allowed. No allowance is to be made for impact. Only one lane is to be loaded with HB type loading, all other lanes being considered as occupied by one-third full lane HA loading only if it's presence results in worst effect. The plan view of Type HB loading is shown in Fig. 2.5. These loadings are followed in Malaysia, Sri Lanka, Kenya and Zambia.

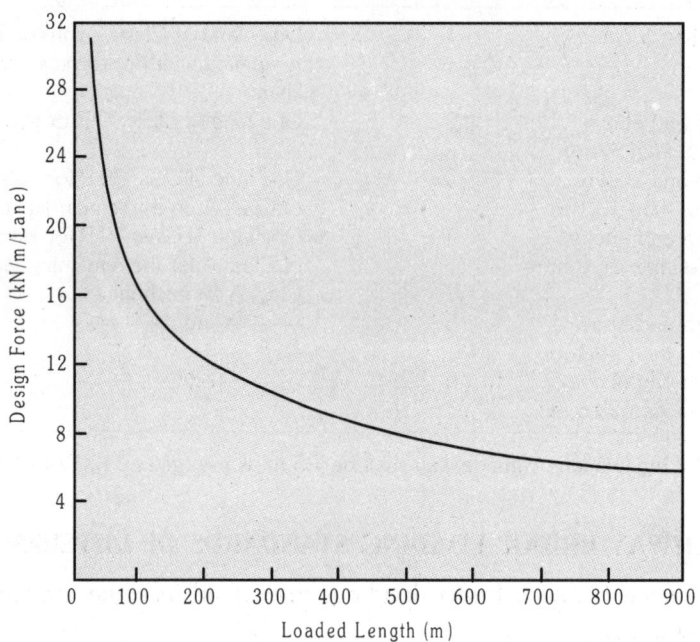

Fig. 2.4: Loading Curve for Type HA-Loading (British Standard Loading-U.K.)

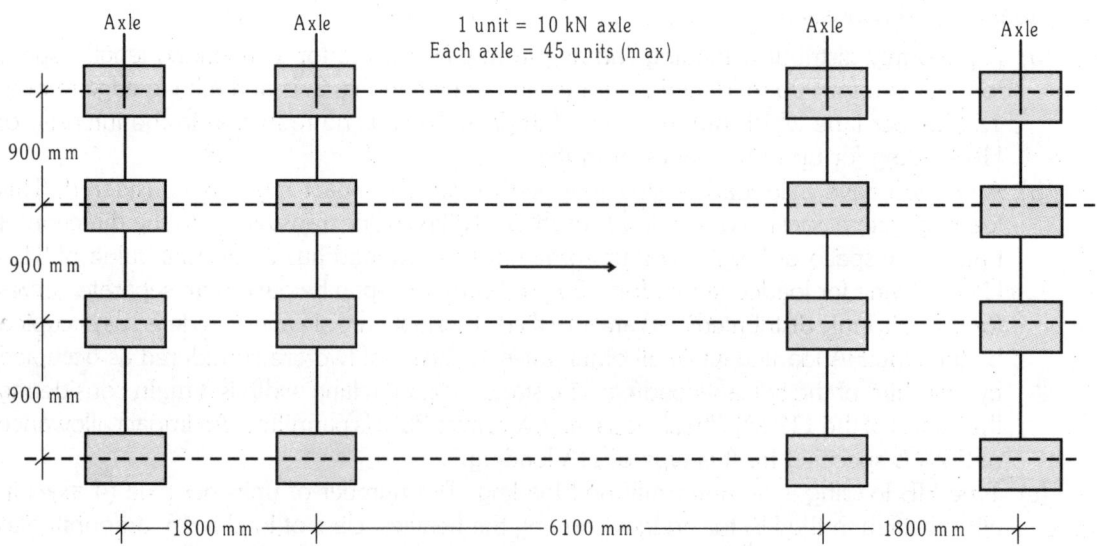

Fig. 2.5: Plan View of HB Loading (British Standard Loading-U.K.)

2. **AASHTO (American) Loading:** The American Association of Highway and Transportation Officials (AASHTO), has prescribed the heaviest loading designated as HS20-44. This type of loading is followed in USA, Australia, Bangladesh, Canada, Ethiopia, Philippines and Turkey. This loading comprises of a heavy tractor truck with a semi-trailer of a total load of 320.3 kN or the corresponding lane loading. The lane loading consists of a uniformly distributed load of intensity 9.3 kN/m together with a knife edge load of 80 kN for bending moment and 115.7 kN for shear force computations. In addition, impact effect is to be added for both the cases as prescribed by the AASHTO recommendations. While designing bridges, both the truck and lane loading should be considered and the one which gives the worst effect is to be adopted.

When truck loading is used, only one truck is considered for each traffic lane for the whole of it's length. Also there is no reduction in load intensity for up to two lanes of traffic loaded.

In USA, Canada and Australia, AASHTO loadings comprise of standard HS20-44 truck or HS20-44 lane loading (shown in Fig. 2.6) are used.

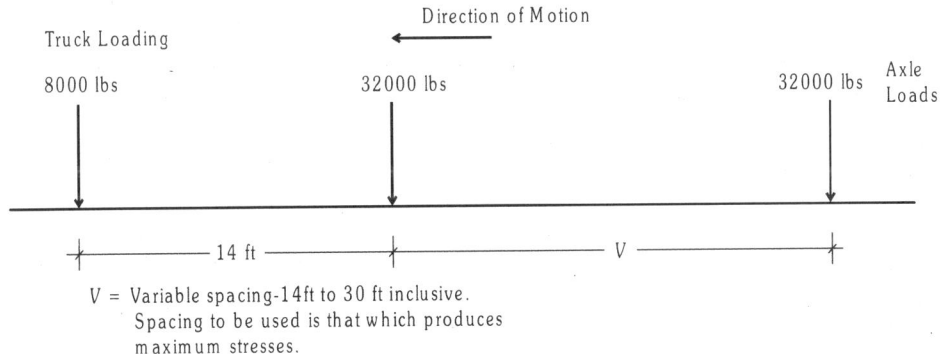

(a) Standard HS 20-44 Truck (U.S.A.)

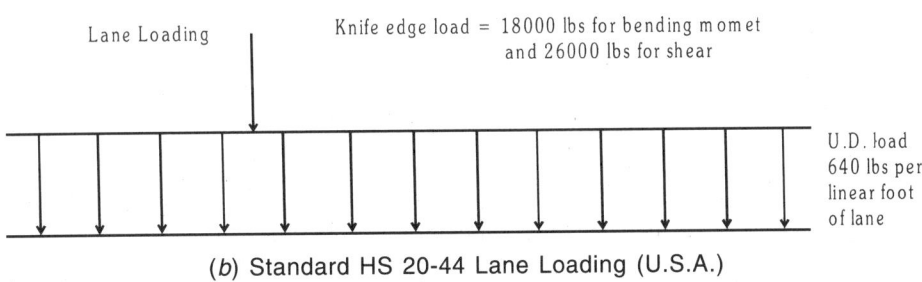

(b) Standard HS 20-44 Lane Loading (U.S.A.)

Fig. 2.6: AASHTO Loading (U.S.A.)

3. **French Highway Loadings:** The highway loadings prescribed in France are categorized as system A and system B loads. System A loading is expressed in an empirical form given by

$$\text{U.d Load } A(L) = \left\{ 230 + \frac{3600}{L+12} \right\} \text{ kg/m}^2$$

where L = Loaded length in metres.

The uniformly distributed load A(L) obtained from the above formula is to be multiplied by a coefficient a_1 whose value is 1.0 up to two lanes and then reduces with an increase in the number of lanes. Also $a_1 \cdot A(L)$ should be not less than (400–0.2 L) kg/m².

In the case of class 1 roads, if the lane width is different from the standard lane width of 3.50 m, the value of A(L) is to be multiplied by the coefficient a_2 so as to keep the total load per linear metre of lane unaltered for any loaded length.

System B loads are grouped as B_c and B_t where B_c comprises of truck loading with wheel loads (as shown in Fig. 2.7(a)) and B_t is made up of tandem axle loading (as shown in Fig. 2.7(b)). System A loading is inclusive of impact. For system B loading, impact factor is expressed as an empirical expression depending on the length of the element in metres, the permanent weight of the bridge and the maximum load of the truck.

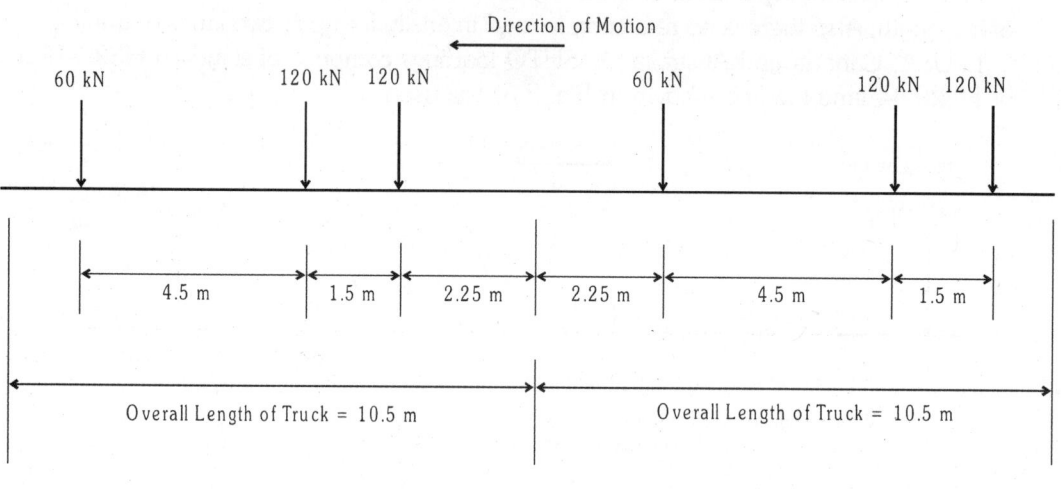

(a) System B_c Truck Loading

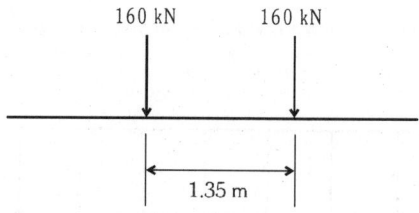

(b) Systme B_t Tandem Axle Loading

Fig. 2.7: Highway Loadings of France

4. **Highway Loadings of Germany:** Two types of loadings are specified for highway bridge design in Germany. The Class 60 loading consists of a 600 kN truck together with a uniformly distributed load of 5 kN/m² in the portion of the lane not occupied by the truck. The substitute uniformly distributed load for the 60 t truck is 33.3 kN/m². The standard design lane width is 3.0 m. There is no reduction in intensity of load for up to two lanes of traffic. For the area outside the main lanes, a uniformly distributed load of 3 kN/m² is specified (Refer Fig. 2.8). Impact factor is expressed by a formula depending upon the governing length.

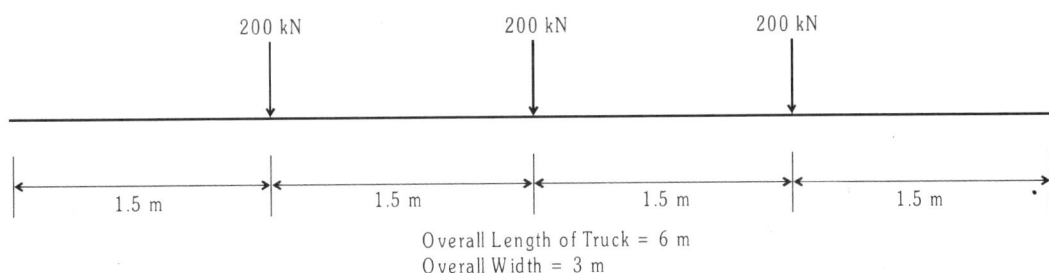

Fig. 2.8: Highway Loadings of Germany

5. **Highway Loadings of Japan:** The L-20 loading consisting of a knife edge line load P of 5000 kg/m and a uniformly distributed load p which has the following values specified for a lane width of 5.5 m or less.

   For $L < 80$ m, $p = 350$ kg/m^2

   For $L > 80$ m, $p = (430 - L) \geq 300$ kg/m^2

In the case of bridges with a width of more than 5.5 m, the values of P and p are to be reduced by one-half on the portion of the road way in excess of 5.5 m. The composition of L-20 loading used in Japan is shown in Fig. 2.9. Impact factor depending upon the length of the element expressed as an empirical formula should be applied for the live loads.

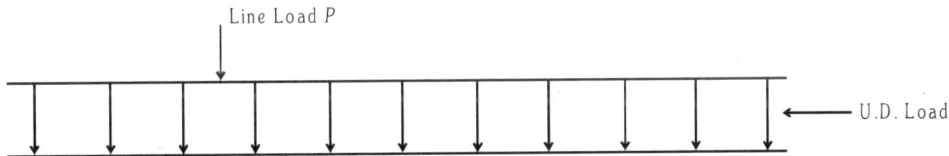

Fig. 2.9: Highway Loadings of Japan

6. **Highway Loadings of New Zealand:** The design load prescribed per lane is the same as that of AASHTO loading comprising HS20-44 truck. Alternately H20-S16-T16 truck loading (Fig. 2.10) is also specified. Between the two, the loading which gives the worst effects should be considered. The standard design lane is 3 m. The impact allowance is the same as that given by the AASHTO without specifying any upper limit.

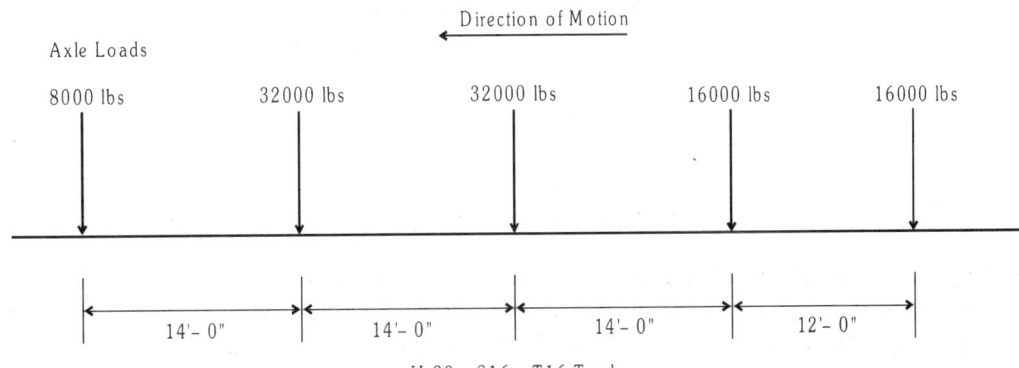

Fig. 2.10: Highway Loadings of New Zealand

7. **Highway Loadings of Sweden:** Two different types of highway loading are specified in Sweden. The first type consists of uniformly distributed load of 'p' t/m which depends on the loaded length and a concentrated axle load of 14 t [Fig. 2.11(a)]. The values of p are specified as follows:

$$p = 2.4 \text{ t/m for } L < 10 \text{ m}$$

$$p = 2.4\frac{1.3(L-10)}{80} \text{ t/m for } 10 < L < 90 \text{ m}$$

$$p = 1.1 \text{ t/m for } L > 90 \text{ m}$$

where L = loaded length in metres.
Design lane width = 3.0 m

The second type comprises of single truck loading with axles loads spaced as shown in Fig. 2.11 (b). Two lane bridges are designed with lane loading in both the lanes or only with single truck loading which gives the worst results. For continuous structures there is a separate loading consists of two axle loads and a uniformly distributed load.

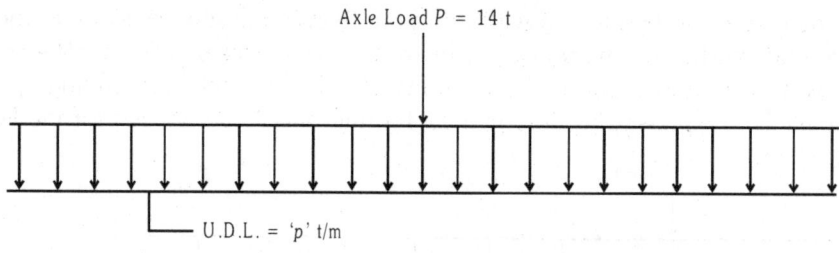

(a) Composition of Lane Loading

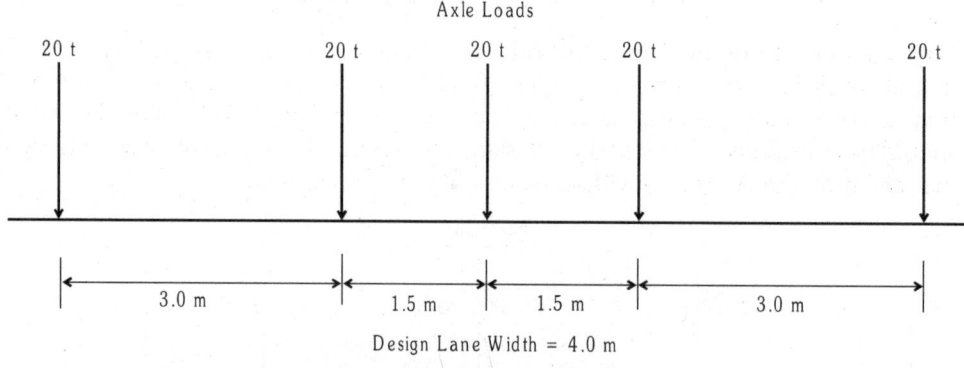

(b) 100 t Single Truck Loading

Fig. 2.11: Highway Loadings of Sweden

8. **Highway Loadings of Austria:** Austrian loadings consist of tracked and truck loadings. The tracked loading with the overall length of the caterpillar being 6 m and overall width 3 m is 60 t as shown in Fig. 2.12 (a). No allowance for impact is to be made for uniformly distributed load. The truck loading with axle loads is shown in Fig 2.12 (b). The axle loads may be increased by 40% for impact effects. A uniformly distributed load of 0.5 t/m^2 is

prescribed for the portion of the lane not occupied by the truck. The specifications also give the following equivalent weights of the caterpillar and truck loading which are to be used for the design of spans exceeding 30 m.

60 t caterpillar---3.33 t/m²
25 t truck---------1.67 t/m²

Impact allowances varying from zero to 40% corresponding to spans of 70 m to zero respectively. Separate impact factors are recommended for concrete and steel bridges as a function of the span and direct and indirect loads on the main girder.

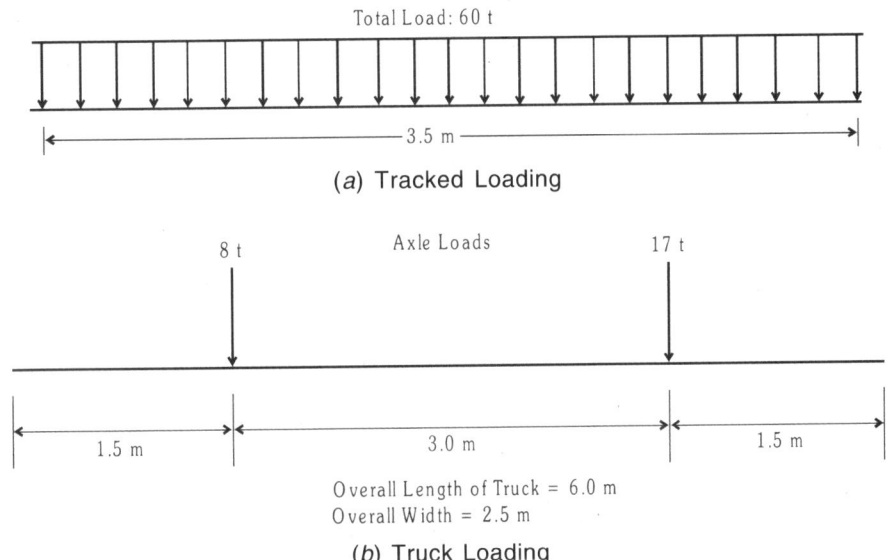

(a) Tracked Loading

(b) Truck Loading

Fig. 2.12: Highway Loadings of Austria

9. **Highway Loadings of Belgium**: Loadings specified in Belgium are categorized as normal and heavy truck loading. The normal truck load of 32 t can have different combinations as shown in Fig. 2.13 (a). Heavy truck loading of 60 t comprising 3 axles is shown in Fig 2.13 (b). Impact factor is expressed by a formula depending upon span, speed of vehicle, moving loads and dead ways and static deflection due to dead weight.

10. **Highway Loadings of Italy**: Highway bridges of Category 1 should be designed for different trains of military loads prescribed in the standards. Three different types of axles (Fig. 2.14) are to be used and the worst effect should be taken for design. The total load of trains consists of 32 t, 61.5 t and 74.5 t. The width of the three types of loadings is 3.5 m. Alternately, an equivalent uniformly distributed load having different values for bending moment and shear is also specified for the design of highway bridges. Impact factor specified by a formula depend upon the span length of the bridge.

11. **Highway Loadings of Netherlands**: The highest class of loading in Netherlands is the Class 60 loading consisting of a 600 kN vehicle of three axles of 200 kN each plus a uniformly distributed load of 4 kN/m² (Fig. 2.15). The impact factor depending upon the span length is expressed by an empirical formula depending upon the span length.

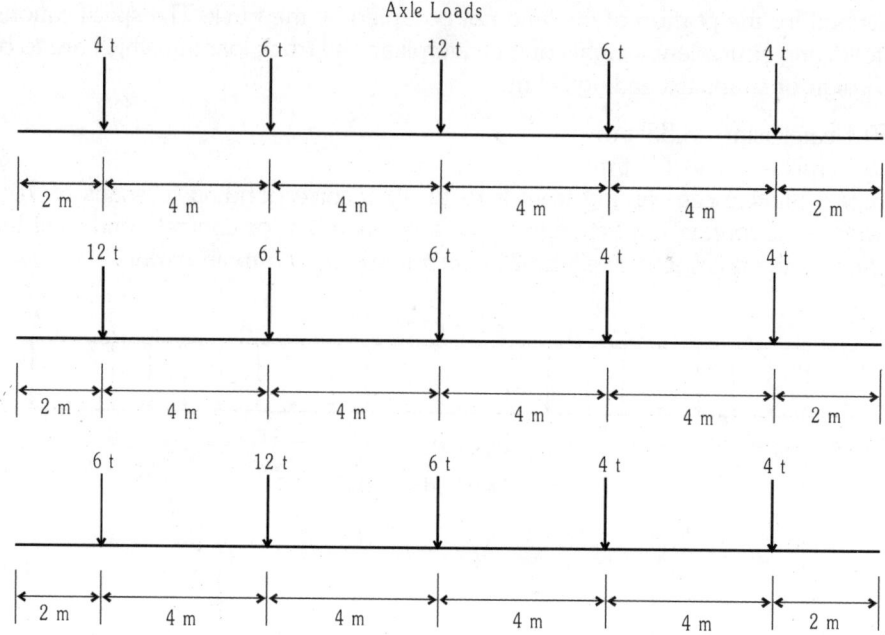

(a) Different Combinations of 32 t Normal Truck

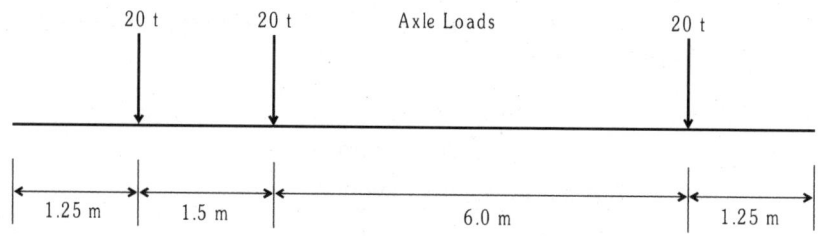

(b) 60 t Heavy Truck

Fig. 2.13: Highway Loadings of Belgium

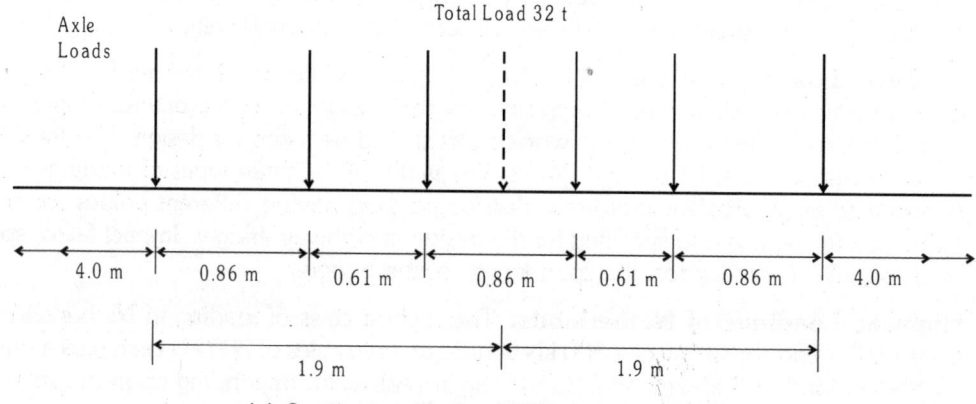

(a) Continuous Train of Military Load of 32 t

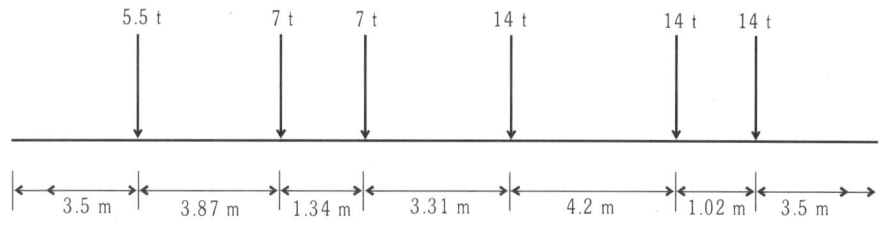

(b) Continuous Train of Military Load of 61.5 t

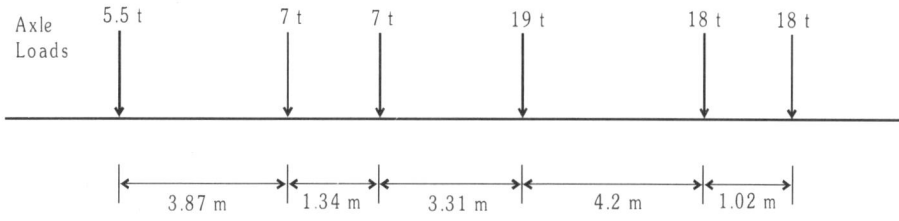

(c) Single Military Load of 74.5 t

Fig. 2.14: Highway Loadings of Italy

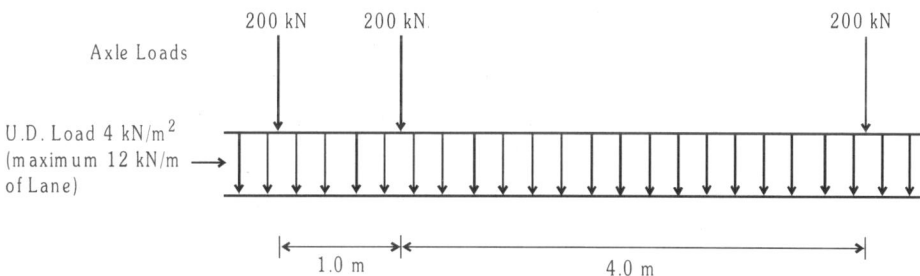

Fig. 2.15: Highway Loadings of Netherlands

12. **Highway Loadings of Norway:** For the design of Class 1 type bridges, the loadings specified consists of an equivalent lane loading per lane together with a knife edge load 'A' and a uniformly distributed load 'p' (Fig. 2.16). The variables 'A' and 'p' are expressed as

$$A = 12 + \left(\frac{8x}{L}\right) t$$

$$p = 0.5 + \left(\frac{35}{L+5}\right) \text{t/m of lane}$$

L = Actual loaded length of lane in metres

x = Distance of the knife edge load from the center of span

The loadings specified above are normally considered for lane widths varying from 3.0 to 3.75 m. For two lane bridges, the full equivalent loading is assured on both the lanes. Besides the lane loading, the structure is designed for a local loading of two axles each of 13 ft. It is assumed that 38.5% impact is to be added to the heaviest axle and it is unnecessary to add any impact to the remaining axles. The values of knife edge load, 'A' and uniformly distributed load 'p' are inclusive of impact calculated on this basis. An impact of 38.5% is to be added to the axle loads.

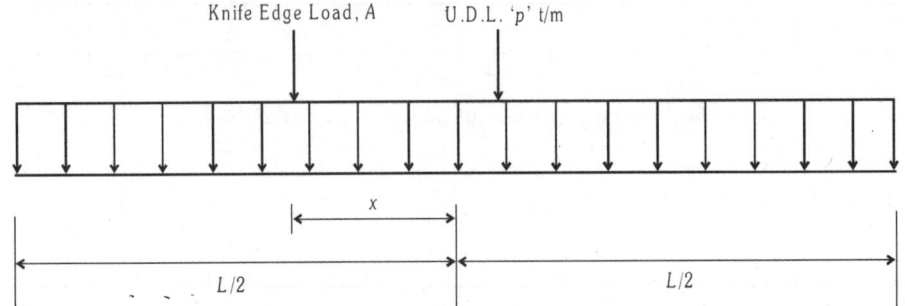

Fig. 2.16: Highway Loadings of Norway

13. **Highway Loadings of USSR:** Three different types of wheeled vehicle loadings are specified in USSR (Fig. 2.17 a, b, c). These are designated as N-30, N-10 and NK-80 loading. Also a tracked loading designated as NG-60 caterpillar loading is specified (Fig. 2.17 d). The following conditions are also prescribed:

(i) The design load should be either (a) or (b) whichever produces the maximum effect.
(ii) Impact allowance as per AASHTO specification
(iii) For more than two lanes loaded, reduction factor on live load effect as per AASHTO specification.

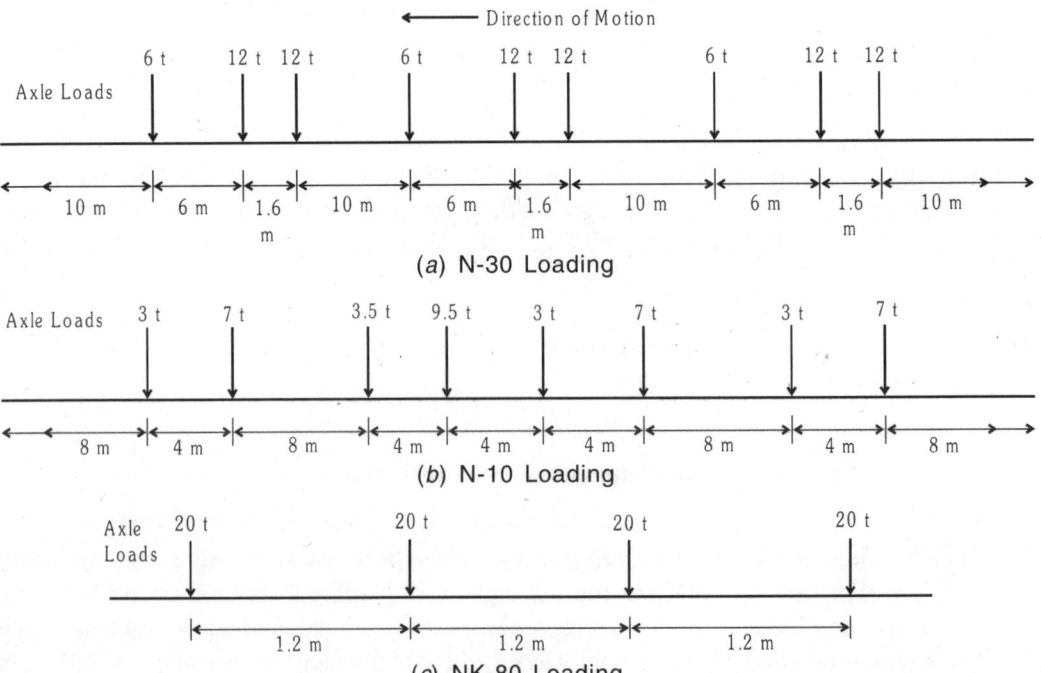

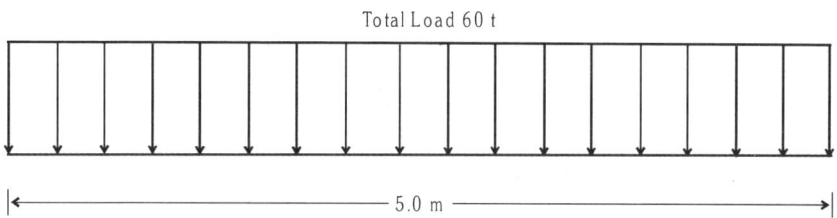

(*d*) NG-60 Caterpillar Loading

Fig. 2.17: Highway Loadings of USSR

14. **Highway Loadings of Saudi Arabia:** Highway bridge loadings of Saudi Arabia comprise of both heavy truck wheeled vehicle loading and uniformly distributed load. The truck of three axle loads comprise of 80, 260 and 260 kN; totaling 600 kN with only one truck allowed per lane in the longitudinal direction.

The uniformly distributed load is of intensity 20 kN/m per lane width with a concentrated live load per lane width of 150 kN for moment computation and 220 kN for shear force calculations. The axle loads and uniformly distributed load are shown in Fig. 2.18 (*a*) and (*b*) respectively.

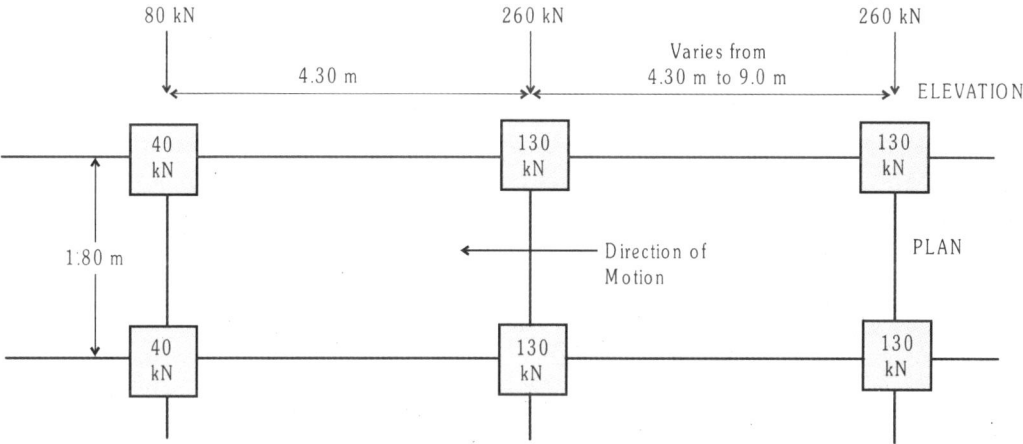

(*a*) 600 kN (Each Lane can be loaded by a truck but only one Truck/Lane Longitudinally)

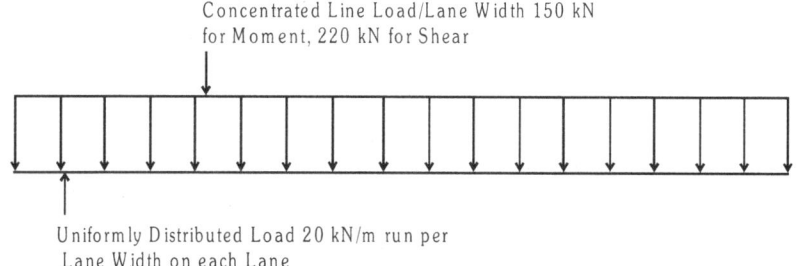

(*b*) Uniformly Distributed Load

Fig. 2.18: Highway Loadings of Saudi Arabia

2.4 IMPACT FACTORS

Impact factors are generally applied to the moving wheel or distributed loads to enhance their magnitude to include their dynamic effects on the bridge deck. The impact allowance is generally expressed as a fraction of the applied live load and is expressed as an empirical expression involving constants and the span length of the bridge deck. The impact factor is always inversely proportional to the length of the span and is different for reinforced concrete and steel bridges.

The impact factors to be considered for different types of live loads of various countries are as follows:

1. **Indian Standard Loadings:**
    (a) IRC Class A loading
    The impact allowance is expressed as a fraction of the applied load and is computed by the expression

    $$I = \frac{A}{(B+L)}$$

    where I = Impact factor fraction
    A = Constant having the value of 4.5 for reinforced concrete bridges and 9.0 for steel bridges.
    B = Constant having a value of 6.0 for reinforced concrete bridges and 13.5 for steel bridges.
    L = Span in metres

    For spans less than 3 m, the impact factor is 0.5 for R.C. bridges and 0.545 for steel bridges. When the span exceeds 45 m, the impact factor taken is 0.088 for R.C. bridges and 0.154 for steel bridges. The impact percentages for highway bridges can also be directly obtained from the curves of IRC code[5] given in Fig. 2.19.

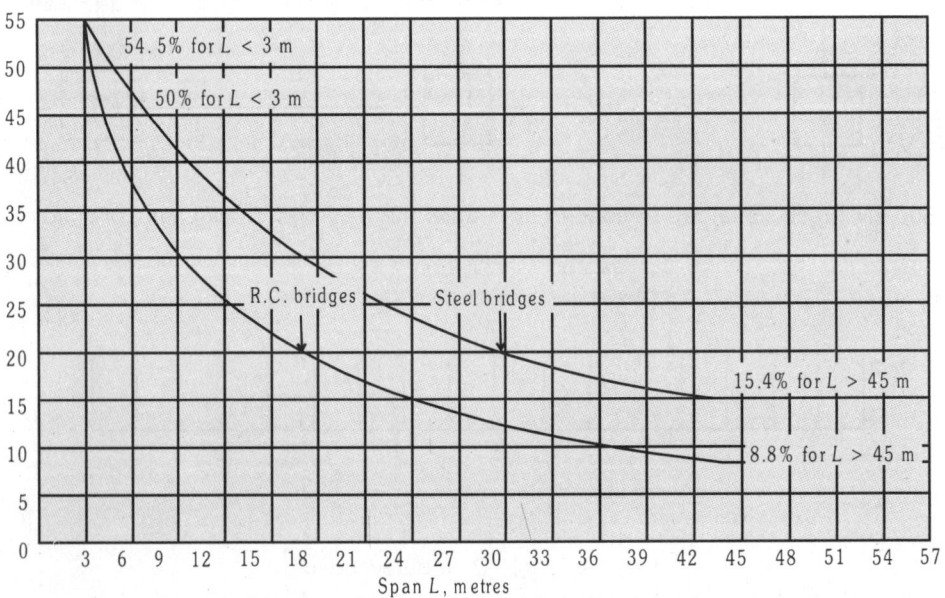

Fig. 2.19: Impact Percentages for Highway Bridges (IRC: 6-2000)

Bridge Loading Standards

(b) IRC Class AA or 70 R loadings.
   For spans less than 9 m
   (i) For tracked vehicle-25% of the span up to 5 m linearly reduced to 10% for span of 9 m
   (ii) For wheeled vehicle-25%
   For spans of 9 m and more,
      For tracked vehicle- for R.C.Bridges, 10% up to a span of 40 m and in accordance with Fig. 2.19 for spans exceeding 40 m. For steel bridges, 25% for spans up to 23 m and as per Fig. 2.19 for spans exceeding 23 m.

2. **British Standard Loadings:**
   (a) For type HA loadings, an impact allowance of 25% on the heaviest axle in the train of vehicles from which the loading has been derived is provided.
   (b) For type HB loading, no impact allowance has been provided since this type is considered as an abnormal loading.

3. **AASHTO Loadings:**

$$\text{Impact allowance} = \frac{50}{L + 25}$$

   where L = Length in feet of the portion of the span to produce the maximum stress in the member subject to a maximum of 30%. For shear due to truck loads, the span (L) is taken as the loaded part of the span from the point being considered to the reaction, except for cantilever arms where the impact allowance is 30%.

4. **French Highway Loadings:** System A loading is inclusive of impact.
   For system B loads, the impact factor '∂' is expressed by the formula,

$$\partial = 1 + \frac{0.4}{(1 + 0.2L)} + \frac{0.6}{1 + 4(G/S)}$$

   where L = Length of the element in metres
   G = Permanent weight of the bridge
   S = Maximum load of the truck

5. **German Highway Loading:** The impact factor ϕ which has to be applied to the live load is given by the relation

$$\phi = 1.4 - 0.008 \, L_\phi \geq 1.0$$

   where L_ϕ = Governing length in metres

6. **Highway Loading of Japan:** The impact allowance 'I' is determined by the impact formula expressed as,

$$I = \left(\frac{20}{50 + L}\right)$$

   where L = Length of the element in metres.

7. **Loadings of New Zealand:** The impact allowance is the same as that specified by the AASHTO but there is however no upper limit is prescribed.

8. **Loading of Sweden:** An impact allowance of 40% is to be applied for the axle loads. However no allowance for impact is to be made for the uniformly distributed load and for the single truck loading.

9. **Loadings of Austria:** The following impact factors are specified for concrete and steel bridges.
   (i) Concrete bridges

Impact Factor for Different Spans

Span Length (m)	Direct Loaded Main Girder	Indirect Loaded Main Girder
0	1.40	1.40
10	1.30	1.25
30	1.20	1.10
50	1.10	1.00
70	1.00	1.00

Impact Factor for Floor Slab = 1.4

(ii) Steel bridges

Impact Factor for Different Spans

Span Length (m)	Lane-I	Lane-II
2	1.64	1.32
6	1.61	1.20
10	1.30	1.15
20	1.18	1.09
40	1.10	1.05
60	1.07	1.03
80	1.05	1.02
100	1.04	1.02

10. **Loadings of Belgium:** The impact factor ϕ is given by the relation

$$\phi = 1 + \frac{0.377v}{\sqrt{L/\alpha}}\sqrt{1 + \frac{2Q}{P}}$$

where v = speed in km/hour (always greater than 60)
L = span length in metres
$\alpha = (L/f_s)$
f_s = static deflection in metres due to dead weight
Q = moving loads on the bridge deck in tons
P = dead weight of the bridge in tons

11. **Loadings of Italy:** The impact factor ϕ expressed as a function of span length L is given by the formula:

$$\phi = 1 + \frac{(100-L)^2}{100(250-L)}$$

This relation is applicable for spans up to 100 m.
For spans exceeding 100 m, the value of φ is unity.

12. **Loadings of Netherlands:** The impact coefficient 'S' for bridges carrying normal traffic is is given by the formula,

$$S = 1 + \frac{40}{(100 + L)}$$ where L is the span in metres.

13. **Loadings of Norway:** Impact factor of 38.5% is prescribed for the heaviest axle load and it is unnecessary to add any impact to the remaining axles. The value of the knife edge load and the uniformly distributed load are inclusive of impact calculated on this basis.

14. **Loadings of Saudi Arabia:** The impact allowance is similar to that specified in AASHTO specifications. For more than two lanes loaded, a reduction factor on live load effect as per AASHTO specifications should be applied.

2.5 COMPARATIVE ANALYSIS OF HIGHWAY LOADING STANDARDS

(a) **Type of Loads:** A critical comparative analysis of highway bridge loadings prescribed in different countries has been presented by Thomas[7]. Many countries specify the same uniformly distributed load for flexure and shear except Italy which does not specify knife edge load with uniformly distributed load. Many countries specify uniformly distributed loads for the full width of the traffic lane while Sweden specifies it as a strip load in two strips of 0.6 m each running for the entire loaded length. Alternatively Sweden allows the uniformly distributed load to be applied uniformly over a width of 2.4 m. Except the HA type group, all countries which have an equivalent uniformly distributed load system, have at least an alternative truck loading which has to be considered in the design. Designs based on HA type loading should generally be checked for HB type loading.

In Italy separate civil and military loadings are specified and prominent bridges are invariably designed for the heavier military loading. Many of the countries do not explicitly specify military loadings, however IRC Class 70 R of India, the caterpillar of Austria and the NK-80 loading of Russia are based on the loading of military vehicles.

(b) **Lane Width:** The lane width in most countries is in the range of 2.5 to 4.0 m with 3 m as the most common value. However Japan prescribes a wider lane width of 5.5 m. In countries like India, Pakistan and Russia, where there is no standard lane loading; only the minimum widths of carriage way for different number of lanes are specified.

(c) **Impact Allowance:** In most of the countries, impact is related to the loaded length (span length) although the exact relationship varies considerably from country to country. Some of the countries like India and Austria specify different impact factors for concrete and steel bridges. Higher impact factors specified for steel bridges are due to the lighter structures being subjected to a larger dynamic effect.

In countries like Italy and West Germany, impact is ignored for span lengths exceeding 100 and 50 m respectively. In comparison, countries like India and Australia specify certain minimum values of impact. Belgium and France relate the impact factor to the dead load of the bridge. The impact formula of Belgium is further complicated by the inclusion of speed of the vehicle on the bridge deck.

(d) Magnitude of Loads, Bending Moment and Shear Forces: A comparative analysis of loadings specified in different countries has been reported by Thomas[7]. British loading of Type HB is the most severe followed by IRC loadings for both single and two lane simply supported spans. The loadings of different countries vary considerably both qualitatively and quantitatively.

Raina[6] has prescribed a critical comparison of bending moments and shear forces developed in simply supported bridge decks due to the live loads prescribed in various countries for spans ranging from 5 to 100 m. Tables 2.2 to 2.5 show the variation of maximum bending moment and shear force (inclusive of impact) for salient values of simply supported predominant spans of 5, 25, 50, 75 and 100 m.

Based on the comparative analysis, the following general observations are noteworthy.

1. The maximum bending moments for a single lane loading occurs due to the British loadings of HB type for the span range of 5 to 100 m. However the maximum shear force develops for HB type loading for spans from 5 to 75 m and IRC loadings yield the highest shear force for 100 m span.
2. For two lanes, the West German Class 60 loading results in the maximum bending moment and shear force for all the spans from 5 to100 m.
3. For both single and double lanes, the AASHTO loading gives the least bending moment and shear force for all spans with magnitudes nearly half that given by the German loading.
4. The British type HA and French loadings are almost similar in effect for spans up to about 50 m and beyond that, the latter yields slightly higher values of moments and shear forces for both single and double lanes.
5. The New Zealand loading is marginally heavier than that of AASHTO loading for spans up to 50 m, beyond that, it gives the same values of bending moment, but in the case of shear force, it results in higher values for all spans.

The comparative analysis indicates the wide qualitative and quantitative variation in the loadings of different countries. Analysis indicates that the equivalent uniformly distributed load system appears to be the most acceptable since it is simpler for application. Due to this factor many countries have adopted the HA and AASHTO loadings for the design of bridge decks.

A critical survey of the impact factors specified by the various countries indicates the basic differences in their approach in assessing the dynamic effect of live loads on the bridge deck. Raina[6] has indicated that the impact formula prescribed by some countries is unnecessarily complicated since the effect of live load on the bridge deck is comparatively less than that of dead load for span lengths exceeding 25 m. However there is need for qualitative research in this field to investigate the behavior of the bridge structure under dynamic loads and the resulting data will be helpful in evolving rational design procedures. From considerations of simplicity of loading and ease of its application in design, type HA loading appears to be the most favourauble among the various load systems.

Bridge Loading Standards

There is significant variation in the type of highway loads specified in the standards of various countries. Raina[6] has made a critical analysis of the bending moments and shear forces developed due to the standard loadings specified by the various countries for single and double lane traffic, covering spans in the range of 5 to 100 m as shown in Tables 2.2 to 2.5. For the range of spans covered, the extreme variations in the bending moment and shear forces developed due to AASHTO loadings is only half of that due to German loadings. The loadings of the various other countries generally lie in between the American and German loadings. In view of these wide variations, there is a need for a systematic survey of vehicular loads on bridges for rationalization of Highway Bridge loading standards.

Table 2.2: Maximum Bending Moments for one Lane Simply Supported Spans of Bridge Decks due to Loads of Various Countries

Maximum Bending Moment Inclusive of Impact (kN.m)

Span (m)	U.K. Type HA Loads	U.K. Type HB Loads	India IRC Loads	Germany Loads	Japan Class-60 Loads	Sweden L-20	France	North America AASHTO HS:20-44 Loads	New Zealand
5	243	756	687	672	551	450	390	231	231
25	3156	7862	5680	4952	4083	5050	3290	2022	2485
50	8656	19029	12496	10830	11344	11300	8870	4597	5738
75	15117	30251	19683	19877	21682	17550	15848	9155	9155
100	22875	41487	33580	31268	33494	23800	24106	15184	15184

Table 2.3: Maximum Shear Forces for one Lane Simply Supported Spans of Bridge Decks Due to Loads of Various Countries

Maximum Shear Force Inclusive of Impact (kN)

Span (m)	U.K. Type HA Loads	U.K. Type HB Loads	India IRC Loads	Germany Loads	Japan Class-60 Loads	Sweden L-20	France	North America AASHTO HS:20-44 Loads	New Zealand
5	199	738	549	571	441	420	359	212	212
25	505	1451	978	807	653	820	569	369	447
50	693	1625	1090	872	908	910	710	410	501
75	806	1684	1477	1064	1156	940	845	529	606
100	915	1713	1776	1254	1340	955	964	647	758

2.6 INDIAN RAILWAY BRIDGE LOADING STANDARDS

Railway bridge loadings[8] should conform to the specifications of the Indian Railway Standards (IRS) prescribed by the Ministry of Railways, Government of India. The various loads to be used are specified in the IRS Bridge rules. Specific recommendations are available for the design of steel, RCC, PSC, masonry and plain concrete arch bridges in the relevant bridge codes.

Table 2.4: Maximum Bending Moments for Two Lane Simply Supported Spans of Bridge Decks Due to Loads of Various Countries

Maximum Bending Moment Inclusive of Impact (kN.m)

Span (m)	U.K. Type HA Loads	U.K. Type HB Loads	India IRC Loads	Germany Class-60 Loads	Japan L-20 Loads	Sweden	France	North America AASHTO HS:20-44 Loads	New Zealand
5	488	838	687	1224	827	640	780	462	462
25	6312	8914	5680	9904	6125	5819	6580	4044	4970
50	17132	21914	12496	21660	17016	15838	17740	9194	11476
75	30234	35290	23486	39754	32523	26250	31696	18310	18310
100	45750	49112	42880	62536	50241	37300	48212	30368	30368

Table 2.5: Maximum Shear Forces for Two Lane Simply Supported Spans of Bridge Decks Due to Loads of Various Countries

Maximum Shear Force Inclusive of Impact (kN)

Span (m)	U.K. Type HA Loads	U.K. Type HB Loads	India IRC Loads	Germany Class-60 Loads	Japan L-20 Loads	Sweden	France	North America AASHTO HS:20-44 Loads	New Zealand
5	398	804	594	1142	661	512	718	424	424
25	1010	1619	978	1614	980	931	1138	738	894
50	1386	1856	1174	1744	1361	1267	1420	820	1002
75	1612	1952	1572	2128	1735	1400	1690	1058	1212
100	1830	2018	1996	2508	2010	1492	1928	1294	1516

The railway tracks are classified according to the importance of traffic as main and branch lines. The three types of gauges used in the Indian Railways are:

1. Broad gauge (BG): 1676 mm (5'-6")
2. Metre Gauge (MG): 1000 (3'-3.375")
3. Narrow gauge (NG): 762 mm (2'- 6")

At present, the Indian Railways have adopted the unigauge policy with the broad gauge as the standard gauge throughout the country. Consequently many important old lines are being converted into broad gauge.

The various loads and forces to be considered in the design of bridge members are:

1. Dead and live loads
2. Dynamic effects
3. Cetrifugal force due to curvature of track
4. Temperature and frictional effects
5. Racking force
6. Wind and earthquake forces

Bridge Loading Standards

IRS Bridge Rules recommends the use of equivalent uniformly distributed loads (EUDL) on each track and also the coefficient of dynamic augment (CDA) for spans varying from 1 to 130 m for both BG and MG as shown in Tables 2.6 and 2.7.

The equivalent loads specified for the computation of bending moment and shear forces can directly be used in place of the various wheel loads of the rolling stock. Hence except in the case of special bridges like the Rigid Frame, Balanced cantilever and Suspension bridges, the designer can directly use the equivalent loads in place of the basic wheel loads. The Impact Factor (CDA) listed in the Tables are for single track spans of BG and MG based on the relation:

$$CDA = 0.15 + \frac{8}{(6+L)} < 1.0 \quad \text{where } L = \text{span}$$

For main girders of double track spans, the value specified above is multiplied by a reduction factor of 0.72.

Table 2.6: E.U.D.L., C.D.A. and Longitudinal Loads for Modified B.G. Loading

Span (m)	Total E.U.D.L For B.M (kN)	Total E.U.D.L. For S.F. (kN)	C.D.A. (I.F.)	Tractive Effort (kN)	Braking Force (kN)
1	490	490	1.000	81	62
2	490	519	1.000	164	123
3	490	662	1.000	245	184
4	596	778	0.950	245	184
5	741	888	0.877	245	184
6	838	985	0.817	245	185
7	911	1068	0.765	327	221
8	981	1154	0.721	409	276
9	1040	1265	0.683	409	276
10	1101	1377	0.650	490	331
12	1377	1589	0.594	490	331
15	1631	1801	0.531	490	368
20	1964	2168	0.458	735	496
25	2356	2586	0.408	735	565
30	2727	2997	0.372	981	662
40	3498	3815	0.324	981	816
50	4253	4630	0.293	981	978
60	5051	5442	0.271	981	1140
70	5831	6254	0.255	981	1301
80	6603	7065	0.243	981	1463
90	7391	7876	0.233	981	1625
100	8201	8686	0.225	981	1787
110	9011	9496	0.219	981	1949
120	9820	10306	0.213	981	2110
130	10630	11115	0.209	981	2272

Table 2.7: E.U.D.L., C.D.A. and Longitudinal Loads for Modified M.G. Loading

Span (m)	Total E.U.D.L. For B.M. (kN)	Total E.U.D.L. For S.F. (kN)	C.D.A. (I.F.)	Tractive Effort (kN)	Braking Force (kN)
1	314	314	1.000	89	57
2	314	365	1.000	118	78
3	326	452	1.000	118	118
4	429	536	0.950	157	118
5	501	616	0.877	157	124
6	581	685	0.817	157	124
7	644	755	0.765	176	135
8	714	819	0.721	209	157
9	774	871	0.683	262	169
10	828	934	0.650	262	198
12	953	1061	0.594	314	235
15	1138	1252	0.531	353	253
20	1421	1532	0.458	471	353
25	1677	1833	0.408	523	401
30	1991	2144	0.372	628	486
40	2589	2748	0.324	628	594
50	3099	3269	0.293	628	702
60	3625	3819	0.271	628	810
70	4178	4372	0.255	628	918
80	4727	4922	0.243	628	1026
90	5274	5470	0.233	628	1134
100	5822	6017	0.225	628	1242
110	6365	6562	0.219	628	1349
120	6908	7106	0.213	628	1457
130	7451	7649	0.209	628	1565

REFERENCES

1. ROWE, R.E., *Concrete Bridge Design* (CR Books, London, 1962), John Wiley and Sons, New York, 1963.
2. SENI, A., Comparison of Live Loads used in High Way Bridge Design in North America with those used in Western Europe, Second International Symposium on Concrete Bridge Design, Chicago, April 1969, American Concrete Institute, Detroit, 1971, pp. 1-34.
3. RAJAGOPALAN, K.S., Comparison of Loads around the World for Design of High Way Bridges, International Symposium on Concrete Bridge Design, Chicago, April 1969, American Concrete Institute, Detroit, 1971, pp. 35-48.
4. GALAMBOS, C.F., International Road Federation in Depth Study on Fatigue, Fracture and Stress Corrosion problems of High Way Bridges, World Survey of Current Research and Development on Roads and Road Transport, International Road Federation, Washington D.C., 1972, pp. 325-365.
5. IRC: 6- 2000, *Standard Specifications and Code of Practice for Road Bridges*. Section II, Loads and Stresses (Fourth Revision), Indian Roads Congress, New Delhi, 2000, pp. 1-61.
6. RAINA, V.K., *Concrete Bridge Practice, Analysis, Design and Economics*. Tata McGraw Hill Publishing Co. Ltd, New Delhi, 1991, pp. 9-25.
7. THOMAS, P.K., A Comparative Study of High Way Bridge Loadings in Different Countries, U.K. Transport and Road Research Laboratory, Supplementary Report, 135 UC, Crowthorne, 1975, p. 47.
8. Indian Railway Standard Code of Practice for the Design of Steel and Wrought Iron bridges carrying rail, road or pedestrian traffic. Government of India, Ministry of Railways, 1962, p. 87.

MATERIALS FOR PRESTRESSED CONCRETE BRIDGES

3.1 GRADES OF CONCRETE

Prestressed concrete used for bridge construction should have a compressive strength at a reasonably early age so that it can withstand pre compression imparted by high tensile tendons. Generally high strength concrete is mandatory for prestressed concrete work. Many of the early researchers who used ordinary concrete for prestressed concrete work noticed that sufficient prestress was not retained in concrete for effective use in structural elements mainly due to the loss of prestress and the low magnitude of prestress. Hence it is universally accepted that concrete of grades less than M-35 are not generally used for prestressed concrete work.

According to Indian Roads Congress specifications, IRC: 18-2000[1], the minimum prescribed characteristic compressive strength of concrete should be not less than 35 N/mm^2. The code also stipulates that for prestressed concrete construction, only "**Design Mix Concrete**" should be used. Also the concrete mix should be designed as per the Indian Standard Code IS: 10262-1982[2] which sets out the guide lines for concrete mix design. The minimum grade of concrete and the corresponding minimum cement content and maximum water/cement ratio for moderate and severe conditions of exposure for prestressed concrete bridges specified in IRC: 18-2000 are compiled in Table 3.1.

3.2 HIGH STRENGTH CONCRETE MIXES

The success of the revolutionary material 'Prestressed Concrete' is mainly attributed to the development of high strength concrete according to Freyssinet[3] who struggled for several years during 1930's to develop the new material. The development of vibration techniques for the production of high strength concrete and the invention of the double acting jack for stressing high tensile steel wires are considered to be the most significant contribution made by Freyssinet between 1928 and 1933.

Recent developments in the technology of cement production has resulted in high quality cements with superior chemical composition capable of producing high strength concrete in the range of 30 to 80 N/mm^2 without recourse to unusual materials or processing and without facing any significant technical difficulties. Experimental investigations by Erntroy and Shaclock[4] have indicated that in high strength concrete mixes, workability, type and maximum size of aggregate and the targeted strength of concrete influence the selection of water/cement ratio. It is a well-established fact that crushed rock aggregates being angular in shape generally produce stronger concretes at the same age in comparison with gravel aggregate.

High strength concrete mixes can be designed by using any of the following well established methods:

1. Erntroy and Shacklock's empirical method[4]
2. British D.O.E. method based on the work of Teychenne, Franklin and Erntroy[5] which has replaced the traditional Road Note No. 4[6] mix design procedure.
3. American Concrete Institute's mix design procedure for No Slump Concrete[7]
4. Indian Standard Code method[2].

Table 3.1: Minimum Grade of Concrete, Cement Content and Maximum Water/Cement Ratio for Post Tensioned P.S.C. Bridges (IRC: 21-2000)

Minimum Grade of Concrete	Type of Exposure	Minimum Cement Content for all Exposure conditions (kg/m³)	Maximum Water/Cement Ratio
M-35	Severe	400	0.40
M-40	Moderate	400	0.40

Notes:
1. *Conditions of Exposure*

    Severe: Marine environment, alternate wetting and drying due to sea spray, alternate wetting and drying combined with freezing, buried in soil (having corrosive effect), members in contact with water where the velocity of flow and the bed material are likely to cause erosion of concrete.

    Moderate: Conditions other than severe.

2. The minimum cement content is based on 20 mm size aggregates. For larger size aggregates, it may be reduced suitably by not more than 10%. Similarly for smaller aggregates, it may be suitably increased by not more than 10%.

The use of these various methods for designing high strength concrete mixes has been demonstrated through a number of examples in a separate monograph[8] by the author. The exact specifications with regard to the acceptance criteria for concrete generally vary from one code to the other. The British Code BS: 8110-1985[9] stipulates that not more than 5 per cent of the test results should fall below the 28 day characteristic compressive strength while the Indian Roads Congress specifications IRC: 21-2000 prescribes that the concrete is deemed to comply with the strength requirement if

(a) the mean strength determined from any group of four consecutive samples (average of three specimens) should exceed the specified characteristic compressive strength by 3 N/mm².

(b) strength of any sample is not less than the specified characteristic compressive strength minus 3 N/mm².

The durability of concrete is ensured by prescribing the minimum cement content and maximum water/cement ratio for different exposure conditions as shown in Table 3.1.

Materials for Prestressed Concrete Bridges

A comparative analysis of the Indian, British and American methods of high strength concrete mix design has been presented by Krishna Raju and Krishna Reddy[10] based on experimental investigations. M-45 grade concrete was designed and tested according to the standard procedure resulting in the following observations:

1. The British and American methods resulted in higher water/cement ratio compared to the Indian Standard Method. The water/cement ratio varied from 0.30 for I.S. to 0.44 for the British method.
2. The aggregate/cement ratio used in the Indian, American and British methods were 2.7, 3.4 and 5.3 respectively.
3. The Indian method resulted in the highest cement content of 591 kg/m^3 of concrete while the British method used the least cement content of 356 kg/m^3.
4. The A.C.I. method of mix design resulted in concrete of compressive strength very nearly equal to the specified characteristic compressive strength with the most economical cement content in comparison with other methods.

In these investigations, the cement used was of the grade C-53, but now the market is flooded with better quality and high grade cements like Birla Super, L&T, A.C.C., Coramendel, Ultra Tech, Gujarat Ambuja and various other brands of cement. Most of these cements more than satisfy the various Indian Standard Code specifications[11, 12] and hence the design of concrete mixes of characteristic strength in the range of M-40 to M-60 does not pose any serious problems. Using good quality aggregates and modern cements, even the nominal mix proportions of 1 : 1.5 : 3 or 1 : 1 : 2 with controlled water/cement ratios in the range of 0.35 to 0.4 and with proper compaction using vibrators can result in concrete having characteristic compressive strength exceeding M-40.

The use of light weight aggregate for prestressed concrete construction is well established since 1955 with the main advantage of reduction in the self weight of the structure. The light weight criterion becomes important especially in long span structures where dead load forms the major portion of the total design load on the structure or when the self weight of the member is a factor to be considered in transportation and erection as in the case of precast concrete construction. Teychenne[13] has developed empirical graphs that relates the important parameters of light weight concrete. ACI Standard[14] provides a generally applicable method for selecting mix proportions of light weight concrete using different types of light weight aggregates. An excellent survey of the effective utilization of light weight concrete in prestressed concrete structures is reported by Gerwick[15].

The modulus of elasticity of concrete generally used for computations of short term and long term deflections of post tensioned prestressed concrete beams as recommended in IRC: 18-2000 is expressed as,

$$E_c = 5000\sqrt{f_{ck}} \text{ N/mm}^2$$

where f_{ck} = characteristic compressive strength of concrete expressed in N/mm^2. The code also recommends varying creep strain in concrete depending upon the maturity of concrete at the time of stressing as a percentage of the characteristic compressive strength of concrete. The loss of stress in steel due to shrinkage of concrete is to be estimated from the values of strain due to residual shrinkage which varies from 4.3×10^{-4} for concrete at the age of 3 days to a value of 1.5×10^{-4} for concrete at the age of 90 days.

3.3 HIGH TENSILE STEEL

(a) **Types of High Tensile Steel:** Concrete is precompressed using high tensile steel available in the form of wires, bars and strands. The higher tensile strength is generally achieved by marginally increasing the carbon content in the steel in comparison with mild steel. High tensile steel usually contains 0.6 to 0.85 per cent carbon, 0.7 to 1 per cent manganese, 0.05 per cent sulphur and phosphorus with traces of silica. The high carbon steel ingots are hot rolled into rods and cold drawn through a series of dies to reduce the diameter and increase the tensile strength.

The cold drawing process decreases the durability of the wires. Hence they are subsequently tempered to improve their properties. Tempering or ageing or stress relieving by heat treatment of the wires at 150 to 420° C enhances the tensile strength. The cold drawn stress relieved wires are generally available in nominal sizes of 2.5, 3, 4, 5, 7 and 8 mm diameter and their properties should conform to the Indian Standard Specification IS: 1785-1983 (Part-I) compiled in Table 3.2[16].

Table 3.2: Tensile Strength and Elongation Characteristics of Cold Drawn Stress Relieved Wires (IS: 1785-1983 Part-I)

Nominal Diameter (mm)	Tensile Strength (Minimum) N/mm²	Elongation (Per cent)
2.50	2010	2.5
3.00	1865	2.5
4.00	1715	3.0
5.00	1570	4.0
7.00	1470	4.0
8.00	1375	4.0

The bond characteristics of the hard drawn steel wires can be improved by indentations and crimping process. The deformed wires are preferred for pre tensioned elements due to their superior bond characteristics. The specifications for indented wires covered in IS: 6003-1983[17] are compiled in Table 3.3.

Table 3.3: Mechanical Properties of High Tensile Indented Wires (IS: 6003-1983)

Nominal Diameter (mm)	Tensile Strength (Minimum) N/mm²	Elongation (Per cent)
5.00	1570	4.00
4.00	1715	3.00
3.00	1865	2.5

High tensile wires of small diameter ranging from 2 to 5 mm are mostly used in the form of strands comprising two, three or seven wires. The helical form of twisted wires in the strand substantially improves the bond characteristics of the tendons. Two or three ply strands are made up of 2 mm and 3 mm diameter individual wires, while the 7 ply strands are twisted using wires of 2 to 5 mm diameter. The nominal diameter of 7 ply strands vary from 6.3 mm to 15.2 mm. The mechanical properties of strands covered in the Indian Standard, IS: 6006-1983 is shown in Table 3.4.

Materials for Prestressed Concrete Bridges

Table 3.4: Mechanical Properties of Prestressing Strands (IS: 6006-1983)

Class	Nominal Diameter (7-ply) (mm)	Area A_p (mm²)	Weight (kg/m)	Ultimate Tensile Strength (f_p)(N/mm²)	$0.8 f_p A_p$ (kN)	$f_p A_p$ (kN)
1	6.3	23.2	0.182	1723	31.97	39.97
	7.9	37.4	0.294	1723	51.55	64.40
	9.5	51.6	0.405	1723	71.12	88.90
	11.1	69.7	0.548	1723	96.07	120.09
	12.7	92.9	0.730	1723	128.05	160.06
	15.2	139.4	1.094	1723	192.14	240.18
2	9.5	54.8	0.432	1862	81.63	102.03
	11.1	74.2	0.582	1862	110.52	138.16
	12.7	98.7	0.775	1862	147.02	183.77
	15.2	140.0	1.102	1862	208.54	260.88

The high tensile steel bars commonly employed in prestressing are manufactured in nominal sizes of 10, 12, 16, 20, 22, 25, 28, and 32 mm diameter. Larger diameter bars with threaded ends are used as straight tendons in conjunction with nuts serving as anchorages. The ultimate tensile strength of the bars does not vary appreciably with the diameter, since the high strength of the bars is due to alloying rather than the cold working as in the case of wires. The mechanical properties of high tensile steel bars covered in the Indian Standard Code specification IS: 2090-1983[19] is shown in Table 3.5.

Table 3.5: Mechanical Properties of High Tensile Steel Bars (IS: 2090-1983)

Nominal Diameter (mm)	Area (A_p) (mm²)	Weight (kg/m)	Ultimate Tensile Strength (N/mm²)	$0.8 f_p A_p$ (kN)	$f_p A_p$ (kN)
10	79	0.62	980	61.93	77.42
12	113	0.89	980	88.59	110.74
16	201	1.58	980	157.58	196.98
20	314	2.47	980	246.17	307.72
22	380	2.98	980	297.92	372.40
25	491	3.85	980	384.94	481.18
28	616	4.83	980	482.94	603.68
32	804	6.31	980	630.33	787.92

The IRC: 18-2000 also permits the use of uncoated stress relieved low relaxation steel conforming to the Indian Roads Congress Code IRC: 14268-1995[20] in post tensioned prestressed concrete road bridges. The British Standard Codes BS: 2691, BS: 4486 and BS: 3671 present specifications for the use of high tensile wires, bars and strands respectively. The British standards provide for the use of 19 wire strands in addition to the earlier 7 wire strands[21].

(b) Tensile Strength and Deformation Characteristics of High Tensile Steel: The characteristic tensile strength of plain hard drawn steel wire varies with the diameter. The tensile strength decreases with increase in the diameter of the wires. In contrast to mild steel, high tensile steel wires do not exhibit any well defined yield point and it is necessary to refer to the 0.2 per cent proof stress corresponding to the specified permanent strain. It is prescribed that the 0.2 per cent proof stress for high tensile steel wires and bars should be not less than 85 and 80 per cent respectively of the minimum specified tensile strength. Extensibility of the tendons near the ultimate stress is also an important requirement to facilitate progressive failure of the prestressed concrete structural element. To prevent the possibility of brittle fracture, the Indian Standard Code prescribes a minimum percentage elongation varying from 2.5% for wires to 10% for bars. Prestressed concrete bridge decks requiring large prestressing forces are prestressed using strands. The mechanical properties of uncoated stress relieved strand specified in IS: 6006-1983 is compiled in Table 3.4.

Relaxation of stress in steel is required for the computation of losses in prestress. The IRC: 18-2000 code recommends the relaxation loss as a percentage of the initial stress which varies from 0.5 to 0.8 times the characteristic tensile strength of the steel. The relaxation loss varies from zero to 9% corresponding to the initial stress varying from $0.5f_p$ to $0.8f_p$.

The loss of stress due to friction between the tendons and the duct or sheath depends upon the steel stress at jacking end, the length and curvature of the cable. Different values are recommended in the code for the values of the coefficient of friction and the wobble coefficient per unit length depending upon the type of duct or sheath and the high tensile steel which may be in the form of wires or strands. For design purposes, the nominal value of modulus of elasticity of steel recommended is compiled in Table 3.6.

Table 3.6: Modulus of Elasticity of Steel (E_s) (IRC: 18-2000)

Sl.No.	Type of Steel	Modulus of Elasticity (N/mm²)
1	Plain hard drawn wires (conforming to IS: 1785 & IS: 6003)	2.1×10^5
2	High tensile steel bars rolled or heat treated (conforming to IS: 2090)	2.0×10^5
3	High tensile strands (conforming to IS: 6006)	1.95×10^5

3.4. UNTENSIONED STEEL OR SUPPLEMENTARY REINFORCEMENT

Reinforcement used as untensioned steel in the form of longitudinal bars and stirrups comprise the following:

1. Mild steel and medium tensile steel bars conforming to IS: 432 (Part-I)[22]
2. High strength deformed steel bars conforming to IS: 1786[23]
3. Hard drawn steel wire fabric conforming to IS: 1566[24]

Supplementary reinforcements are required in prestressed concrete beams and slabs to safeguard against shrinkage cracks and for resisting shear forces. Normally stirrups and hanger bars are made up of untensioned reinforcement. They are also provided as anchorage zone reinforcement at the end blocks of post tensioned prestressed concrete beams.

The minimum quantity of reinforcement prescribed in the vertical direction in beams and box girders should be not less than 0.3 per cent of the cross-sectional area of the rib/web in plan for mild steel and 0.18 per cent for HYSD bars respectively. The reinforcements should be uniformly spaced along the length of the web. The longitudinal reinforcements provided should be not less than 0.25 and 0.15 per cent of the gross cross-sectional area of the section for mild steel and HYSD bars respectively where the specified grade of concrete is less than M-45. For M-45 grade or more, the reinforcement percentage is increased to 0.3 per cent and 0.18 per cent respectively. For cantilever slabs, the minimum reinforcement comprising 4 HYSD bars of 16 mm diameter or 6 mild steel bars of 16 mm diameter should be provided parallel to the free edge at 150 mm spacing at the tip divided equally between the top and bottom surfaces.

The maximum size of reinforcement is restricted to 40 mm and the minimum diameter of any type of secondary reinforcement should be not less than 8 mm and the longitudinal reinforcing bars in columns should be not less than 12 mm. The diameter of reinforcements in slabs should be limited to one-tenth the depth of the slab and the diameter of shear reinforcement in beam-webs including cranked bars should be limited to one-eighth the thickness of the web. The horizontal distance between two parallel reinforcing bars should be not less than the greatest of the following three dimensions:

1. The diameter of the bar if the diameters are equal
2. The diameter of the largest bar if the diameters are unequal and
3. 10 mm more than the nominal size of the coarse aggregate used in concrete.

3.5. PERMISSIBLE STRESSES IN CONCRETE

The maximum permissible stresses in high strength concrete at the stages of transfer and service loads are compiled in Table 3.7. However concrete should have attained a minimum compressive strength of 20 N/mm² before any prestress is applied.

Table 3.7: Permissible Stresses in Concrete (IRC: 18-2000)

Loading Stage	Compressive Stresses	Tensile Stresses
At Transfer	Not to exceed 0.5 f_{cj} which Shall not be more than 20 N/mm², where f_{cj} = concrete strength at the stage of loading subject to a maximum of characteristic compressive strength (f_{ck})	Not to exceed one-tenth of the permissible compressive stress
At Service loads	Not to exceed 0.33 f_{ck}	No tensile stresses permitted

In the case of precast segmental elements joined by prestressing, the stresses in the extreme fibres of concrete during service loads should be compressive and the minimum compressive stress in the extreme fibres should be not less than 5 per cent of the maximum permanent stress that may be developed in the same section.

The properties of concrete and the basic permissible stresses in concrete of different grades prescribed in IRC: 21-2000 is compiled in Table 3.8.

Table 3.8: Properties and Basic Permissible Stresses in Concrete (IRC: 21-2000)

Properties/Permissible Stresses	M 15	M 20	M 25	M 30	M 35	M 40	M 45	M 50	M 55	M 60
1. Modulus of Elasticity (GPa)	26	27.5	29	30.5	31.5	32.5	33.5	35	36	37
2. Permissible Direct Compressive Stress (σ_{co}) (N/mm^2)	3.75	5.00	6.25	7.50	8.75	10.0	11.25	12.5	13.75	15
3. Permissible Flexural Compressive Stress (σ_{cb}) (N/mm^2)	5.00	6.67	8.33	10.0	11.67	13.33	15.0	16.67	18.3	20

The maximum permissible stress immediately behind the anchorages in adequately reinforced end blocks according to IS: 1343 Code may be computed by the equation:

$$f_b = 0.48 f_{ci} \sqrt{\frac{A_{br}}{A_{pun}}} \text{ or } 0.8 f_{ci} \text{ whichever is smaller}$$

where f_b = the permissible unit bearing stress
 A_{br} = the bearing area
 A_{pun} = punching area

However, the bearing stress is permissible only subject to the following conditions:

1. The projection of concrete beyond the bearing area of the anchorage should be at least 50 mm or one-fourth of the depth of the bearing area whichever is more all round the anchorage.

2. Where embedded anchorages are provided, the reinforcement details, concrete strength, cover and other dimensions shall conform to the manufacturer's specifications/specialist literature.

3. The pressure operating on the anchorage shall be taken before allowing for losses due to creep and shrinkage of concrete and relaxation of steel but after allowing for losses due to elastic shortening and seating of anchorage.

3.6 PERMISSIBLE STRESSES IN STEEL

The IRC: 18-2000 code prescribes that the maximum jack pressure should not exceed 90% of the 0.1 per cent proof stress of the prestressing steel. In addition it stipulates that 0.1 per cent proof stress should be taken as equal to 85% of the minimum ultimate tensile strength.

The permissible stress in untensioned reinforcement comprising mild steel and HYSD bars as prescribed in IRC: 21-2000 are compiled in Table 3.9.

Materials for Prestressed Concrete Bridges

Table 3.9: Permissible Stresses in Reinforcing Bars (IRC: 21-2000)

Bar Grade	Type of Stress in Steel Reinforcement	Permissible Stress (N/mm^2)
Fe-240	Tension in Flexure, Shear or Combined bending	125
Fe-415		200
Fe-500	Tension in Flexure or Combined bending	240
	Tension in Shear	200
Fe-240		115
Fe-415	Direct Compression	170
Fe-500		205
Fe-240		95
Fe-415	Tension in Helical	95
Fe-500	reinforcement	95

3.7. ANCHORAGES AND SHEATHING DUCTS

(a) Anchorages: In the case of post tensioned prestressed concrete members, proprietary anchorages are used to stress the tendons and anchor them against concrete using anchoring devices. The various types of post tensioning anchorages currently in use are explained in detail by the author in a separate monograph[24]. The most widely used anchorages are the Freyssinet, BBRV, Le-McCall, Dywidag, Gifford-Udall, Baur-Leonhardt systems. Raina[25] has presented the various types of anchorages along with an exhaustive data and has critically analyzed the suitability of these anchorages in a given particular situation.

(b) Sheathing Ducts: In post-tensioned members, tendons are threaded in sheathing ducts or cables which are positioned to conform to the predesigned profile before concreting the member. The sheathing ducts are generally made of either mild steel or high density polyethylene (HDPE). The mild steel sheathing ducts are made of cold rolled and cold annealed (CRCA) mild steel sheet with a thickness not less than 0.3 mm, 0.4 mm and 0.5 mm for sheathing ducts having internal diameters up to 50 mm, 75 mm and 90 mm respectively. For larger diameter ducts, thickness of sheathing should be based on the recommendations of the manufacturers and suppliers.

Corrugated HDPE sheathing ducts are made from high density polyethylene, with more than 2 per cent carbon black to provide resistance to ultraviolet degradation. The thickness of the wall should be not less than 2.3 ± 0.3 mm as manufactured and 1.5 mm after loss in the compression test for duct size up to 160 mm. The ducts should be corrugated on both sides and transmit the full tendon force to the surrounding concrete over a length not greater than 40 times the duct diameter.

The ducts can be joined by using any one or more of the following methods:

1. Screwed together with male and female threads
2. Joining with thick walled HDPE shrink couplers with glue
3. Welding with electro fusion couplers.

The sheathing ducts should conform to the requirements specified in Appendix I A of IRC: 18-2000. The sheathing ducts should also conform to the various tests such as

1. Workability test
2. Transverse load rating test
3. Tension load test
4. Water loss test
5. Bond test
6. Compression test.

The details of these various tests are outlined in Appendix I A and I B of IRC: 18-2000.

REFERENCES

1. IRC: 18-2000, Design Criteria for Prestressed Concrete Road Bridges (Post-Tensioned Concrete), 2nd Revision. The Indian Roads Congress, New Delhi 1985, pp. 1-44.
2. IS: 10262-1982, Indian Standard Guide Lines for Concrete Mix Design, Indian Standards Institution, New Delhi, 1983, pp. 1-21.
3. Freyssinet, E, The Birth of Prestressing, Cement and Concrete Association. *Translation No.* CJ.59, London, 1956, p. 44.
4. Erntroy, H.C., and B.W. Shacklock., Design of High Strength Concrete Mixes, Proceedings of a Symposium on Mix design and Quality control of Concrete, Cement and Concrete Association, London, May 1954, pp. 55-65.
5. Teychenne, D.C., R.E. Franklin and H.C. Erntroy, Design of Normal Concrete Mixes, Department of Environment, London, H.M.S.O, 1975, pp. 11-31.
6. Road Research Laboratory, Design of Concrete Mixes, *D.S.I.R. Road Note No.*4, London, H.M.S.O, 1950, pp. 1-14.
7. ACI Committee-211, Recommended Practice for selecting proportions for No-Slump Concrete (ACI 211.3-75), *Journal of the American Concrete Institute*, Vol. 71, No. 4, 1974, pp. 153-170.
8. Krishna Raju, N., *Design of Concrete Mixes*, Fourth Edition, C.B.S. Publishers and Distributors, New Delhi, 2002, pp. 1-316.
9. BS: 8110-1985, British Standard Code of Practice for the Structural Use of Concrete (Part 1&2), British Standards Institution, London,1985, pp. 2-3 to 2-7.
10. Krishna Raju, N and Krishna Reddy, Y., A Critical Review of the Indian, British and American Methods of Concrete Mix Design. *The Indian Concrete Journal*, Vol.63, No. 4, April 1989, pp. 196-201.
11. IS: 269-1967, Indian Standard Specifications for Ordinary, Rapid Hardening and Low Heat Portland Cement, Indian Standards Institution, New Delhi.
12. IS: 516-1959, Indian Standard Code of Practice, Methods of Test for Strength of Concrete, Indian Standards Institution, New Delhi, 1959.
13. Teychenne, D.C., Structural Concrete made with Light Weight Aggregates. *Concrete*, Vol.1, No. 4, April 1967, pp. 111-112.
14. ACI Standard Recommended Practice for Selecting Proportions of Structural Light Weight Concrete, ACI: 211.2-69, *ACI Manual of Concrete Practice*, Detroit, Michigan, Part-I, 1970, pp. 211-15 to 32.
15. Gerwick, B.C., *Construction of Prestressed Concrete Structures*. Wiley Inter Science, New York, 1971.
16. IS: 1785 (Part-I)-1983, Indian Standard Specifications for Plain Hard Drawn Steel Wire for Prestressed Concrete, Part-I, Cold drawn stress relieved wires (II revision), I.S.I., New Delhi, 1983.
17. IS: 6003-1983, Indian Standard Specifications for Indented Stress Relieved Wire for Prestressed Concrete (First Revision), Indian Standards Institution, New Delhi (First Reprint), April 1983.
18. IS: 6006-1983, Indian Standard Specifications for Uncoated Stress Relieved Strand for Prestressed Concrete (First Revision), Indian Standards Institution, New Delhi, October 1989.

19. IS: 2090-1983, Indian Standard Specifications for High Tensile Steel Bars used In Prestressed Concrete (First revision), Indian Standards Institution, New Delhi, (First reprint), October 1988.
20. IS: 14268-1995, Indian Standard Specifications for Uncoated Stress Relieved Low Relaxation Seven Ply Strand for Prestressed Concrete, Bureau of Indian Standards New Delhi, 1995.
21. Bate, S.C.C. and E.W. Bennett, *Design of Prestressed Concrete*. Surrey University Press, International Text Book Company Limited, London, 1976, pp. 13-16.
22. IS; 432-1966, Indian Standard Specifications for Mild Steel and Medium Tensile Steel Bars and Hard Drawn Steel Wire for Concrete Reinforcement, Indian Standards Institution, New Delhi, 1966, Part-1, p. 12.
23. IS: 1786-1985, Indian Standard Specifications for High Strength Deformed Steel Bars and Wires for Concrete Reinforcement (Third Revision), Bureau of Indian Standards, New Delhi, 1985.
24. IS: 1566-1982, Indian Standard Specifications for Hard Drawn Steel Wire Fabric for Concrete Reinforcement (Second Revision), Bureau of Indian Standards, New Delhi, 1982.
25. Krishna Raju, N., *Prestressed Concrete*, (Fourth Edition). Tata McGraw Hill Publising Co, Ltd, New Delhi, 2007, pp. 79-98.
26. Raina, V.K., *Concrete Bridge Practice, Analysis, Design and Economics*. Tata McGraw Hill Publishing Co. Ltd., New Delhi, 1991, pp. 41-59.

4
Limit State Design of Reinforced and Prestressed Concrete Bridge Deck Sections

4.1 DESIGN PHILOSOPHY OF REINFORCED AND PRESTRESSED CONCRETE STRUCTURES

The elastic method of design based on permissible stresses in concrete and steel, pioneered by Thaddeus Hyatt, Roennen and Coignet[1] during the last part of the 19th century formed the basis of design for most of the structural concrete members up to 1950. The inadequacy of the working load or permissible stress or elastic method of design in predicting the ultimate strength[2] of structures paved the way for ultimate load or the load factor method of design[3]. However the inadequacies of both the elastic and ultimate load methods of design were realized, paving the way for a new philosophy of design termed the '**Limit State Approach**'[4, 5] which was incorporated in the Russian Code as early as 1954[6].

Basically, the limit state design[7] is a method of designing structures based on statistical concept of safety and the associated statistical probability of failure. The limit state design philosophy recognizes the need to provide structures which are serviceable at working loads and have the desired load factor against collapse. The reader may refer other publications[8, 9] for an exhaustive treatment of the evolution of design methods for structural concrete members[10].

4.2 ELASTIC DESIGN COEFFICIENTS FOR REINFORCED CONCRETE SECTIONS

In the elastic design of slabs in Tee beam bridges, while the beam is prestressed, the slab between the girders is normally of reinforced concrete designed as per the permissible stresses specified in IRC: 21-2000 (Refer Table 3.8). The grade of concrete used can vary from M-20 to M-40 and the reinforcement generally used being HYSD bars.

Based on these permissible stresses, the design constants to be used for computation of effective depth 'd' of the structural element and the area of steel 'A_{st}' in the tension zone along with the neutral axis depth factor 'n', lever arm factor 'j' and the moment factor 'Q' expressed as a function of the permissible compressive stress 'σ_{cb}' in concrete as given by the following expressions are compiled in Table 4.1.

$$n = \left[\cfrac{1}{1+\left(\cfrac{\sigma_{st}}{m\sigma_{cb}}\right)}\right]$$

$$j = \left[1 - \frac{n}{3}\right]$$

$$Q = 0.5\sigma_{cb}nj$$

Table 4.1: Elastic Design Coefficients

Grade of Concrete and Steel	m	σ_{cb} (N/mm^2)	σ_{st} (N/mm^2)	n	j	Q
M-20 Fe-415	10	6.7	200	0.25	0.91	0.762
M-25 Fe-415	10	8.3	200	0.29	0.90	1.100
M-30 Fe-415	10	10.0	200	0.333	0.889	1.480
M-35 & 40 Fe-415	10	11.5	200	0.365	0.879	1.844

4.3 FLEXURAL AND SHEAR STRENGTH OF REINFORCED CONCRETE SECTIONS

The basis of design specified in Clause 304.2.1 of IRC: 21-2000 regarding reinforced concrete road bridges confines to the principles of elastic theory and no recommendations are specified for computation of the ultimate strength of reinforced concrete sections in flexure and shear. However the design equations recommended for shear stress computations in this code are the same as that specified in IS: 456-2000.[12, 13]

The design shear stress 'τ_v' at any cross-section of the beam or slab of uniform depth is calculated by the equation,

$$\tau_v = \left(\frac{V}{bd}\right)$$

where V = the design shear force across the section
b = breadth of the member which for flanged sections should be taken as the breadth of the web, and
d = effective depth of the section

The code also specifies that for obtaining maximum shear stress, the section at a distance equal to the effective depth from the face of the support should be considered in the computations.

When the shear stress 'τ_v' exceeds the permissible shear stress 'τ_c' in beams compiled in Table 4.2, shear reinforcements in the form of vertical links, inclined stirrups or bent up bars should be designed.

When shear reinforcements are provided, the shear stress in beams should not exceed the value of shear stress '$\tau_{c,\,max}$' shown in Table 4.3. In the case of solid slabs, the permissible design shear stress in concrete is computed as '$K\tau_c$' where 'K' is a factor having the values shown in Table 4.4. In the case of slabs, the value of shear stress should not exceed half the value of '$\tau_{c,\,max}$' shown in Table 4.3.

Table 4.2: Permissible Shear Stress in Concrete (IRC: 21-2000)

$\left(\dfrac{100A_s}{bd}\right)$	Permissible Shear Stress in Concrete τ_c N/mm² Grade of Concrete				
	M-20	M-25	M-30	M-35	M-40 & above
(1)	(2)	(3)	(4)	(5)	(6)
0.15	0.18	0.19	0.20	0.20	0.20
0.25	0.22	0.23	0.23	0.23	0.23
0.50	0.30	0.31	0.31	0.31	0.32
0.75	0.35	0.36	0.37	0.37	0.38
1.00	0.39	0.40	0.41	0.42	0.42
1.25	0.42	0.44	0.45	0.45	0.46
1.50	0.45	0.46	0.48	0.49	0.49
1.75	0.47	0.49	0.50	0.52	0.52
2.00	0.49	0.51	0.53	0.54	0.55
2.25	0.51	0.53	0.55	0.56	0.57
2.50	0.51	0.55	0.57	0.58	0.60
2.75	0.51	0.56	0.58	0.60	0.62
3 & above	0.51	0.57	0.60	0.62	0.63

Table 4.3: Maximum Shear Stress ($\tau_{c,\,max}$) in Concrete (N/mm²)

Concrete Grade	M-20	M-25	M-30	M-35	M-40 & above
$\tau_{c,\,max}$ (N/mm²)	1.8	1.9	2.2	2.3	2.5

Table 4.4: Values of K for Solid Slabs

Overall Depth of Slab (mm)	300 or more	275	250	225	200	175	150 or less
K	1.00	1.05	1.10	1.15	1.20	1.25	1.30

Stirrups are not generally used in slabs due to practical difficulties and shear stresses are limited to be within permissible limits either by increasing the depth of the slab or the longitudinal reinforcement or both, so as to avoid the use of vertical link reinforcements.

Control of cracking in slabs is ensured by restricting the spacing of main bars to 150 mm or less and the diameter of the bars not to exceed 25 mm. Also the diameter of the reinforcements used in slabs should not exceed one-tenth of the depth of the slab.

4.4 DESIGN OF PRESTRESSED CONCRETE SECTIONS FOR SERVICE LOADS

Prestressed concrete sections subjected to flexural moments should satisfy the limits specified for permissible stresses at the stage of transfer of prestress at service loads. Expressions for the minimum section modulus required, prestressing force and the corresponding eccentricity derived from the first principles in prestressed concrete text books[14, 15] have been reproduced here using the standard notations recommended in the Indian Standard Code IS: 1343.[16]

Minimum section modulus is expressed as,

$$Z_b \geq \left[\dfrac{(M_g + M_q) - \eta M_g}{f_{br}}\right] \qquad \ldots(4.1)$$

where M_g = Dead load bending moment
M_q = Live load bending moment
Z_b = Section modulus of bottom fibre of structural element
f_{br} = Range of stress at bottom fibre = $(\eta f_{ct} - f_{tw})$
f_{ct} = Permissible compressive stress in concrete at transfer of prestress
f_{tw} = Permissible tensile stress in concrete under service loads

The equation for minimum prestressing force is expressed as

$$P = \left[\frac{A(f_t Z_t + f_b Z_b)}{Z_t + Z_b}\right] \quad \ldots(4.2)$$

where P = Minimum prestressing force
A = Cross-sectional area of concrete section
Z_t & Z_b = Section modulus of top and bottom fibres of concrete section

Also
$$f_t = \left[f_{tt} - \frac{M_g}{Z_t}\right] \quad \ldots(4.3)$$

$$f_b = \left[\frac{f_{tw}}{\eta} + \frac{(M_g + M_q)}{\eta Z_b}\right] \quad \ldots(4.4)$$

The eccentricity corresponding to the minimum prestressing force is expressed as

$$e = \left[\frac{Z_t Z_b (f_b - f_t)}{A(f_t Z_t + f_b Z_b)}\right] \quad \ldots(4.5)$$

where f_t = Prestress in concrete at top of section
f_b = Prestress in concrete at bottom of section
e = Corresponding eccentricity of prestressing force

In the case of slabs and beams where the live load moment (M_q) is equal to or larger in magnitude than the dead load moment (M_g), the theoretical eccentricity computed using Eq. (4.5) lies within the selected concrete section. However, in cases where the dead load moment is considerably larger than the applied or live load moment (which is normal in the case of long span members), the theoretical eccentricity (e) computed from Eq. (4.5) lies below the soffit of the section, i.e. outside the section and consequently it is impracticable to house the prestressing cables outside the section.

Hence, in such cases, the theoretical eccentricity is reduced so that the prestressing force lies within the section at the lowest practicable position. Consequently, the magnitude of the prestressing force should be increased so that the prestress at the soffit is unaltered. There will be a reduction in the negative prestress at top at the transfer stage and an increase in the compressive stress at the top fibre under service loads.

In the case of long span bridge girders, the self weight moment is very large and the minimum stress at transfer developed at the top fibre will be compressive in nature due to the limitation of position of prestressing force within the cross-section. The required prestressing force P acting at known eccentricity e, which can develop the required prestress f_b at the bottom fibre is computed using the equation,

$$P = \left[\frac{Af_b Z_b}{Z_b + Ae}\right] \qquad ...(4.6)$$

where A and Z_b are the cross-sectional area and the section modulus of the cross-section actually provided.

The prestress along the length of the beam is generally adjusted by varying the eccentricity of the prestressing force. This practice is employed in post tensioned beams by using curved cables.

In the case of prismatic members, the permissible tendon zone is controlled by the following equations.

$$e \leq \left[\frac{Z_b f_{ct}}{P} - \frac{Z_b}{A} + \frac{M_g}{P}\right] \qquad ...(4.7)$$

$$e \geq \left[\frac{Z_b f_{tw}}{\eta P} - \frac{Z_b}{A} + \frac{M_g + M_q}{\eta P}\right] \qquad ...(4.8)$$

In long span bridge girders, there will be number of cables and at each section, the profile of the cables should be adjusted in such a way that the centroid of all the cables lies within the permissible tendon zone defined by the equations 4.7 and 4.8.

4.5 FLEXURAL AND SHEAR STRENGTH OF PRESTRESSED CONCRETE SECTIONS

The critical sections of a bridge deck designed for service loads should satisfy the limit state of collapse both in flexure and shear. According to IRC: 18-2000, prestressed concrete bridge sections should be checked for failure conditions. The ultimate load of the section under different exposure conditions should satisfy the following criteria:

$U = (1.25\ G + 2.5\ SG + 2.5\ Q)$ under moderate exposure conditions

$U = (1.5\ G + 2\ SG + 2.5\ Q)$ under severe exposure conditions

where
U = ultimate load
G = permanent dead load
SG = super imposed dead load
Q = live load including impact

The super imposed dead load includes the dead load of foot path, hand rails, wearing course, utility services, kerbs, etc.

(a) **Flexural Strength:** The code recommends separate equations for computation of the ultimate flexural strength of sections failing by
1. Yielding of steel (under reinforced sections)
2. Crushing of concrete (over reinforced sections)

The ultimate moment of resistance of sections failing by yielding of steel is expressed as,

$M_{us} = (0.9\ d\ A_p\ f_p)$

where
d = effective depth
A_p = area of high tensile steel
f_p = ultimate tensile strength for steel without definite yield point or yield stress or stress at 4 per cent elongation whichever is higher for steel with a definite yield point.

Any supplementary untensioned reinforcement used is considered as contributing to the steel section and the effective area is calculated $(A_s f_y / f_p)$

where A_s = area of untensioned reinforcement

f_y = yield stress

The ultimate moment of resistance of sections failing by crushing of concrete is calculated by the equation,

$$M_{uc} = (0.176 \, b d^2 f_{ck}) \text{ for rectangular sections and}$$

$$M_{uc} = [0.176 \, b d^2 f_{ck} + (2/3) 0.8 (B_f - b)(d - 0.5 \, t) \, t f_{ck}] \text{ for flanged sections}$$

where b = width of rectangular section or web of a flanged section

B_f = effective width of flange

t = thickness of flange

(b) Shear Strength: The Indian Roads Congress Code IRC: 18-2000 specifies separate equations for computing the shear strength of sections uncracked in flexure (support sections) and sections cracked in flexure (span section). The empirical equation recommended for sections uncracked in flexure is the same as that specified in IS: 1343 code.

(i) Sections Uncracked in Flexure: The ultimate shear strength of sections uncracked in flexure (normally support sections) is governed by the occurrence of maximum principal tensile stress at the centroidal axis of the section leading to diagonal tension cracks and is expressed as a function of the characteristic compressive strength of concrete by the relation $f_t = 0.24 \sqrt{f_{ck}}$.

The ultimate shear strength V_{co} is given by the relation:

$$V_{co} = 0.67 \, bh \sqrt{f_t^2 + 0.8 f_{cp} f_t} + \eta P \sin \theta$$

where b = width of rectangular section or width of rib in flanged section

h = overall depth of the section

f_t = tensile strength of concrete corresponding to maximum principal tensile stress

f_{ck} = characteristic compressive strength of concrete

f_{cp} = compressive stress at centroidal axis due to prestress

η = loss ratio

P = prestressing force

θ = inclination of the cable to the horizontal axis

The support sections free from flexure are generally checked for shear strength using this expression and suitably designed to resist the shear forces.

(ii) Sections Cracked in Flexure: The ultimate shear strength of sections cracked in flexure (V_{cf}) is calculated using the empirical relation:

$$V_{cf} = 0.037 \, bd \sqrt{f_{ck}} + \left(\frac{M_o}{M} \right) V$$

where d = distance from the extreme compression face to the centroid of the tendons
M_o = cracking moment at the section considered and is calculated as

$$\left(0.37\sqrt{f_{ck}} + 0.8 f_{pt}\right)\frac{I}{y}$$

f_{pt} = stress in concrete due to prestress only at the extreme tensile fibre

I = second moment of area of the section

y = distance of the extreme tensile fibre from the centroid of the concrete section

V and M are the shear force and corresponding bending moment at the section considered due to the ultimate loads.

In the case of inclined tendons at the section cracked in flexure, the vertical component of the prestressing force is ignored as per the specifications of IRC: 18-2000 clause 14.1.3.

4.6 TORSIONAL RESISTANCE OF PRESTRESSED CONCRETE SECTIONS

Prestressed concrete spine beams of continuous bridge decks used in flyovers are commonly subjected to torsion and shear. Torsion does not usually decide the dimensions of members. However torsion design should be carried out as a check following the flexural design of the members. Where torsional resistance or stiffness of members is not considered in the analysis of the structure, no specific computations for torsion is necessary. However, in structures where torsional forces are significant as in the case of curved continuous girders and circular girders, the structure should be checked for torsional resistance according to the provisions made in clause 14.2 of IRC: 18-2000 code and suitable reinforcements should be designed.

The structural behavior of prestressed concrete members subjected to combined bending moment, torsion and shear is complex depending upon the combination of different force components and the reader may refer to the various theories proposed by Collins[17], Warner[18] and the design recommendations of the British Code[19] and the comprehensive treatise on torsion by Mattock[20] for an exhaustive understanding of the problems of torsion.

The IRC: 18-2000 code based on the British Code BS: 8110-1985 prescribes that the calculations for torsion are required only for ultimate loads and the torsional shear stress should be calculated assuming a plastic stress distribution across the section The torsional shear stress τ_t developed due to the torsional moment T_u in rectangular and flanged sections is computed using the following equations:

(a) Rectangular Section

$$\tau_t = \frac{2T_u}{\left(h_{min}^2\right)\left[h_{max} - \dfrac{h_{min}}{3}\right]}$$

where T_u = ultimate torsional moment

h_{min} = smaller dimension of the section

h_{max} = larger dimension of the section

(b) Flanged Section: In the case of T, L and I sections, they are divided into component rectangles for purposes of computations. The torque shared by each rectangle is computed as

$$T_{ur} = \frac{T_u(h_{max} \times h_{min}^3)}{\Sigma(h_{max} \times h_{min}^3)}$$

(c) Box Section: In the case of closed box sections, the area enclosed by the centre line of the member forming the box section is used for computation of the torsional shear stress which is computed using the relation:

$$\tau_t = \frac{T_u}{2At_i}$$

where t_i = wall thickness of the member where the stress is required
A = the area enclosed by the centre line of the box section members

The torsional shear stress (τ_t) computed from the above relations, should not exceed the maximum torsional shear stress (τ_{tu}) compiled in Table 4.5 for different grades of concrete varying from M-30 to M-60.

Table 4.5: Maximum and Minimum Torsional Shear Stress (IRC: 18-2000) Grade of Concrete (N/mm²)

Torsional Shear Stress	30	35	40	45	50	55	60
$\tau_{t, min}$	0.37	0.40	0.42	0.42	0.42	0.42	0.42
τ_{tu}	4.10	4.45	4.75	5.03	5.30	5.56	5.81

In the case of small sections where the larger dimension (y_1) of the link reinforcement is less than 550 mm, the torsional shear stress τ_t should not exceed the value ($\tau_{tu} \, y_1/550$). If either of these conditions is not satisfied, the overall dimensions of the cross-sections are inadequate and the section has to be redesigned.

If the computed torsional shear stress exceeds the value of $\tau_{t, min}$ shown in Table 4.5, suitable torsion reinforcements should be designed to resist the ultimate torsional moment. This reinforcement is in addition to any requirement for shear and flexure which occurs simultaneously with the torsion and is computed as:

$$\left(\frac{A_{sv}}{S_v}\right) \geq \left[\frac{T_u}{(0.8 x_1 y_1)(0.87 f_{yv})}\right]$$

$$\text{and } A_s \geq \left(\frac{A_{sv}}{S_v}\right)(x_1 + y_1)\left(\frac{f_{yv}}{f_y}\right)$$

where T_u = the ultimate torsional moment
A_{sv} = the area of the legs of closed links at the section
A_s = the area of longitudinal torsion reinforcement
f_{yv} = the characteristic strength of links (not to exceed 415 N/mm²)

f_y = the characteristic strength of longitudinal reinforcement
(not to exceed 415 N/mm^2)
S_v = the spacing of links
x_1 = the smaller centre to centre dimension of the links
y_1 = the larger centre to centre dimensions of the links.

If the computed torsional shear stress (τ_t) is less than $(\tau_{t,\ min})$, then nominal links are designed satisfying the equation,

$$S_v = \left[\frac{A_{sv} 0.87 f_y}{0.4b}\right]$$

Torsion reinforcement should consist of longitudinal steel bars and effectively closed links. This reinforcement should be provided in addition to that required for flexure and shear. To prevent failure due to improper detailing, the closed links should have minimum cover and pitch with the smallest of the following:

(a) $(x_1 + y_1)/4$
(b) 16 times the longitudinal corner bar
(c) 200 mm

The longitudinal reinforcement should consist of bars at each corner of the links and the diameter of the corner bars should be not less than the diameter of the links.

4.7 FORCES IN END BLOCKS

In the case of post tensioned prestressed concrete members, larger forces from cables are transmitted to concrete concentrated over a small area through the anchorages. The concentrated force at the end of the member develops bursting tension in the concrete over a length equal to the depth of the structural member which constitutes the end block. To prevent splitting of concrete in these zones, suitable reinforcements are to be designed in the end blocks.

According to clause 17.2 of IRC: 18-2000, the bursting tensile force in the end block depends upon the anchorage force, size of the end block and the type of anchoring device. The ratio of the bursting tension to the force in the tendon is computed from the Table 4.6 for values of ratio (y_{po}/y_o) varying from 0.3 to 0.7.

Table 4.6: Design Bursting Tensile Force in End Blocks (IRC: 18-2000)

(y_{po}/y_o)	0.3	0.4	0.5	0.6	0.7
(F_{bst}/P_k)	0.23	0.20	0.17	0.14	0.11

where P_k = force in the tendons or cables
F_{bst} = bursting tensile force
$2y_o$ = side of the end block
$2y_{po}$ = side of the loaded area

The bursting tensile force is distributed over a region extending from 0.2 y_o to 2 y_o from the loaded area of the end block and suitable reinforcements are designed to resist the tensile force and distributed in the end block zone generally in the form of a mesh with reinforcements in the longitudinal and transverse directions.

Limit State Design of Reinforced and Prestressed Concrete Bridge Deck Sections

4.8 DESIGN EXAMPLES

1. A prestressed concrete slab deck of a bridge is 400 mm thick with an effective span of 8 m. The service load bending moment due to IRC loads is computed as 3200 kNm/m at centre of span section. If the compressive stress permissible at transfer is 16 N/mm² and tensile stresses are not permitted, check the adequacy of the section and estimate the minimum prestressing force and the corresponding eccentricity at mid span section. Assume loss ratio as 0.8.

   (i) Data
   Effective span = L = 8 m
   Thickness of slab = 400 mm
   Service live load bending moment = M_q = 3200 kNm/m
   Loss ratio = η = 0.8

   (ii) Permissible stresses
   f_{ct} = 16 N/mm²
   $f_{tt} = f_{tw} = 0$

   (iii) Dead load bending moment
   $M_g = (0.125 \times 1 \times 0.4 \times 24 \times 8^2) = 76.8$ kNm/m

   (iv) Check for Minimum Section Modulus

   $$Z \geq \left[\frac{M_q + (1-\eta)M_g}{f_{br}}\right]$$

   $$\geq \left[\frac{(3200 \times 10^6) + (1-0.8)76.8 \times 10^6}{(0.8 \times 16 - 0)}\right]$$

   $$\geq (26.2 \times 10^6) \text{ mm}^3$$

   Section modulus provided = $\left(\frac{1000 \times 400^2}{6}\right) = (26.66 \times 10^6)$ mm³ > (26.2×10^6) mm³

   Hence the section modulus provided is adequate to resist the service load bending moment without exceeding the permissible stresses.

   (v) Minimum Prestressing Force

   $$f_t = \left[f_{tt} - \frac{M_g}{Z_t}\right] = \left[0 - \frac{76.8 \times 10^6}{26.66 \times 10^6}\right] = -2.88 \text{ N/mm}^2$$

   $$f_b = \left[\frac{f_{tw}}{\eta} - \frac{M_g + M_q}{\eta Z_b}\right] = \left[0 + \frac{(76.8 + 3200)10^6}{(0.8 \times 26.66 \times 10^6)}\right] = 18.6 \text{ N/mm}^2$$

   $$\therefore P = \left[\frac{A(f_t Z_t + f_b Z_b)}{Z_t + Z_b}\right]$$

   $$= \left[\frac{(1000 \times 400) \times 26.66 \times 10^6 (18.60 - 2.88)}{(2 \times 26.66 \times 10^6)}\right]$$

$$= (3144 \times 10^3) \text{ N}$$
$$= 3144 \text{ kN}$$

(vi) Eccentricity of the prestressing force

$$e = \left[\frac{Z_t Z_b (f_b - f_t)}{A(f_t Z_t + f_b Z_b)} \right]$$

$$= \left[\frac{(26.66)^2 \times 10^{12}(18.6 + 2.88)}{(1000 \times 400)(26.66 \times 10^6)(18.6 - 2.88)} \right] = 91.07 \text{ mm}$$

2. An unsymmetrical concrete I-section has been selected for a National Highway Bridge to support IRC Class AA loads over an effective span of 20 m. The section properties, permissible stresses and design bending moment at centre of span section are as follows:

$A = (596 \times 10^3) \text{ mm}^2$ $f_{ct} = 16 \text{ N/mm}^2$
$I = (1407 \times 10^8) \text{ mm}^4$ $f_{tt} = f_{tw} = 0$
$y_t = 546 \text{ mm and } y_b = 854 \text{ mm}$ $\eta = 0.8$
$Z_t = (2.57 \times 10^8) \text{ mm}^3 \text{ and}$
$Z_b = (1.65 \times 10^8) \text{ mm}^3$
$M_g = 1073 \text{ kNm}$
$M_q = 1295 \text{ kNm}$

Check for the adequacy of the section and design the prestressing force to resist the design bending moments.

(i) Prestress at Top and Bottom of Section

$$f_t = \left[f_{tt} - \frac{M_g}{Z_t} \right] = \left[0 - \frac{1073 \times 10^6}{2.57 \times 10^8} \right] = -4.17 \text{ N/mm}^2$$

$$f_b = \left[f_{tw} + \frac{M_g + M_q}{\eta Z_b} \right] = \left[0 + \frac{(1073 + 1295) \times 10^6}{0.8 \times 1.65 \times 10^8} \right] = 17.93 \text{ N/mm}^2$$

(ii) Eccentricity of Prestressing Force: Since the dead load bending moments are almost similar in magnitude to that of live load bending moments, eccentricity is first checked for its position to be accommodated within the section with suitable cover requirements.

$$e = \frac{Z_t Z_b (f_b - f_t)}{A(f_t Z_t + f_b Z_b)}$$

$$= \frac{(2.57 \times 1.65) \times 10^8 [17.93 + 4.17]}{(596 \times 10^3) \left[(-4.17 \times 2.57 \times 10^8) + (17.93 \times 1.65 \times 10^8) \right]}$$

$$= 833 \text{ mm}$$

Since, $y_b = 854$ mm and minimum effective cover required is 150 mm, the computed eccentricity corresponding to minimum prestressing force is impracticable. Hence provide the maximum possible eccentricity as

$$e = [854 - 154] = 700 \text{ mm}$$

(iii) Modified Prestressing Force

$$P = \left[\frac{Af_b Z_b}{Z_b + Ae}\right]$$

$$= \left[\frac{596 \times 10^3 \times 17.93 \times 1.65 \times 10^8}{(1.65 \times 10^8) + (596 \times 10^3 \times 700)}\right]$$

$= (3028 \times 10^3)$ N
$= 3028$ kN

3. A tee beam having an effective width of flange (B_f) = 1400 mm and thickness of flange (t) = 150 mm, width of rib = 200 mm and overall depth (h) = 1400 mm is designed as a main bridge girder. The design moments at the critical section are as follows:
B.M. due to permanent dead load of girder (G) = 1122 kNm
B.M. due to superimposed dead load (SG) = 180 kNm
B.M. due to live load (Q) = 1500 kNm

The tee beam is prestressed by six high tensile cables each with effective area of tendons equal to 603 mm² at an effective depth (d) = 1250 mm.
Characteristic compressive strength of concrete (f_{ck}) = 40 N/mm²
Characteristic tensile strength of tendons (f_p) = 1500 N/mm²

Estimate the ultimate flexural strength of the section and check its adequacy as per IRC: 18-2000 code.

(i) Ulimate flexural strength according to IRC: 18-2000
$M_u = 1.25G + 2.0SG + 2.5Q$
$= 1.25(1200) + 2(180) + 2.5(1500)$
$= 5610$ kNm

(ii) Ultimate flexural strength of tee section
  (a) Failure by yielding of steel
  $M_{us} = (0.9 \, d \, A_p \, f_p)$
  $= (0.9 \times 1250 \times 6 \times 603 \times 1500)$
  $= (6105 \times 10^3)$ Nm
  $= 6105$ kNm

  (b) Failure by crushing of concrete
  $M_{uc} = (0.176 \, b \, d^2 \, f_{ck}) + [0.67 \times 0.8 \, (B_f - b)(d - 0.5 \, t) \, t \, f_{ck}]$
  $= (0.176 \times 200 \times 1250^2 \times 40)$
  $\qquad + [0.67 \times 0.8 \times (1400 - 200)(1250 - 75) \, 150 \times 40]$
  $= (6666 \times 10^6)$ Nmm
  $= 6666$ kNm

The ultimate flexural strength of the section is the least of (a) and (b)
Hence $M_u = 6105$ kNm

Required M_u = 5610 kNm
Hence the section designed is adequate to resist the design moments.

4. A prestressed concrete bridge beam is of flanged section having a web width of 200 mm and overall depth of the girder being 1400 mm. At the support section, the ultimate shear force due to dead and live loads is computed as 1000 kN. The cross-sectional area of beam section is 600×10^3 mm² and the girder is concentrically prestressed at the support section with an effective prestressing force of 3000 kN. f_{ck} = 40 N/mm², the vertical component of the prestressing force at support is 680 kN. Estimate the shear resistance of the support section and check its adequacy to resist the design shear force.

(i) Data

$A = 600 \times 10^3$ mm² b = 200 mm
h = 1400 mm f_{ck} = 40 N/mm²
P = 3000 kN

Design ultimate shear force V_u = 1000 kN
Vertical component of prestressing force at support P_v = 680 kN

(ii) Compressive prestress and principal tensile stress

$$f_{cp} = \left[\frac{P}{A}\right] = \left[\frac{3000 \times 10^3}{600 \times 10^3}\right] = 5 \text{ N/mm}^2$$

$$f_t = 0.24\sqrt{f_{ck}} = 0.24\sqrt{40} = 1.52 \text{ N/mm}^2$$

(iii) Shear strength of support section
Ultimate shear resistance of uncracked section is given by

$$V_{co} = 0.67bh\sqrt{f_t^2 + 0.8f_{cp}f_t} + P_v$$

$= [0.67 \times 200 \times 1400\sqrt{1.52^2 + 0.8 \times 5 \times 1.52} + (680 \times 10^3)$

$= (1223 \times 10^3)$ N

$= 1223$ kN

(iv) Check for safety against shear force

V_u = 1000 kN
V_{co} = 1223 kN

Since $V_{co} > V_u$ the support section is safe against shear failure.
However minimum shear reinforcements as prescribed in IRC: 18-2000 should be provided.

5. A prestressed concrete beam of hollow rectangular box girder section has overall dimensions of 2200 mm by 1500 mm with a uniform wall thickness of 200 mm. The section is subjected to an ultimate torque of 5000 kNm. The grade of concrete is 45 N/mm². Estimate the maximum torsional shear stress in the section and compare it with the maximum torsional shear stress permitted in the IRC: 18-2000 code and check the adequacy of the section to resist the torque.

(i) Data: Overall dimensions of hollow box girder section = 2200 mm by 1500 mm, Thickness of walls = t_i = 200 mm, Area enclosed by centre line = A = (2000 × 1300) = 26 × 10^5 mm^2, Ultimate torque on section = T_u = 5000 kNm.

(ii) Maximum torsional shear stress

$$\tau_t = \frac{T_u}{2At_i} = \left[\frac{5000 \times 10^6}{2 \times 26 \times 10^5 \times 200}\right] = 4.8 \text{ N/mm}^2$$

(iii) Permissible torsional shear stress: Maximum permissible torsional shear stress from Table 4.5 for M-45 grade concrete is τ_{tu} = 5.03 N/mm^2.

Since $\tau_t < \tau_{tu}$, the section is adequate to resist the torsional moment.

6. A prestressed concrete bridge deck has an unsymmetrical I-section beam spanning over 20 m. The top flange is 1200 mm wide by 200 mm thick. The bottom flange is 500 mm wide by 400 mm thick. The web is 200 mm wide by 900 mm deep. The section is subjected to an ultimate torsional moment of 200 kNm. Estimate the maximum torsional shear stress in the section and check the adequacy of the section as per IRC: 18-2000 code specifications assuming the grade of concrete as M-40.

(i) Data

	h_{min} (mm)	h_{max} (mm)
Top Flange	200	1200
Bottom Flange	400	500
Web	200	900

Grade of concrete (M-40): Hence f_{ck} = 40 N/mm^2
Ultimate torsional moment = T_u = 200 kNm

(ii) Torsion shared by web and flanges

$$T_{ur} = \frac{T_u(h_{max} \times h_{min}^3)}{\Sigma(h_{max} \times h_{min}^3)}$$

$$\Sigma(h_{max} \times h_{min}^3) = \left[(1200 \times 200^3) + (500 \times 400^3) + (900 \times 200^3)\right] = (488 \times 10^8) \text{ mm}^4$$

$$T_{ur}(\text{Top Flange}) = \frac{T_u(1200 \times 200^3)}{488 \times 10^8} = 0.20 T_u$$

$$T_{ur}(\text{Bottom Flange}) = \frac{T_u(500 \times 400^3)}{488 \times 10^8} = 0.65 T_u$$

$$T_{ur}(\text{Web}) = \frac{T_u(900 \times 200^3)}{488 \times 10^8} = 0.15 T_u$$

(iii) Maximum torsional shear stress (τ_t)

$$\tau_t = \left\{ \frac{2T_u}{(h_{min}^2)\left[h_{max} - \dfrac{h_{min}}{3}\right]} \right\}$$

$$\tau_t (\text{Top flange}) = \left\{ \frac{2(0.20T_u)}{200^2\left[1200 - \dfrac{200}{3}\right]} \right\} = 0.0088 \times 10^{-6} \, T_u$$

$$\tau_t (\text{Bottom flange}) = \left\{ \frac{2(0.65T_u)}{400^2\left[500 - \dfrac{400}{3}\right]} \right\} = 0.0223 \times 10^{-6} \, T_u$$

$$\tau_t (\text{Web}) = \left\{ \frac{2(0.15T_u)}{200^2\left[900 - \dfrac{200}{3}\right]} \right\} = 0.0088 \times 10^{-6} \, T_u$$

Maximum torsional shear stress occurs in the bottom flange and is computed as
$$\tau_t = (0.0223 \times 10^{-6}) \, T_u = (0.0223 \times 10 - 6 \times 200 \times 10^6) = 4.46 \text{ N/mm}^2$$

(iv) Check for adequacy of the section: Referring to Table 4.5, the maximum torsional shear stress corresponding to M-40 grade concrete is $\tau_{tu} = 4.75 \text{ N/mm}^2$.

Since $\tau_t < \tau_{tu}$, the section can safely resist the ultimate torsional moment.

7. The end block of a prestressed concrete beam 200 mm wide by 300 mm deep has two Freyssinet anchorages (100 mm diameter) with their centres at 75 mm from top and bottom of the beam. The force transmitted by each anchorage being 200 kN. Estimate the bursting tension developed in the end block.

Anchorage force = P_k = 200 kN
Anchorage diameter = 100 mm

Equivalent side of square = $2y_{po} = \sqrt{\dfrac{\pi \times 100^2}{4}} = 89$ mm

Side of the surrounding prism = 150 mm

$\therefore$ Ratio $\left[\dfrac{y_{po}}{y_o}\right] = \left[\dfrac{89}{150}\right] = 0.593 \cong 0.6$

Referring to Table 4.6 and interpolating, the bursting tension expressed as a ratio is

$$\left(\frac{F_{bst}}{P_k}\right) = 0.14$$

$\therefore$ $F_{bst} = (0.14 \times 200) = 28$ kN

REFERENCES

1. RAINA, V.K, *Concrete Bridge Practice, Analysis, Design and Economics*. Tata McGraw Hill Publishers, New Delhi, 1991, pp. 587-617.
2. HOGNESTAD, E. HANSEN, N.W. and McHENRY, D., Concrete stress distribution in ultimate strength design. *Journal of the American Concrete Institute*, Vol. 52, Dec. 1955, pp. 455-479.
3. EVANS, R.H., The plastic theories for the ultimate strength of reinforced concrete beams. *Journal of the Institution of Civil Engineers*, London, Dec. 1943, pp. 98-121.
4. ROWE, R.E. CRANSTON, W.B. and BEST, B.C., New concepts in the design of structural concrete. *Structural Engineer*, Vol. 43, 1965, pp. 339-403.
5. BATE, S.C.C., Why limit state design? *Concrete*, London, March 1968, pp. 103-108.
6. MURASHEV, V. SIGALOV, E. BAIKOV, V., *Design of Reinforced Concrete Structures*. Mir Publishers, Moscow, 1968.
7. KRISHNA RAJU, N., Limit State Design for Structural Concrete, Proceedings of the Institution of Engineers (India), Vol. 51, Jan.1971, pp.138-143.
8. BENNETT, E.W., *Structural Concrete Elements*. Chapman and Hall, London, 1973, pp. 149-170.
9. KRISHNA RAJU, N., Limit State design of prestressed concrete members, engineering design, special issue on prestressed concrete. *Journal of National Design and Research Forum*, Institution of Engineers (India), April 1979, pp. 16-21.
10. FIB-CEB; Joint Committee Practical Recommendations for the Design and Construction of Prestressed Concrete Structures, June 1966.
11. KRISHNA RAJU, N., *Design of Bridges*, Oxford and IBH Publishers (Third Edition) New Delhi, 1998, p. 26.
12. IS: 456-2000 *Indian Standard Code of Practice for Plain and Reinforced Concrete*, (Fourth Edition), BIS, New Delhi, July 2000, pp. 100.
13. KRISHNA RAJU, N. and PRANESH, R. N., *Reinforced Concrete Design (IS:456-2000) Principles* and *Practice*. New Age International Publishers, New Delhi, 2003, pp. 609.
14. EVANS, R.H. and BENNETT, E.W., *Prestressed Concrete*. Chapman and Hall, London, 1958, pp. 115-118.
15. KRISHNA RAJU. N., *Prestressed Concrete* (Fourth Edition), Tata McGraw Hill Publishers, New Delhi, 2007, pp.
16. IS: 1343 Indian Standard Code of Practice for Prestressed Concrete (under revision) BIS, New Delhi, 2007.
17. COLLINS, M.P. and MITCHELL, D., Shear and torsion design of prestressed and non prestressed concrete beams. *Journal of Prestressed Concrete Institute*, Vol. 25, 1980, pp. 32-100.
18. WARNER, R.F. and FAULKES, K.A., *Prestressed Concrete*. Pitman, Australia 1978, pp. 175-182.
19. BS; 8110 (Part-1):1985, British Standard Code of Practice for Structural Use of Concrete, British Standards Institution, London, 1985, pp. 415-416.
20. MATTOCK, A.H., *How to Design for Torsion, Torsion of Structural Concrete*. Special Publication SP-18, American Concrete Institute, 1968, pp. 469-495.

EXERCISES

1. The deck slab of a prestressed concrete bridge is 500 mm thick with parallel post tensioned cables. The effective span is 10 m. The working live load bending moment due to I.R.C. loading is computed as 280 kN·m/m at centre of span section. The safe permissible compressive stress at transfer is 15 N/mm^2 and the structure is to be designed as Class 1 type without any tensile stress. Check the adequacy of the section and estimate the minimum prestressing force and the corresponding eccentricity assuming a loss ratio of 0.85. For the section designed estimate the ultimate flexural strength assuming M-40 grade concrete and the tensile strength of tendons as 1600 N/mm^2.

2. A symmetrical I-section girder has been used for a continuous post tensioned prestressed concrete bridge deck of spans 40 m. The section properties, permissible stresses and design bending moment at the critical support section are as follows:

$A = (880 \times 10^3)$ mm^2 $\qquad M_g = 9400$ kN.m

$I = 0.45$ m^4 $\qquad M_q = 1100$ kN.m

$y_b = y_t = 1$ m $\qquad \eta = 0.85$

$Z_t = Z_b = 0.45$ m^3 $\qquad f_{ct} = 16$ N/mm^2

$h = 2$ m $\qquad f_{tt} = f_{tw} = 0$

Check the adequacy of the section and design the prestressing force and eccentricity to resist the design bending moment.

3. A prestresed concrete bridge deck comprises of tee beam and slab having the following dimensions:

Effective width of flange = 1600 mm

Thickness of flange = 200 mm

Width of rib = 200 mm

Overall depth = 1600 mm

The design moments at the centre of span section are as follows:

Live load moment = 1600 kN.m

Dead load moment = 1200 kN.m

The tee beam is prestressed by 4 high tensile cables each having an effective area of 900 mm^2 located at an effective depth of 1400 mm

Characteristic compressive strength of concrete = 50 N/mm^2

Characteristic tensile strength of tendons = 1600 N/mm^2

Estimate the ultimate flexural strength of the section and check its adequacy as per IRC: 18-2000 code.

4. The web of a prestressed concrete tee beam and slab deck is 250 mm and overall depth of the beam is 1600 mm. At the support section the ultimate shear force due to dead and live loads is estimated as 1200 kN. The cross-sectional area of the tee beam girder is (600×10^3) mm^2.

The girder is concentrically prestressed at support section with an effective prestressing force of 3600 kN. Cube strength of concrete = 40 N/mm^2. The vertical component of the prestressing force at support is 650 kN. Calculate the ultimate shear resistance of the support section and check its adequacy to resist the design ultimate shear force.

5. A continuous prestressed concrete curved bridge deck of a National Highway Comprising a hollow square box girder has overall dimensions of 2 m by 2 m with a uniform wall thickness of 200 mm. The critical section is subjected to an ultimate torque of 4000 kN.m. The grade of concrete used for the girder is M-50. Compute the maximum torsional shear stress developed in the section and check the adequacy of the section to resist the torque according to IRC: 18-2000 code specifications.

Limit State Design of Reinforced and Prestressed Concrete Bridge Deck Sections

6. The cross-section of a prestressed concrete bridge deck is made up of an unsymmetrical I-section having the following sectional dimensions:

   Width and depth of top flange = 1600 mm by 250 mm

   Width and depth of bottom flange = 500 mm by 400 mm

   Thickness of web = 200 mm

   Overall depth of section = 1800 mm

   The section is subjected to an ultimate torsional moment of 400 kN.m.

   Estimate the maximum torsional shear stress in the section and check the safety of the section for resisting torque as per IRC: 18-2000 code specifications assuming M-40 grade concrete.

7. A P.S.C. girder of a highway bridge is of unsymmetrical I-section with the following details:

   Top flange: 1600 mm wide by 250 mm thick

   Bottom flange: 600 mm wide by 400 mm thick

   Web: 200 mm wide by 1500 mm deep

   The beam is reinforced with 7K-15 type Freyssinet cables of total high tensile steel area of 4000 mm^2 located at an effective depth of 1900 mm. If M-50 Grade concrete is used and the characteristic tensile strength of tendons is 1800 N/mm^2, estimate the flexural strength of the section as per IRC: 18-2000 code specifications.

8. The end block of a post tensioned prestressed concrete girder is 800 mm wide by 2000 mm deep. Three Freyssinet cables each carrying a force of 4000 kN is anchored at the end using 400 mm by 400 mm plates spaced at 600 mm centres. Estimate the bursting tension in the end block as per IRC: 18-2000 code recommendations.

5
PRESTRESSED CONCRETE SLAB BRIDGE DECKS

5.1 GENERAL FEATURES

Prestressed concrete slab bridge decks are being increasingly used for the small and medium span ranges. Solid slabs are used in the span range of 10 to 20 m while voided slabs are preferable for longer spans of up to about 40 m. The span/depth ratio for simply supported spans are in the range of 20 to 30 while for continuous spans, the ratio can be as high as 40. The slab decks have high torsional resistance and hence they are ideally suited for curved alignment especially supported on single central columns.

Prestressed concrete slab decks are more economical in comparison with reinforced concrete slab decks for spans exceeding 10 m. Reinforced concrete slab decks are suitable for spans in the range of 5 to 10 m with span/depth ratios in the range of 15 to 20 for simple spans and 20 to 25 for continuous spans. For longer spans, reinforced concrete slab decks are uneconomical and hence prestressed concrete slab decks are preferred.

Prestressed concrete slab decks look neat and simple and are easier to construct when compared to beam and slab deck. The slab decks can either be precast or cast *in situ*. Precast pretensioned or post tensioned slab desk are used for small spans in the range of 10 to 25 m. Span/depth ratios generally used are in the range of 25 to 30. Precast slabs can be cast using simple form work near the work site and field erection is fast using cranes and no false work is required.

Cast *in situ* slab decks are generally post tensioned and require false work to form ducts for threading high tensile cables for post tensioning operations. Maintenance of slab decks is very little except that proper care should be taken in the vicinity of hinges and bearings near the supports. Typical configurations of prestressed concrete solid and cored decks are compiled in Fig. 5.1.

5.2 ANALYSIS OF SLAB DECKS

Prestressed concrete slab decks used for small and medium span bridges are generally spanning in one direction and hence the moments due to dead and live loads are critical in the longitudinal direction, i.e. the direction of the moving loads. Bridge deck slabs simply supported on either side have to be designed for IRC loads[1] specified as Class AA or A depending upon the importance and classification of the bridge. In the case of prestressed concrete tee beam and slab bridges, the deck slab is supported along the longitudinal and lateral directions by main and cross girders. Hence the

Prestressed Concrete Slab Bridge Decks

slabs in such cases have to be analyzed for moments developed in the longitudinal and lateral directions. Analysis of slabs with different support conditions are detailed under the following sub sections.

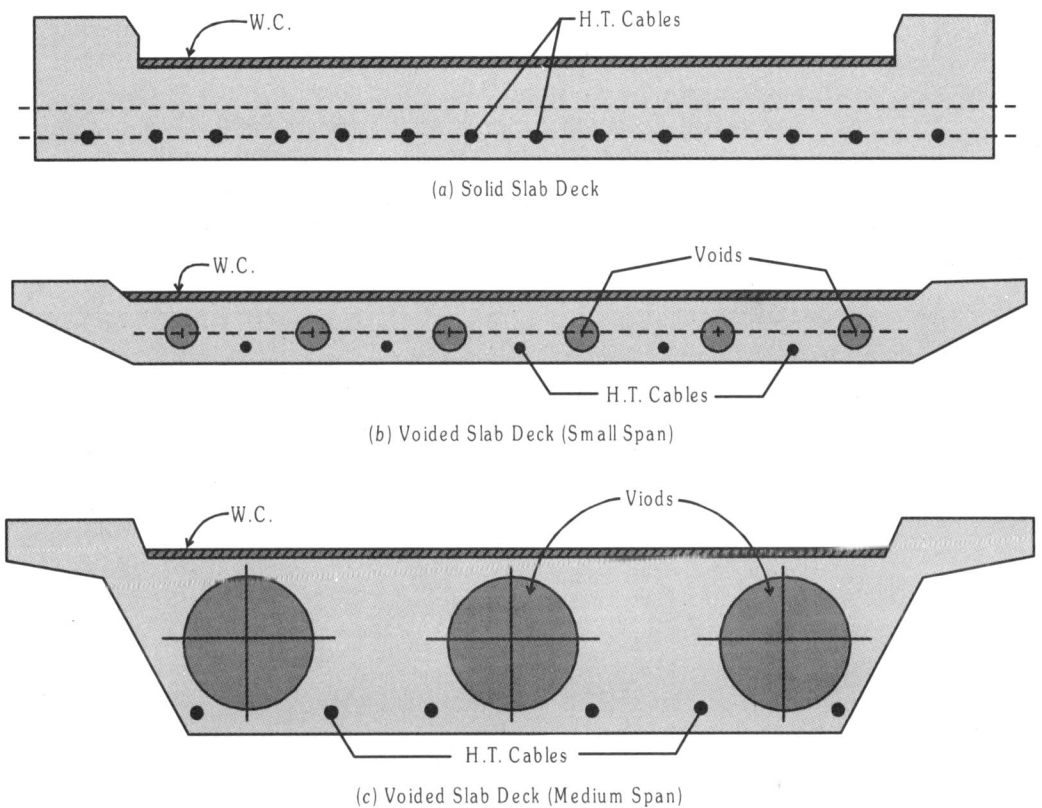

(a) Solid Slab Deck

(b) Voided Slab Deck (Small Span)

(c) Voided Slab Deck (Medium Span)

Fig. 5.1: Prestressed Concrete Slab Deck Configurations

(a) Solid Slabs Spanning in one Direction

1. **Single Concentrated Load:** In the case of slabs spanning in one direction, the dead load moments are directly computed assuming the slab to be simply supported between the bearings. Live loads of vehicles transmitted through wheels are considered as concentrated loads spread over the contact area of the tyres with the deck slab. The bending moment per unit width of slab developed by concentrated loads on solid slabs may be calculated by assuming the width of slab considered as effective in resisting the bending moment due to concentrated loads. In the case of precast slabs, the term 'actual width of slab' should include the actual width of each individual precast element.

   IRC: 21-2000[2] code specifications outline a method of computing the effective width of slab supporting a concentrated wheel load considering various parameters like the span length, dimensions of slab and the concentration area of the wheel load.

   For a single concentrated load, the effective width may be calculated by the equation,

   $$b_e = kx\left[1 - \frac{x}{L}\right] + b_w$$

where b_e = effective width of slab on which the load acts
L = effective span
x = distance of centre of gravity of load from nearer support
b_w = breadth of concentration area of load, i.e. the dimension of the tyre or track contact area over the road surface of the slab in a direction at right angles to the span plus twice the thickness of the wearing coat or surface finish above the structural slab.
K = A constant depending upon the ratio (B/L) where B is the width of the slab

The values of the constant 'K' for different values of the ratio (B/L) are compiled in Table 5.1.

Table 5.1: Values of Constant 'K' (IRC: 21-2000)

B/L	K for Simply Supported Slabs	K for Continuous Slabs	B/L	K for Simply Supported Slabs	K for Continuous Slabs
0.1	0.40	0.40	1.1	2.60	2.28
0.2	0.80	0.80	1.2	2.64	2.36
0.3	1.16	1.16	1.3	2.72	2.40
0.4	1.48	1.44	1.4	2.80	2.48
0.5	1.72	1.68	1.5	2.84	2.48
0.6	1.96	1.84	1.6	2.88	2.52
0.7	2.12	1.96	1.7	2.92	2.56
0.8	2.24	2.08	1.8	2.96	2.60
0.9	2.36	2.16	1.9	3.00	2.60
1.0	2.48	2.24	2.0 & Above	3.00	2.60

The effective width shall not exceed the actual width of the slab. Also in the case of a load near the unsupported edge of a slab, the effective width shall not exceed the above value nor half the above value plus the distance of the load from the unsupported edge.

2. **Two or More Concentrated Loads in Line in the Direction of Span:** When two or more concentrated loads are positioned in a line in the direction of span, the bending moment per unit width of slab shall be calculated separately for each load according to its appropriate effective width of slab as specified under the single concentrated load.

3. **Two or More Concentrated Loads not in Line in the Direction of Span:** In cases where the effective width of slab for one load overlaps the effective width of slab for an adjacent load, the resultant effective width for the two loads equals the sum of the effective widths for each load minus the width of overlap, provided that the slab so designed is tested for the two loads acting separately.

(b) Solid Cantilever Slab

The effective width of dispersion in the direction parallel to the supported edge for a single concentrated load is computed from the equation,

$$b_e = 1.2\,x + b_w$$

where b_e = effective width
  x = distance of the centre of gravity of the concentrated load from the face of the cantilever support.
  b_w = the breadth of the concentration area of the load, i.e. the dimension of the tyre or track contact area over the road surface of the slab in a direction parallel to the supporting edge of the cantilever plus twice the thickness of the wearing coat or surface finish above the structural slab.

The effective width should be limited to one-third the length of the cantilever slab measured parallel to the support. When the concentrated load is placed near one of the two extreme ends of the length of the cantilever slab in the direction parallel to the support, the effective width should not exceed the above value, nor should it exceed half the above value plus the distance of the concentrated load from the nearer extreme end, measured in the direction parallel to the fixed edge.

When two or more loads act on the slab and when the effective width of slab for one load overlaps the effective width of the adjacent load, the resultant effective width for two loads should be taken as the sum of the respective effective widths for each load minus the width of the overlap.

(c) **Dispersion of Loads along the Span**

The effective length of slab in the direction of the span is computed as the sum of the tyre contact area over the wearing surface of slab in the direction of the span and twice the overall depth of the slab inclusive of the thickness of the wearing surface.

If D = depth of the wearing coat
  H = depth of the slab
  x = wheel load contact area along the span
  v = effective length of dispersion along the span

We have the relation,
$$v = x + 2(D + H)$$

(d) **Dispersion of Loads in Slabs Spanning in two Directions**

In bridge decks comprising slab integrally cast with longitudinal and cross girders as in the case of tee beam and slab decks, the moments develop due to wheel loads on the slab both in the longitudinal and transverse directions. These moments are computed by using the design curves developed by M. Pigeaud or Westergaard's method[3, 4]. Pigeaud's method is applicable to rectangular slabs supported freely on all the four sides and the slab should be symmetrically loaded as shown in Fig. 5.2. The following notations are used in calculating the dispersion width and moments due to concentrated wheel loads on slabs.

  L = Long span length
  B = Short span length
  u & v = Dimensions of the load spread after allowing for dispersion through the wearing coat and structural slab
  K = Ratio of short to long span of slab (B/L)
  M_1 = Moment in the short span direction
  M_2 = Moment in the long span direction
  m_1 & m_2 = Coefficients for moments along the short and long spans
  μ = Poisson's ratio for concrete generally assumed as 0.15 as per IRC: 21-2000
  W = Wheel load under consideration

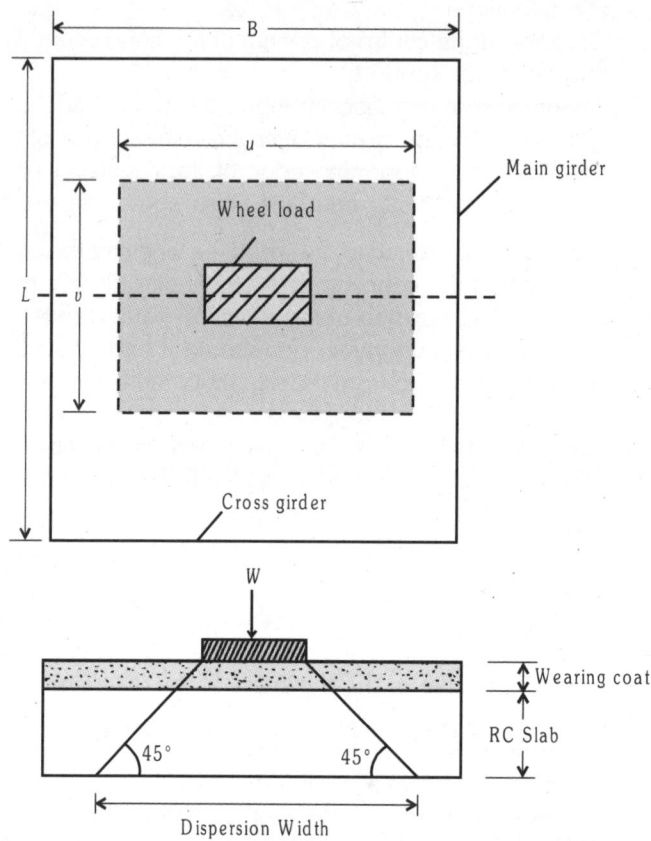

Fig. 5.2: Dispersion of Wheel Load through Wearing Coat and Deck at 45 Degrees

The dispersion of the wheel or track load may be assumed to be at 45 degrees through the wearing coat and structural slab according to Clause 305.16.3 of IRC: 21-2000 specifications. Consequently the effect of contact of wheel or track load in the direction of span length is equal to the dimensions of tyre contact area over the wearing surface of the slab in the direction of the span plus twice the overall depth of the slab inclusive of the thickness of the wearing coat as shown in Fig. 5.2. According to Victor[4], the dispersion is some times assumed to be at 45 degrees through the wearing coat which is flexible and at a steeper angle through the deck slab which is rigid as shown in Fig. 5.3.

The bending moments in the short and long span directions are expressed as

$$M_1 = (m_1 + \mu m_2)W$$
$$M_2 = (m_2 + \mu m_1)W$$

The values of the moment coefficients m_1 and m_2 depend upon the parameters (u/B) and (v/L) and the value of $K = (B/L)$. Figures 5.4 to 5.10 are the Pigeaud's curves used for the estimation of moment coefficients m_1 and m_2 for various values of K ranging from 0.4 to 1.0. Moment coefficients m_1 and m_2 corresponding to K and $(1/K)$ for slabs supporting uniformly distributed load (dead load of slab) are obtained from Fig. 5.11.

Prestressed Concrete Slab Bridge Decks

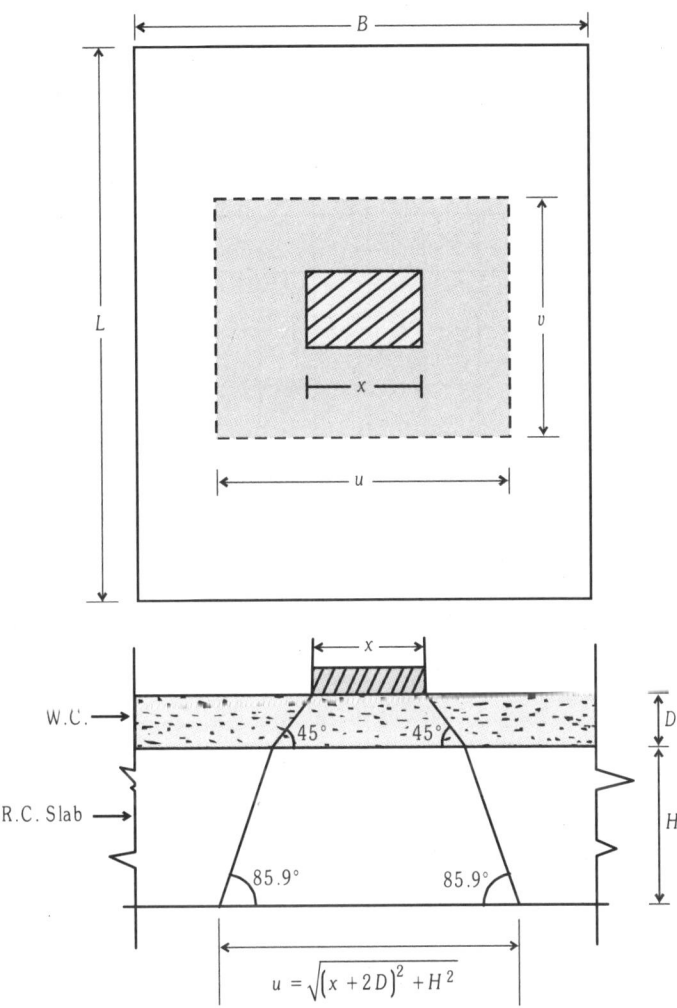

Fig. 5.3: Dispersion of Wheel Load at Different Angles through Wearing Coat and Deck Slab

The effect of different criteria used for determining the dispersion of wheel loads can be examined by a detailed comparative analysis. Hence a comparative study of the bending moments developed in the short and long span directions of a typical deck slab supported on main and cross girders of a tee beam bridge, assuming different types of load distribution through wearing coat and thickness of structural slab is presented for three different types of cases for a wheel load of IRC Class A type loading using the following data:

Data: IRC Class A Wheel Load = W = 57 kN
Wheel contact dimensions = 500 mm by 250 mm
Thickness of wearing coat = D = 80 mm
Thickness of structural slab = H = 200 mm
Dimensions of slab: L = 4 m and B = 2 m
Ratio = K = (B/L) = $(2/4)$ = 0.5
Poisson's Ratio = μ = 0.15

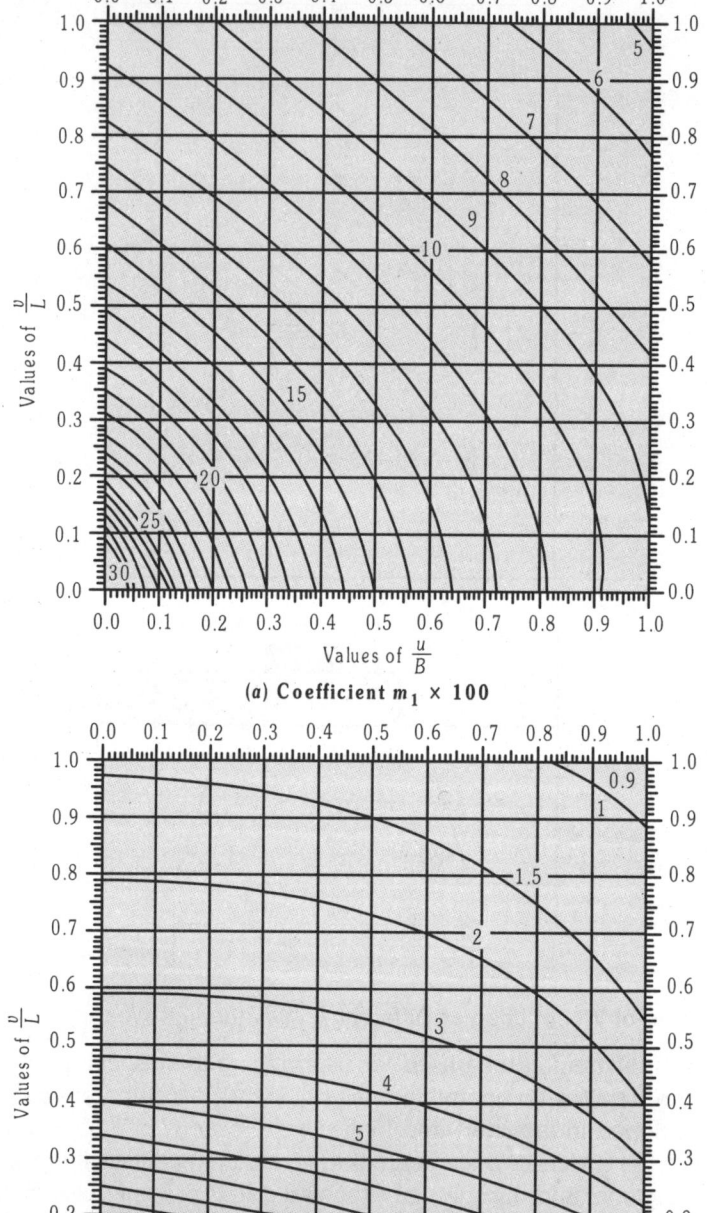

Fig. 5.4: Moment Coeffecients m_1 and m_2 for $K = 0.4$

Prestressed Concrete Slab Bridge Decks

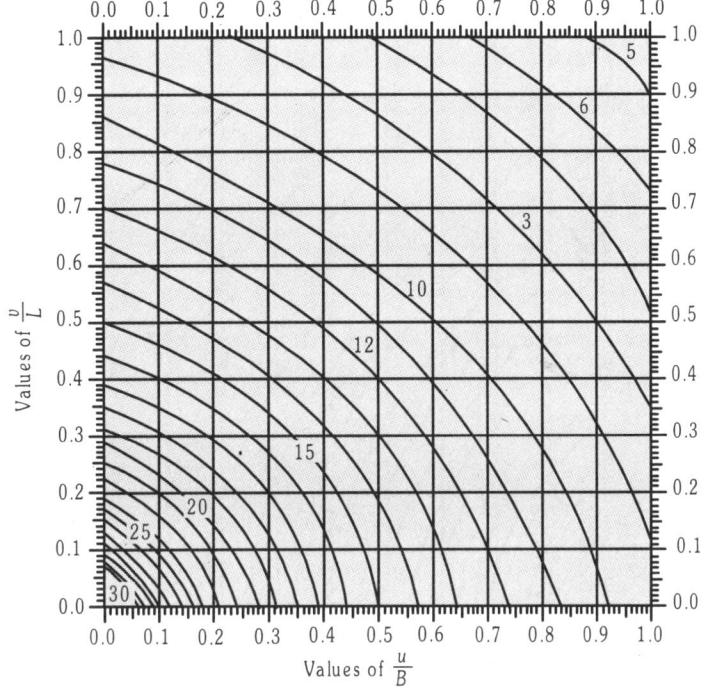

(a) Coefficient $m_1 \times 100$

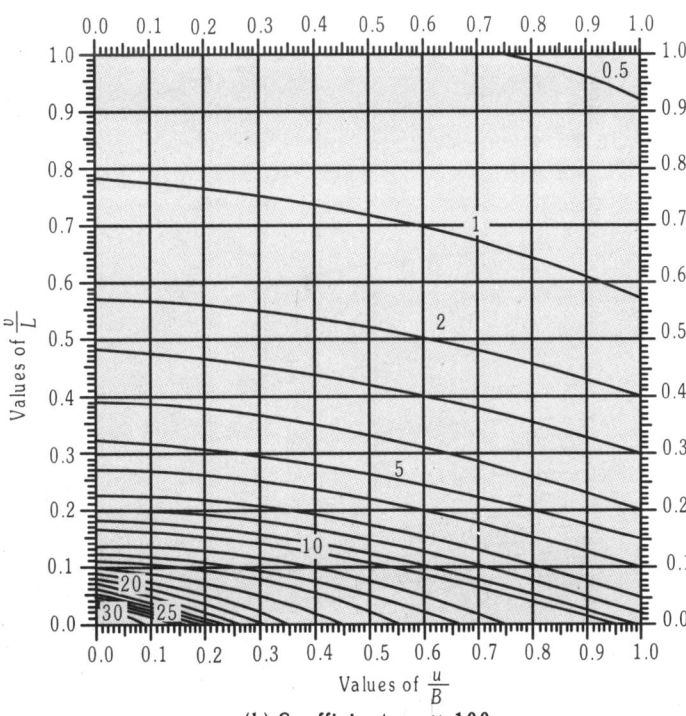

(b) Coefficient $m_2 \times 100$

Fig. 5.5: Moment Coefficients m_1 and m_2 for $K = 0.5$

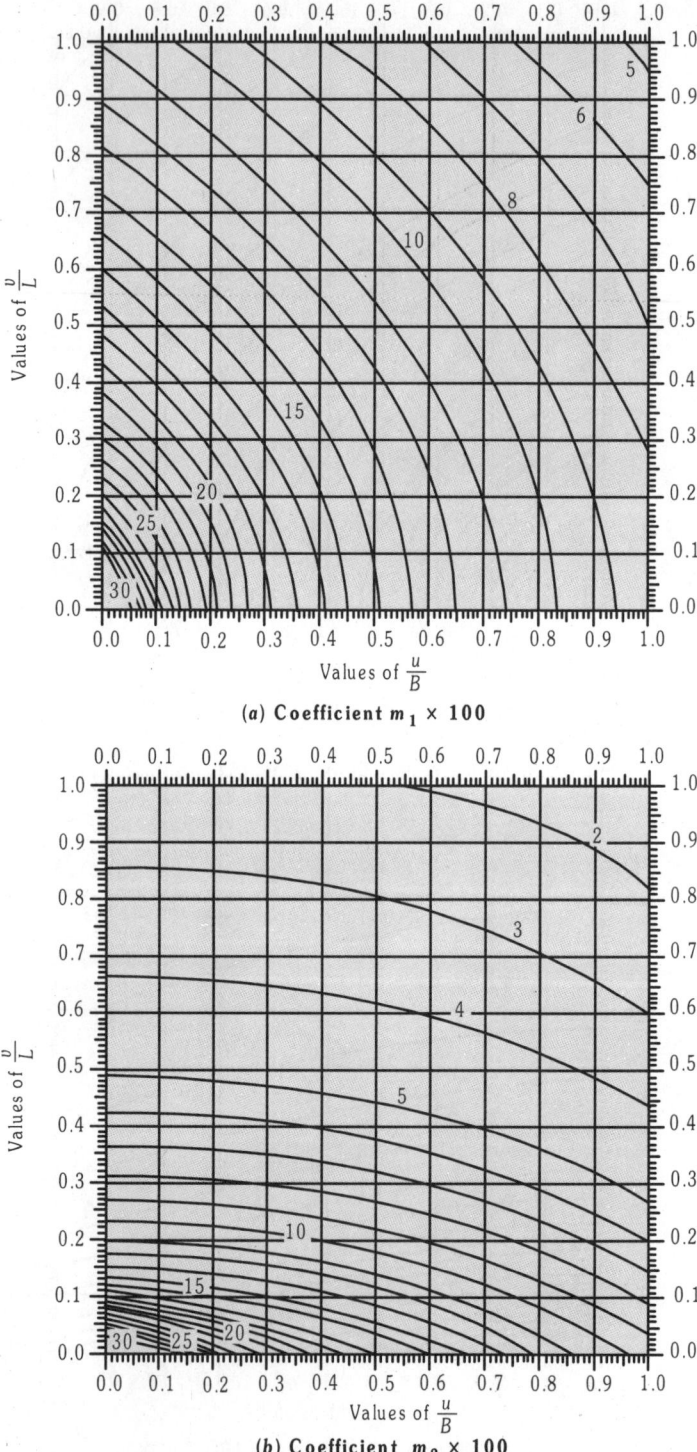

Fig. 5.6: Moment Coefficients m_1 and m_2 for $K = 0.6$

Prestressed Concrete Slab Bridge Decks

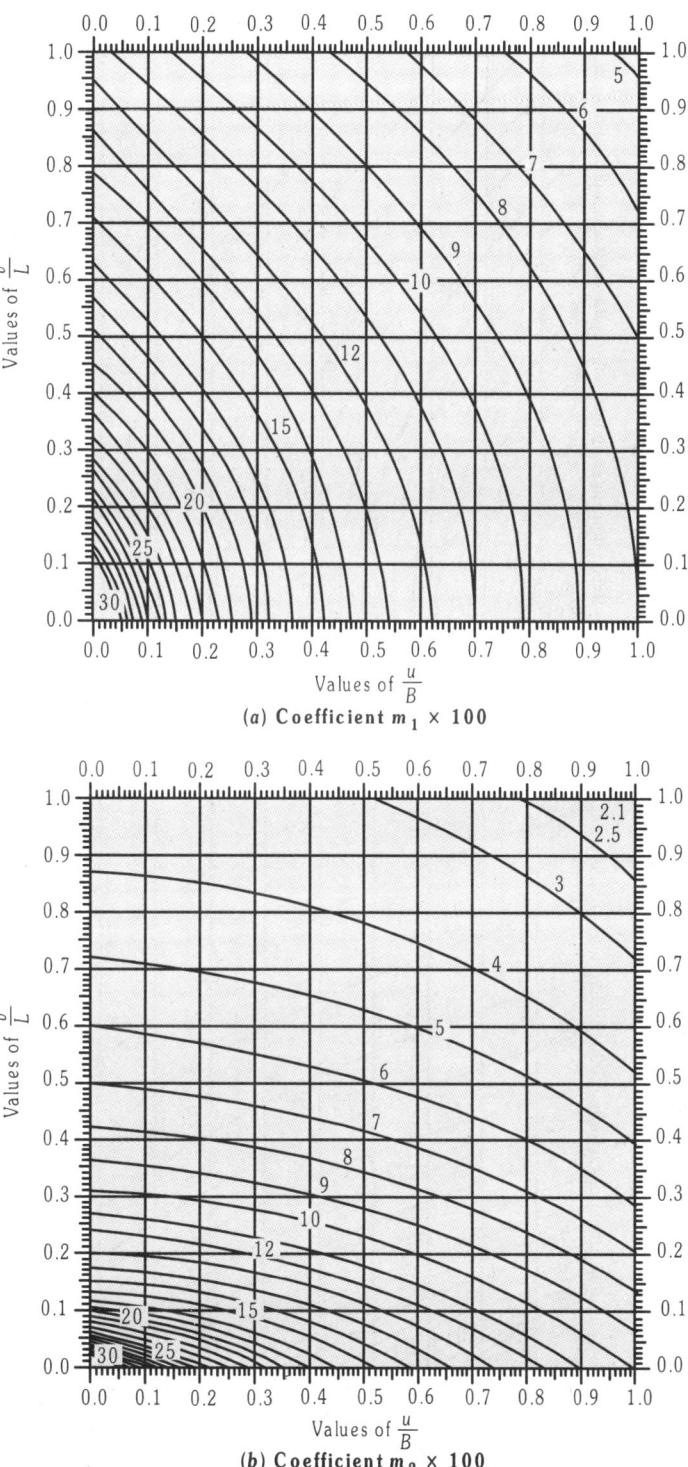

Fig. 5.7: Moment Coefficients m_1 and m_2 for $K = 0.7$

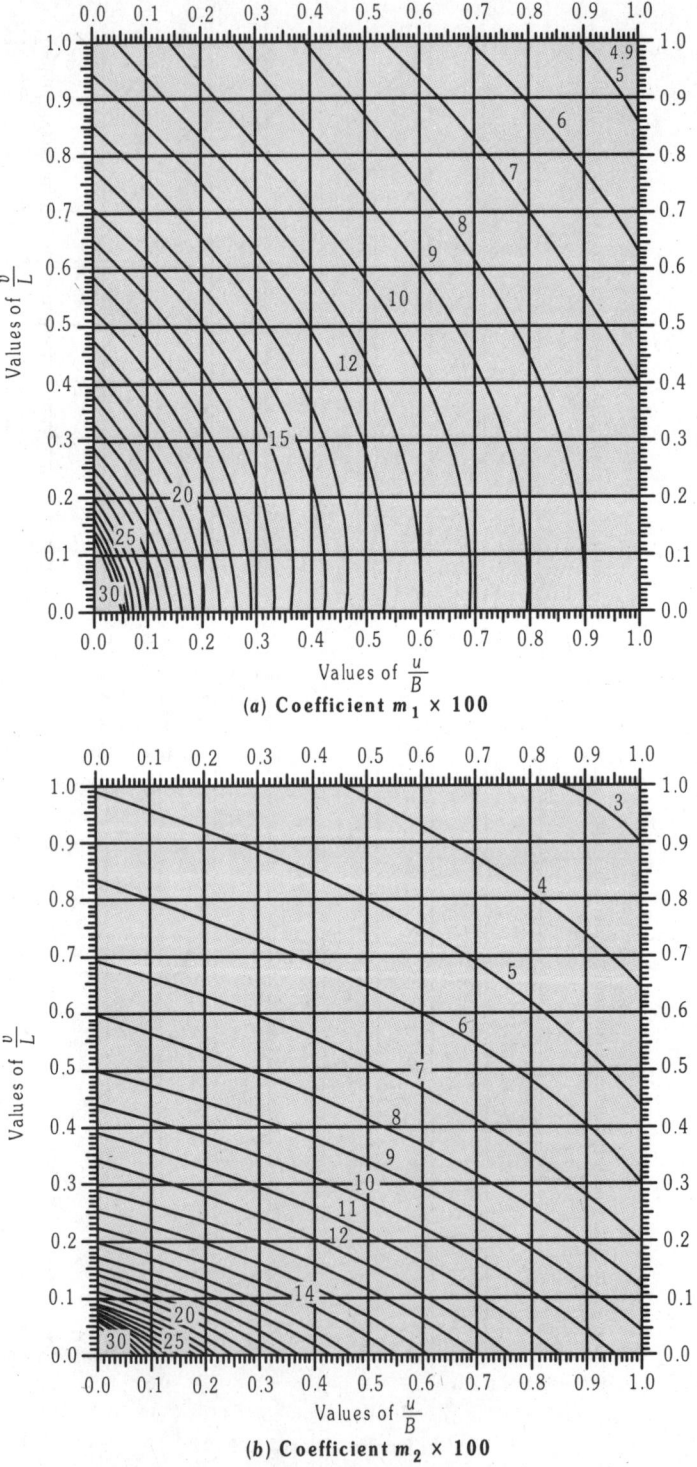

Fig. 5.8: Moment Coefficients m_1 and m_2 for $K = 0.8$

Prestressed Concrete Slab Bridge Decks

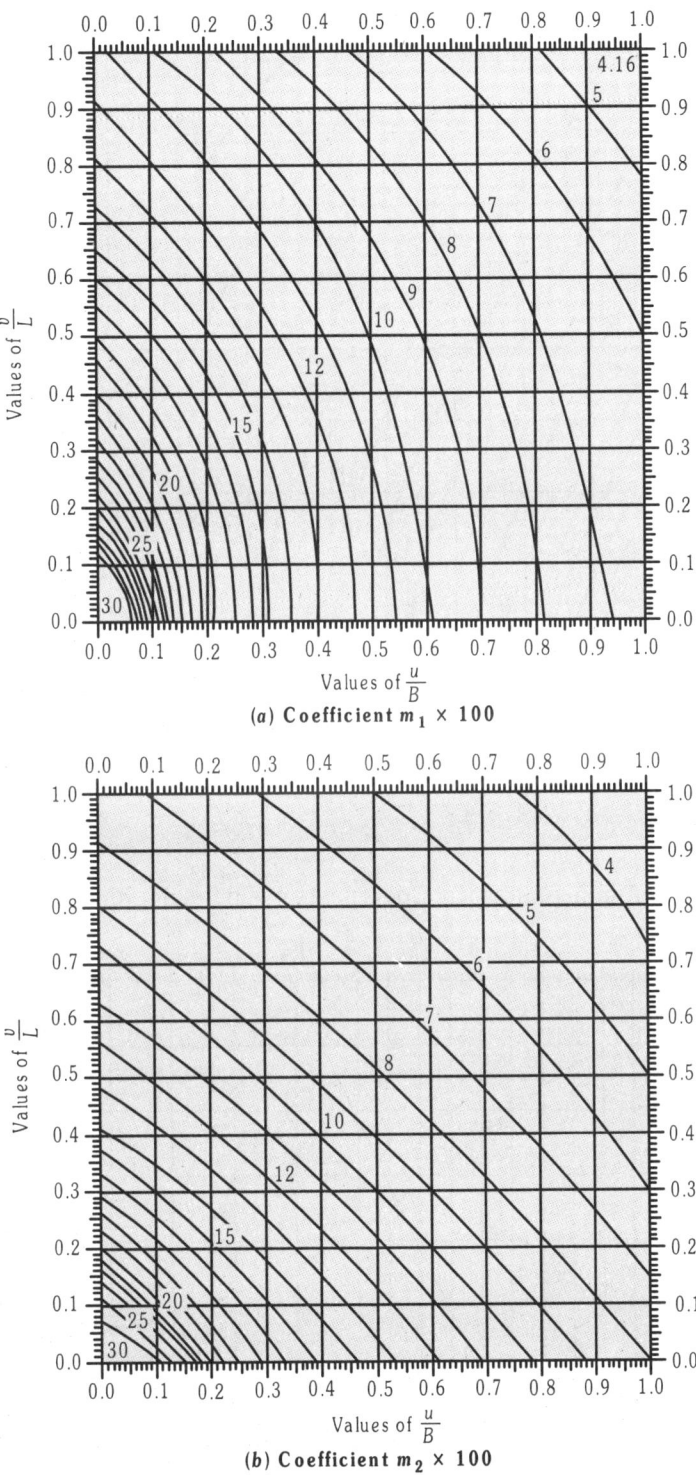

Fig. 5.9: Moment Coefficients m_1 and m_2 for $K = 0.9$

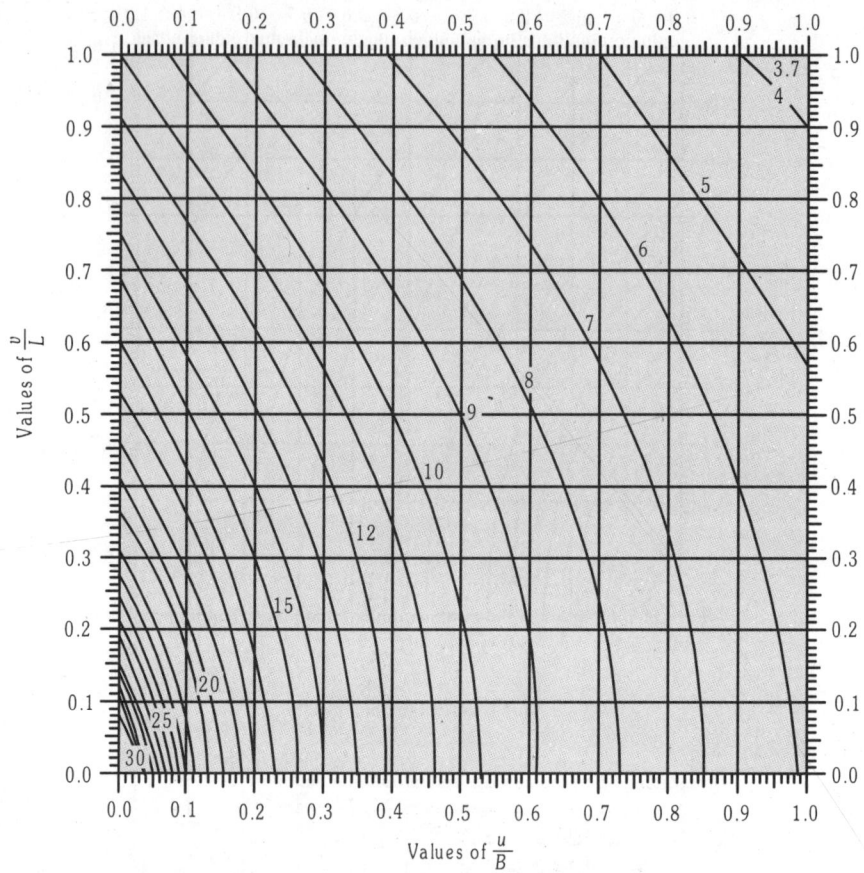

Fig. 5.10: Moment Coefficients m_1 and m_2 for $K = 1.0$

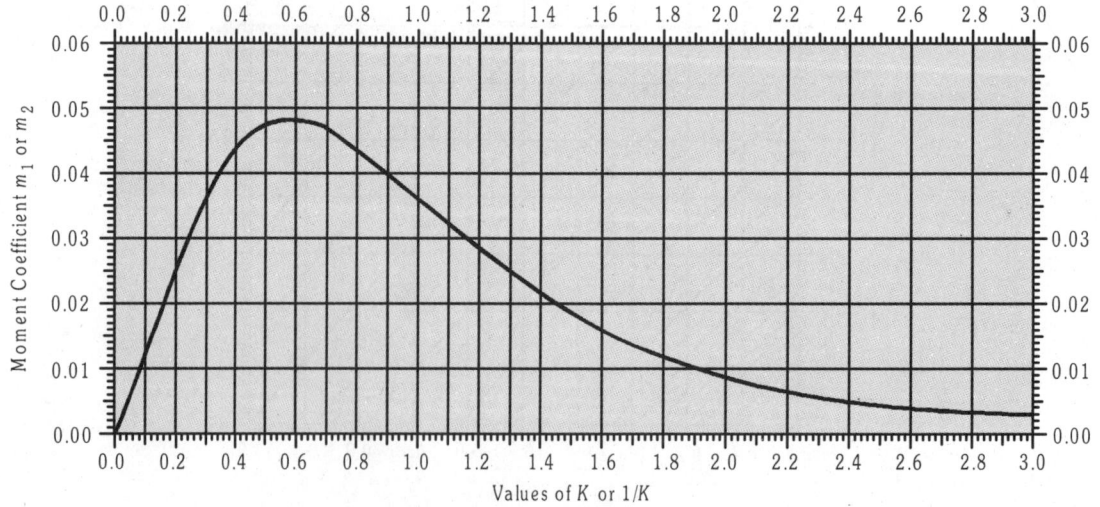

Fig. 5.11: Moment Coefficients for Slabs Completely Loaded with Uniformly Distributed Load, Coefficient is m_1 for K and m_2 for $1/K$

Case 1: Dispersion of wheel load through wearing coat only. Referring to Fig. 5.12,

$u = (0.50 + 2 \times 0.08) = 0.66$ m
$v = (0.25 + 2 \times 0.08) = 0.41$ m
$(u/B) = (0.66/2.0) = 0.33$
$(v/L) = (0.41/4.0) = 0.102$
$K = (B/L) = (2/4) = 0.5$

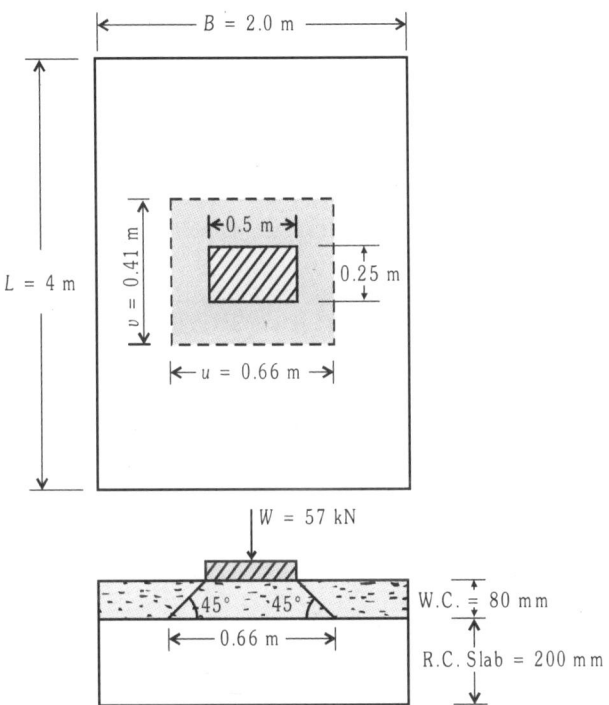

Fig. 5.12: Dispersion of Wheel Load through Wearing Coat

Using Pigeaud's curves corresponding to $K = 0.5$, the moment coefficients are read as $m_1 = 0.18$ and $m_2 = 0.13$

∴ $M_B = W(m_1 + \mu m_2) = 57(0.18 + 0.15 \times 0.13) = 11.37$ kNm
$M_L = W(m_2 + \mu m_1) = 57(0.13 + 0.15 \times 0.18) = 8.94$ kNm

Case 2: Dispersion of wheel load through wearing coat and structural slab at 45 degrees. Referring to Fig. 5.13

$u = (0.50 + 2 \times 0.28) = 1.06$ m
$v = (0.25 + 2 \times 0.28) = 0.81$ m
$(u/B) = (1.06/2.0) = 0.53$
$(v/L) = (0.81/4.0) = 0.202$
$K = (B/L) = (2/4) = 0.5$

Using Pigeaud's curves corresponding to $K = 0.5$, the moment coefficients are read as $m_1 = 0.138$ and $m_2 = 0.080$

∴ $M_B = W(m_1 + \mu m_2) = 57(0.138 + 0.15 \times 0.08) = 8.55$ kNm
$M_L = W(m_2 + \mu m_1) = 57(0.08 + 0.15 \times 0.138) = 5.74$ kNm

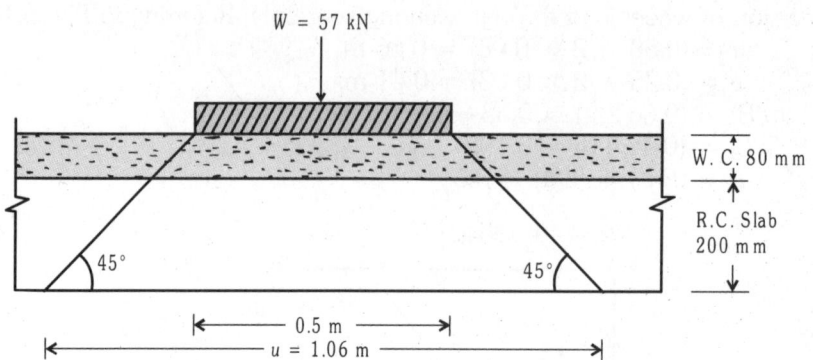

Fig. 5.13: Dispersion of Wheel Load through Wearing Coat and Deck Slab

Case 3: Dispersion of wheel load at 45 degrees through wearing coat and 85.9 degrees through structural slab. Referring to Fig. 5.14

$$u = \sqrt{(x + 2D)^2 + H^2} = \sqrt{(0.5 + 2 \times 0.08)^2 + 0.2^2} = 0.689 \text{ m}$$

$$v = \sqrt{(0.25 + 2 \times 0.08)^2 + 0.2^2} = 0.456 \text{ m}$$

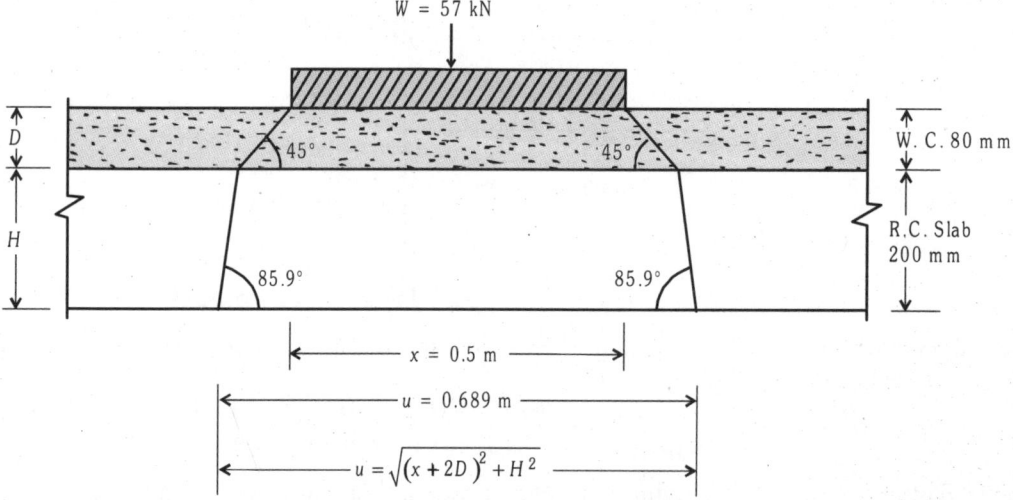

Fig. 5.14: Dispersion of Wheel Load at Different Angles through Wearing Coat and Deck Slab

Using Pigeaud's curves for $K = 0.5$, the moment coefficients are $m_1 = 0.175$ and $m_2 = 0.125$

∴ $M_B = W(m_1 + \mu m_2) = 57(0.175 + 0.15 \times 0.125) = 11.04$ kNm
$M_L = W(m_2 + \mu m_1) = 57(0.125 + 0.15 \times 0.175) = 8.62$ kNm

A comparative study[5, 6] of the moments resulting from different types of dispersion of wheel load indicates that the value of moments are maximum for Case 1 in which the dispersion is assumed at 45 degrees through wearing coat only. If the dispersion is assumed at a steeper angle of 85.9 degrees through the stiffer structural slab, the resulting moments are only 3 per cent less than Case 1. However if the dispersion is assumed at an uniform angle of 45 degrees through both the wearing coat and slab, the resulting moments are the least with a difference of 24 per cent in comparison with the results of Case 1 type load dispersion. Hence it is recommended that Case 1 type of load dispersion yielding maximum moments may as well be used for the design of two way slabs.

5.3 DESIGN AIDS AND TABLES FOR PRESTRESSED CONCRETE BRIDGE DECK SLABS

Pigeaud's method of computing the live load and dead load moments in concrete slabs supported on all sides assumes simple unyielding boundaries and arbitrary continuity coefficients are applied in order to account for the continuity of the slab over supports. This assumption is not valid particularly when the longitudinal and cross girders are cast monolithically with the slab so that the whole deck behaves as an integral structure under loads. In such a case the behaviour of slab can be approximated to that of a slab with fixed boundaries. This factor has been recognized in the German specifications DIN-1075[7] and ONORM B4204[8]. The bridge loadings are generally specified in the form of wheel or tracked loads with the dimensions of loaded area laid down by codes of practice.

Concentrated wheel load pattern does not permit simplified solutions unlike that for uniformly distributed load covering the entire slab. Further, the multiplicity of loading conditions with several wheel loads make the designer's job very tedious and time consuming. Although Pigeaud's curves provide a ready method of analysis for simple boundaries, still a lot of computations will be necessary particularly for eccentric loads.

Based on the work of Ruesch[9], involving slabs with continuous boundaries, design tables have been prepared by Suryaprakasha Rao et al.[10] for computation of bending moments in two way slabs with simply supported and fixed or continuous edges.

The various parameters considered are

1. The width (B) and length (L) of slab panel
2. Type of support at slab edges (Simply supported or fixed or continuous)
3. Type of loading such as Class 70R, Class AA tracked and wheeled vehicles and Class A train and uniformly distributed load.

The design tables are based on the following assumptions:

1. The slab is analyzed as a thin plate using the elastic analysis[11] with different boundary conditions.
2. The dispersion of the wheel loads is taken up to the middle surface of the slab
3. The wearing coat thickness is assumed as 75 mm and that of the structural slab as 150 mm. In case the thickness is greater than 150 mm, the values provided will be slightly on the conservative side.

The bending moments are tabulated for two different cases which are commonly encountered in highway bridge decks. These are (i) All edges are simply supported and (ii) All edges fixed or continuous.

The variables considered are the slab dimensions in the perpendicular directions designated as

L = length of slab in the longitudinal direction (direction of traffic)
B = width of slab in the transverse direction

M_{xc} and M_{yc} are the positive moments developed at the centre of slabs in the principal directions.

M_{xe} and M_{ye} are the negative moments developed at the edges in the principal directions. The range of values of slab dimensions, type of loading and the edge conditions covered in these tables are as follows:

(a) The length 'L' varies from 3 to 8 m with increments of 0.25 m
(b) The width 'B' varies from 1 to 5 m with increments of 0.25 m
(c) The type of loading comprises of IRC Class 70 R, Class AA and Class A
(d) A key chart indicates the plan dimensions of the deck slabs. Chart numbers varying from 1 to 285 cover the entire spectrum of the slab dimensions.
(e) Four sets of tables covering the various types of loading (Three types of live loads and uniformly distributed load due to dead load) and two types of edges, viz. all edges simply supported and all edges fixed or continuous are covered in the design tables with bending moments tabulated for all cases.

Tee beam slab decks generally have slabs continuous at edges enclosed by longitudinal and cross beams and hence only for this type of fixed slabs covering the slab dimensions (Length varying from 3 to 5 m and width from 2 to 2.5 m) commonly used in Indian Highway Bridge decks, bending moments are tabulated.

Table 5.2 gives the live load moments for IRC Class AA tracked loading and Table 5.3 gives the dead load moments for uniformly distributed load of 10 kN/m². These design tables are very useful in the design of slabs in tee beam and slab decks. Since the maximum bending moments in slabs occur for IRC Class AA tracked loading, only this particular case is covered in the following tables.

Table 5.2: Design Tables for Positive and Negative Moments in Slabs with Continuous Edges for IRC Class AA Tracked Loading

| Slab Dimensions | | Positive Moments | | Negative Moments | |
L (m)	B (m)	M_{xc} (kN·m)	M_{yc} (kN·m)	M_{xe} (kN·m)	M_{ye} (kN·m)
3	2	8.54	10.41	−18.52	−19.35
4	2	7.73	11.86	−16.73	−20.22
4	2.5	10.49	15.64	−22.66	−28.11
5	2.5	9.89	16.97	−22.56	−29.03
5	2	7.54	12.18	−16.40	−20.43

Table 5.3: Design Tables for Positive and Negative Bending Moments in Slabs for Uniformly Distributed Load of 10 kN/m²

| Slab Dimensions | | Positive Moments | | Negative Moments | |
L (m)	B (m)	M_{xc} (kN·m)	M_{yc} (kN·m)	M_{xe} (kN·m)	M_{ye} (kN·m)
3	2.0	0.61	1.43	−2.27	−2.99
4	2.0	0.39	1.64	−2.25	−3.28
4	2.5	0.87	2.34	−3.55	−4.83
5	2.5	0.61	2.58	−3.54	−5.16
5	2.0	0.29	1.69	−2.23	−3.33

Prestressed Concrete Slab Bridge Decks

5.4 MINIMUM REINFORCEMENTS IN SLABS

The minimum reinforcement in bridge deck slabs is governed by the specifications in Clause 15.1 of IRC: 18-2000 code[12]. In the case of solid deck slabs, the top and soffit sides of the slabs should be reinforced consisting of a grid formed by layers of bars. The minimum quantity of untensioned reinforcement should be 0.3 per cent and 0.18 per cent of the gross cross-sectional area of the slab for mild steel and HYSD bars respectively. The reinforcement should be equally distributed at the top and bottom of the slab.

In cantilever slabs, minimum reinforcement comprising of 4 HYSD bars of 16 mm diameter or 6 mild steel bars of 16 mm diameter should be provided with minimum spacing at the top divided equally between the top and bottom surface parallel to the support.

5.5 DESIGN EXAMPLE OF PRESTRESSED CONCRETE SLAB DECK FOR IRC STANDARD LOADS

Design a post tensioned prestressed concrete slab bridge deck for a National Highway Crossing to suit the following data:

1. **Data:**
    Clear span = 12 m
    Width of bearing = 500 mm
    Clear width of roadway = 7.5 m
    Foot paths: 1 m on either side
    Kerbs: 400 mm wide
    Thickness of wearing coat = 80 mm
    Live load: IRC Class A or AA tracked vehicle whichever is critical
    Type of structure: Class 1 (No tensile stresses permitted)
    Materials: M-40 Grade concrete and 7 mm diameter high tensile wires with characteristic tensile strength of 1500 N/mm² housed in cables with 12 wires and anchored by Freyssinet anchorages of 150 mm diameter. Adopt loss factor as 0.8
    For supplementary reinforcement, adopt FE-415 grade HYSD bars.
    Compressive strength of concrete at transfer = 35 N/mm².
    The design should conform to the recommendations of the codes IRC: 6-2000[1] and IRC: 18-2000[12]

2. **Permissible Stresses:** The permissible compressive stresses in concrete at transfer and service loads as recommended in IRC: 18-2000 are as follows:
    $f_{ct} = 15$ N/mm² $< 0.5 f_{ci} = (0.5 \times 35) = 17.5$ N/mm²
    $f_{cw} = 12$ N/mm² $< 0.33 f_{ck} = (0.33 \times 40) = 13.2$ N/mm²
    $f_{tt} = f_{tw} = 0$ and $\eta = 0.8$

3. **Depth of Slab and Effective Span:** For highway bridge decks, assume the thickness of slab at 50 mm per metre of span. Overall thickness of the slab = $(12 \times 50) = 600$ mm
    Width of bearing = 500 mm
    Effective span is the least of
    (i) Clear span + effective depth = $12 + (0.6 - 0.1) = 12.5$ m (effective cover = 100 mm)
    (ii) Centre to centre of supports = $(12 + 0.50) = 12.5$ m
    Hence effective span = $L = 12.5$ m
    The cross-section of the deck slab is shown in Fig. 5.15.

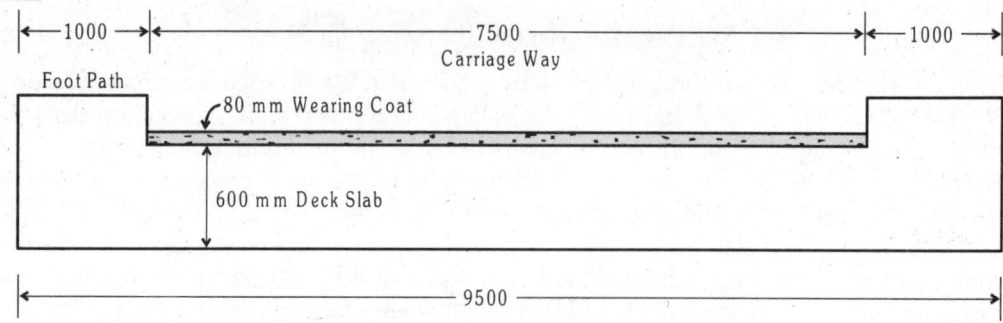

Fig. 5.15: Cross-Section of Deck Slab

4. **Dead Load Bending Moment**
   Dead weight of slab = (0.6 × 24) ... = 14.40 kN/m²
   Dead weight of wearing coat = (0.08 × 22) ... = 1.76 kN/m²

   Average dead weight of foot path = $\left(\dfrac{2 \times 1 \times 0.3 \times 24}{9.5}\right)$ ... = 1.51 kN/m²

   Self weight of parapet railing etc. (Lump sum) ... = 0.33 kN/m²
   Total dead load (g)... = 18.00 kN/m²
   Dead load bending moment (M_g) = (18 × 12.5²)/8 = 352 kNm

5. **Live Load Bending Moment:** Generally the bending moment due to live loads will be maximum for IRC Class AA tracked vehicle. Impact factor for Class AA tracked vehicle is 25 per cent for 5 m span, decreasing linearly to 10 per cent for 9 m span. Interpolating, we have the Impact Factor = 10 per cent for a span of 12.5 m.

   The tracked vehicle is placed symmetrically on the span.
   Effective length of load = [3.6 + 2(0.6 + 0.08)] = 4.96 m
   Referring to Fig. 5.16

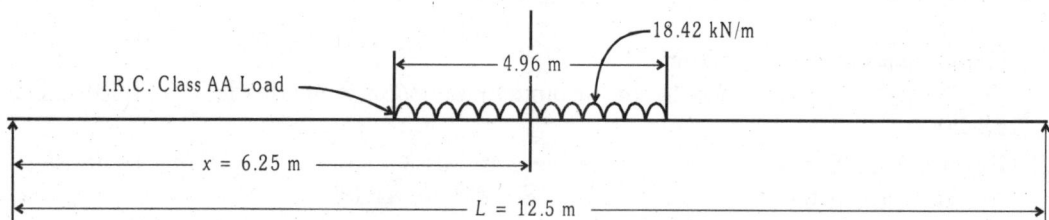

Fig. 5.16: Position of Load for Maximum Bending Moment

Effective width of slab perpendicular to span is expressed by the relation
$$b_e = K x (1 - x/L) + b_w$$
In this case, x = 6.25 m
L = 12.5 m
B = 9.5 m
b_w = (0.5 + 2 × 0.08) = 0.66 m
(B/L) = (9.5/12.5) = 0.76

From Table 5.1, for ratio (B/L) = 0.76, read out the value of K for simply supported span as 2.19.

$$b_e = (2.19 \times 6.25)\left[1 - \frac{6.25}{12.5}\right] + 0.66 = 7.50 \text{ m}$$

The tracked vehicle is placed close to the kerb with the required minimum clearance as shown in Fig. 5.17.

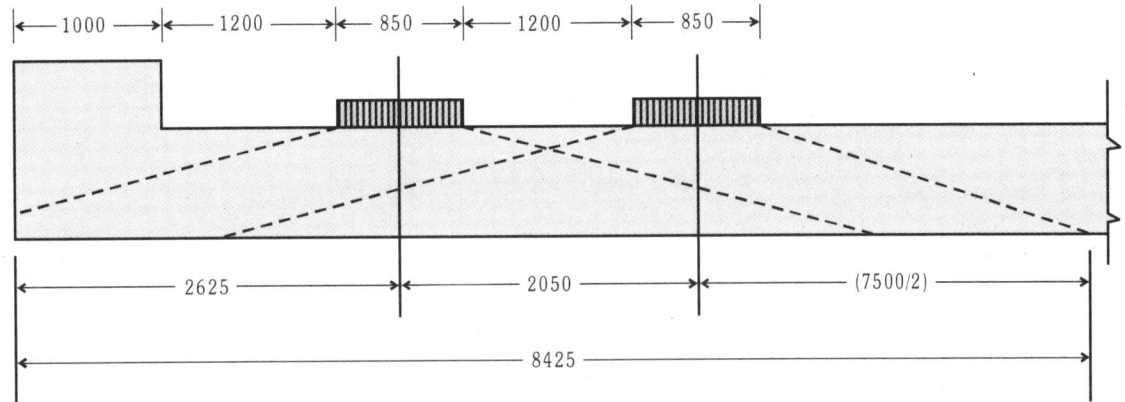

Fig. 5.17: Effective Width of Dispersion for IRC Class AA Tracked Vehicle

Net effective width of dispersion = 8.425 m
Total load of two tracks with impact = (700 × 1.10) = 770 kN

$$\text{Average intensity of load} = \frac{770}{(4.96 \times 8.425)} = 18.42 \text{ kN/m}^2$$

Referring to Fig. 5.16, maximum bending moment due to live load is computed as
$$M_q = [(18.42 \times 4.96)(0.5 \times 6.25)] - [(18.42 \times 4.96)(0.5 \times 0.25 \times 4.96)] = 229 \text{ kNm}$$

6. **Maximum Shear Force Due to Class AA Tracked Vehicle:** For maximum shear force at support section, the IRC Class AA tracked vehicle is arranged near the support as shown in Fig. 5.18.

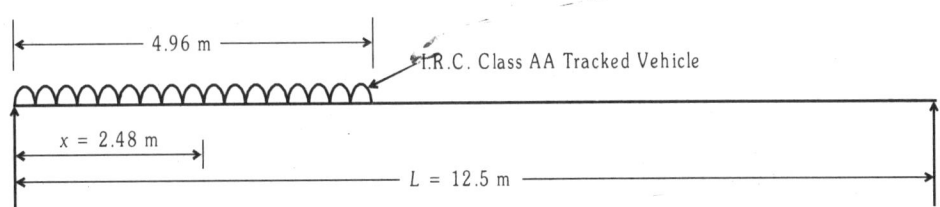

Fig. 5.18: Position of Load for Maximum Shear

Effective width of dispersion is given by the relation,
$$b_e = Kx(1 - x/L) + b_w$$
where $x = 2.48$ m, $L = 12.5$ m, $B = 9.5$ m, $b_w = 1.01$ m and $(B/L) = (9.5/12.5) = 0.76$
From Table 5.1, value of $K = 2.19$

∴ $b_e = [2.19 \times 2.48(1 - 2.48/12.5) + 1.01] = 5.363$ m

Referring to Fig. 5.17,
width of dispersion for two tracks is = [2625 + 2050 + (5363/2)] = 7356 mm

Intensity of load $= \left[\dfrac{770}{(4.96 \times 7.356)}\right] = 21.1$ kN/m^2

Live load shear force $V_A = (21.1 \times 4.96 \times 10.02)/12.5 = 83.9$ kN
Dead load shear force $= V_g = (0.5 \times 18 \times 12.5) = 112.5$ kN
$\therefore$ Total design shear force $= (83.9 + 112.5) = 196.4$ kN

7. Check for Minimum Section Modulus

Dead load moment $M_g = 352$ kNm
Live load moment $M_q = 229$ kNm

Section Modulus $= Z_t = Z_b = Z = \left[\dfrac{1000 \times 600^2}{6}\right] = 60 \times 10^6$ mm^3

The permissible stresses in concrete are

$f_{ct} = 15$ N/mm^2, $f_{cw} = 12$ N/mm^2, $f_{tw} = 0$ and $\eta = 0.8$
$f_{br} = (\eta f_{ct} - f_{tw}) = (0.8 \times 15 - 0) = 12$ N/mm^2

The minimum section modulus is given by

$$Z_b \geq \left[\dfrac{M_q + (1-\eta)M_g}{f_{br}}\right] \geq \left[\dfrac{229 \times 10^6 + (1-0.8)352 \times 10^6}{12}\right]$$

$\geq (24.95 \times 10^6)$ mm^3 $< (60 \times 10^6)$ mm^3

Hence the section selected is adequate to resist the service loads without exceeding the permissible stresses.

8. Prestressing Force:
In the present case the dead load moment (M_g) exceeds the live load moment (M_q) and hence it will not be possible to provide the minimum prestressing force since the corresponding eccentricity computed may be impracticable (eccentricity lying towards the soffit or even lying outside the section).

In such cases where $M_g > M_q$, the prestress at soffit (f_b) is computed and this value is used to compute the modified prestressing force using Eq. 4.6.

$$P = \left[\dfrac{A f_b Z_b}{Z_b + Ae}\right]$$

where $\quad f_b = \left[\dfrac{f_{tw}}{\eta} + \dfrac{(M_g + M_q)}{\eta Z_b}\right] = \left[0 + \dfrac{(352 + 229)10^6}{0.8 \times 60 \times 10^6}\right] = 12.10$ N/mm^2

and $\quad e =$ maximum possible eccentricity with due regard for cover requirements
$= [300 - 100] = 200$ mm

$$P = \left[\dfrac{(1000 \times 600)(12.10 \times 60 \times 10^6)}{(60 \times 10^6) + (1000 \times 600 \times 200)}\right] = (2420 \times 10^3) = 2420 \text{ kN}$$

Using Freyssinet cables containing 12 numbers of 7 mm diameter high tensile wires stressed to 1200 N/mm^2,

Force in each cable $= \left[\dfrac{12 \times 38.5 \times 1200}{1000}\right] = 554$ kN

Prestressed Concrete Slab Bridge Decks

$$\therefore \text{ Spacing of cables } = \left[\frac{1000 \times 554}{2420}\right] = 230 \text{ mm}$$

The cables are arranged in a parabolic profile with a maximum eccentricity of 200 mm at centre of span reducing to zero eccentricity (concentric) at supports and are spaced at 230 mm intervals in the transverse direction.

9. Check for Stresses at Service Loads

$P = 2420$ kN $\qquad M_g = 352$ kNm
$e = 200$ mm $\qquad M_q = 229$ kNm
$A = (1000 \times 600) = 6 \times 10^5 \text{ mm}^2$
$Z = 60 \times 10^6 \text{ mm}^3$
$(P/A) = [(2420 \times 10^3)/(6 \times 10^5)] = 4.03 \text{ N/mm}^2$
$(Pe/Z) = [(2420 \times 10^3 \times 200)/(6 \times 10^5)] = 8.06 \text{ N/mm}^2$
$(M_g/Z) = [(352 \times 10^6)/(60 \times 10^6)] = 5.86 \text{ N/mm}^2$
$(M_q/Z) = [(229 \times 10^6)/(60 \times 10^6)] = 3.81 \text{ N/mm}^2$

–Stresses at Transfer
At top of slab $= (4.03 - 8.06 + 5.86) = 1.83 \text{ N/mm}^2$
At bottom of slab $= (4.03 + 8.06 - 5.86) = 6.23 \text{ N/mm}^2$
Stresses at Working Loads
At top of slab $= 0.8 (4.03 - 8.06) + 5.86 + 3.81 = 6.65 \text{ N/mm}^2$
At bottom of slab $= 0.8 (4.03 + 8.06) - 5.86 - 3.81 = 0.002 \text{ N/mm}^2$
The actual stresses developed are within the safe permissible limits.

10. Check for Ultimate Flexural Strength

Considering 1 m width of slab,
$b = 1000$ mm
$d = 500$ mm
$A_p = [(12 \times 38.5 \times 1000)/230] = 2008 \text{ mm}^2$
$f_p = 1500 \text{ N/mm}^2$

(a) Failure by Yielding of Steel
$M_{us} = (0.9 \, d \, A_p \, f_p)$
$\quad = [0.9 \times 500 \times 2008 \times 1500]$
$\quad = (1355 \times 10^6) \text{ Nmm}$
$\quad = 1355 \text{ kNm}$

(b) Failure by Crushing of Concrete
$M_{uc} = (0.176 \, b \, d^2 \, f_{ck})$
$\quad = (0.176 \times 1000 \times 500^2 \times 40)$
$\quad = (176 \times 10^6) \text{ Nmm}$
$\quad = 1760 \text{ kNm}$

The safe flexural strength is the least of cases (a) and (b)
Hence $M_u = 1355$ kNm
Required Ultimate Flexural Strength $= [1.5 M_g + 2.5 M_q]$
$= [(1.5 \times 352) + (2.5 \times 229)]$
$= 1100$ kNm < 1355 kNm

Hence the section satisfies the limit state of collapse specifications prescribed In IRC: 18-2000[12] Code.

11. **Check for Ultimate Shear Strength**
Ultimate Shear Strength (Required) $= V_u = [1.5 V_g + 2.5 V_q]$
$= [(1.5 \times 112.5) + (2.5 \times 83.9)]$
$= 379$ kN

According to IRC: 18-2000, the ultimate shear resistance of support section uncracked in flexure is given by

$$V_{co} = 0.67bh\sqrt{f_t^2 + 0.8f_{cp} f_t} + \eta P \sin\theta$$

where $b =$ width of slab $= 1000$ mm
$h =$ overall depth $= 600$ mm
$f_t = 0.24\sqrt{f_{ck}} = 0.24\sqrt{40} = 1.51$ N/mm²
$f_{cp} =$ compressive prestress at centroidal axis
$= \left(\dfrac{\eta P}{A}\right) = \left(\dfrac{0.8 \times 2420 \times 10^3}{1000 \times 600}\right) = 3.23$ N/mm²

Eccentricity of cable at centre of span $= e = 200$ mm
The cables are concentric at support section

Slope of cable at support $\theta = \left(\dfrac{4e}{L}\right) = \left(\dfrac{4 \times 200}{12.5 \times 1000}\right) = 0.064$

$V_{co} = (0.67 \times 1000 \times 600)\sqrt{1.51^2 + (0.8 \times 3.23 \times 1.51)}$
$\qquad\qquad +(0.8 \times 2420 \times 10^3 \times 0.064)$
$= (1123 \times 10^3)$ N
$= 1123$ kN

Since the ultimate shear strength required (V_u) is less than 50 per cent of the ultimate shear resistance (V_{co}), no shear reinforcements are required in the slab.

12. **Supplementary Reinforcement:** According to clause 15.4 of IRC: 18-2000 code, for solid slabs, the minimum supplementary reinforcement is given as
$A_s = 0.18$ per cent of the gross cross-sectional area
$= (0.0018 \times 1000 \times 600) = 1080$ mm²

Provide 12 mm diameter FE-415 HYSD bars at 200 mm centres both at top and bottom of the slab in the longitudinal and transverse directions.

13. **Design of End Block Reinforcement:** At the support section, the concentric cables carrying a force of 554 kN are spaced at intervals of 230 mm. The bursting tension is computed using Table 4.6.

Prestressed Concrete Slab Bridge Decks

In the present case using the values of

$P_k = 554$ kN

$2y_o$ = side of the end block = 230 mm

$2y_{po}$ = side of the loaded area = 150 mm

∴ $(y_{po}/y_o) = (150/230) = 0.65$

Interpolating the value of (F_{bst}/P_k) for $(y_{po}/y_o) = 0.65$

$F_{bst} = (0.125 \times 554) = 69.25$ kN

Using 10 mm diameter FE-415 HYSD bars as end block reinforcement

Area of steel = $[(69.25 \times 10^3)/(0.87 \times 415)] = 192$ mm^2

Provide 10 mm diameter bars at 100 mm centres in the vertical and horizontal directions in two planes in the form of mesh at distances of 100 and 200 mm from the anchorages. The longitudinal and cross-sections of the deck slab are shown in Figs. 5.19 and 5.20.

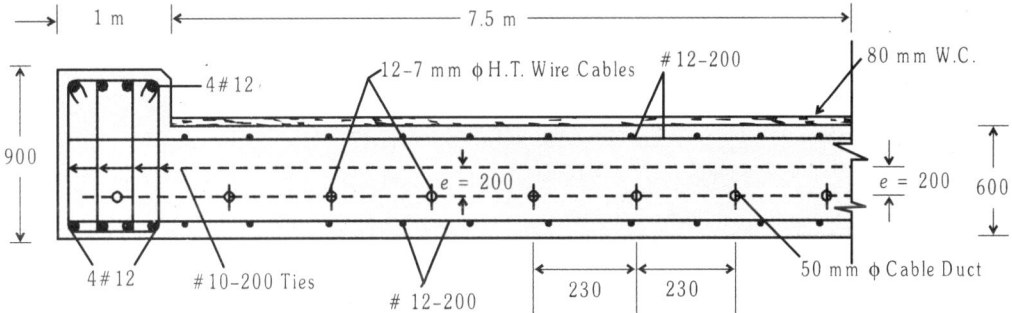

Fig. 5.19: Cross-section of Deck Slab at Centre of Span

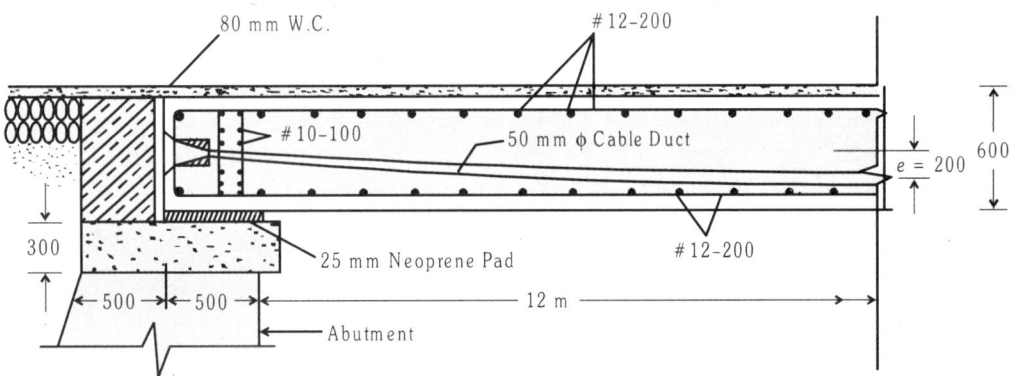

Fig. 5.20: Longitudinal Section of Deck Slab

REFERENCES

1. IRC: 6-2000, Standard Specifications and Code of Practice for Road Bridges Section-II, Loads and Stresses (Fourth Revision), Indian Roads Congress, New Delhi, 2000, pp. 11-16.
2. IRC: 21-2000, Standard Specifications and Code of Practice for Road Bridges Section-III, Cement Concrete (Plain and Reinforced) (Third Revision), Indian Roads Congress, New Delhi, 2000, pp. 49-55.
3. ROWE, R.E., *Concrete Bridge Design*. C.R. Books Ltd, First Edition, London 1962, pp. 336.
4. VICTOR, J.D., *Essentials of Bridge Engineering* (Fifth Edition). Oxford and IBH Publishing Co, New Delhi, 2001, pp. 391-413.
5. KRISHNA RAJU, N., *Design of Bridges* (Third Edition). Oxford and IBH Publishing Co, New Delhi, 1998, pp. 156-169.
6. SARKAR, S, KAPLA, M.S. PRASADA RAO, A.S. and CHHAUDA, J.N., *Hand Book for Prestressed Concrete Bridges*. Structural Engineering Research Centre, Roorkee, 1969, pp. 1-238.
7. DIN 1075-1955, Massisve Bruecken, Berechnungs Grundlagen, German Specifications.
8. ONORM-B4202, *Berechnung und Ausfuehrung der Tragwerke Massisve Strassebruechken*. 1958 (Revised 1975).
9. RUESCH, H., *Berechnungsstafeln fuer rechtwinklige Fahrbahnplatten von Strassenbriecken*. Wilhelm Ernst and Sohn, Berlin, 1965.
10. SURYA PRAKASH RAO, D., TAMHANKAR, M.G., KAPLA, M.S., and AMRIT SINGH., Design Tables for Concrete Bridge Deck Slabs, Structural Engineering Research Centre, Roorkee, Uttaranchal, 1976, pp. 1-70.
11. TIMOSHENKO, S. and WOINOWSKY-KRIEGER, S., *Theory of Plates and Shells*. McGraw Hill Book Co., 1959.
12. IRC: 18-2000, Design Criteria for Prestressed Concrete Road Bridges (Post-Tensioned Concrete) (Third Revision), Indian Roads Congress, New Delhi, 2000, pp. 30-32.

EXERCISES

1. A prestressed concrete slab 450 mm thick with parallel post tensioned cables is provided for a road bridge deck slab of effective span 10 m. The live load analysis indicates an equivalent uniformly distributed live load of 40 kN/m^2.

   The force at transfer in each cable is 400 kN. If the permissible compressive stress in concrete at transfer is 16 N/mm^2, design the slab as Class 1 type structure and determine the spacings of the cables and their eccentricity at mid span section. Assume loss ratio as 0.85.

2. Design a post tensioned prestressed concrete slab bridge deck for a National Highway crossing to suit the following data:
   Clear span = 12 m
   Width of bearing = 500 mm
   Clear width of roadway = 7.5 m
   Foot paths: one metre on either side with 600 mm wide kerbs
   Thickness of wearing coat = 100 mm
   Live load: IRC Class AA tracked vehicle
   Type of structure: Class 1 type
   Materials: M-50 Grade concrete and 8 mm diameter high tensile wires with an ultimate tensile strength of 1375 N/mm^2 housed in cables with 12 wires and anchored by Freyssinet anchorages of 150 mm diameter.

Prestressed Concrete Slab Bridge Decks

Adopt Fe-415 HYSD bars for supplementary reinforcements
Compressive strength of concrete at transfer = 40 N/mm^2
Loss ratio = 0.85
The design should conform to the specifications of IRC Codes 6-2000, 18-2000 and 21-2000. Sketch the details of cables and reinforcements in the slab.

3. A post tensioned prestressed concrete slab bridge deck is to be designed for a National Highway crossing to suit the following data:
Clear width of road way = 15 m
Foot paths of 1 m on either side with 600 mm kerbs
Thickness of wearing coat = 100 mm
Effective span of deck slab = 10 m
Type of structure: Class 1 type
Live loads: IRC Class AA tracked vehicle
Materials: M-40 Grade concrete and 7 mm diameter high tensile wires housed in Freyssinet cables.
Adopt Fe-415 HYSD bars for supplementary reinforcements.
Loss ratio = 0.80
Compressive strength of concrete at transfer = 18 N/mm^2
The design of deck slab should conform to the latest IRC Codes 6-2000 21-2000 and 18-2000.
Sketch the details of cables and reinforcements in the critical longitudinal and cross-sections of the deck slab.

6

PRESTRESSED CONCRETE TEE BEAM AND SLAB BRIDGE DECK

6.1 GENERAL FEATURES

Tee beam and slab bridge decks are by far the most commonly adopted type in the medium span range of 20 to 30 m. Before the advent of prestressed concrete, tee beam and slab decks were made up of reinforced concrete. The structure is so named because the main longitudinal girders spaced at intervals of 2 to 3 m are designed as tee beams integral with the deck slab which is cast monolithically with the beams. Alternatively the tee girders can be precast in a casting yard and transported to the site of construction. Cast *in situ* reinforced concrete slab connects all the girders and ensures structural integrity of bridge deck in resisting the live loads.

Modern tee beam and slab bridge decks comprise of main longitudinal girders together with cross girders at regular intervals along the span to improve the rigidity of the bridge deck and reduce the deflections under heavy rolling loads. Experimental investigations conducted by Rowe et al.[1] and Victor[2] indicate that the structural behaviour of tee beam and slab decks at the limit states of serviceability and collapse shows significant improvement with the use of cross girders at regular intervals along the span.

The main girders house the high tensile cables and these are tensioned after the concrete attains the desired strength. The typical cross-sections of cast *in situ* tee beam and slab bridge decks without and with cross girders and precast tee girders with cast *in situ* slab are shown in Figures 6.1 (a), (b) and (c) respectively.

6.2 STRUCTURAL COMPONENTS OF TEE BEAM AND SLAB BRIDGE DECKS

The main structural components of prestressed concrete tee beam and slab bridge decks are compiled below:

1. Deck Slab
2. Main or Longitudinal Beams
3. Cross Girders or Diaphragms
4. Cantilever Portion
5. Foot Paths, Kerbs and Hand Rails

6. Parapets
7. Wearing Course

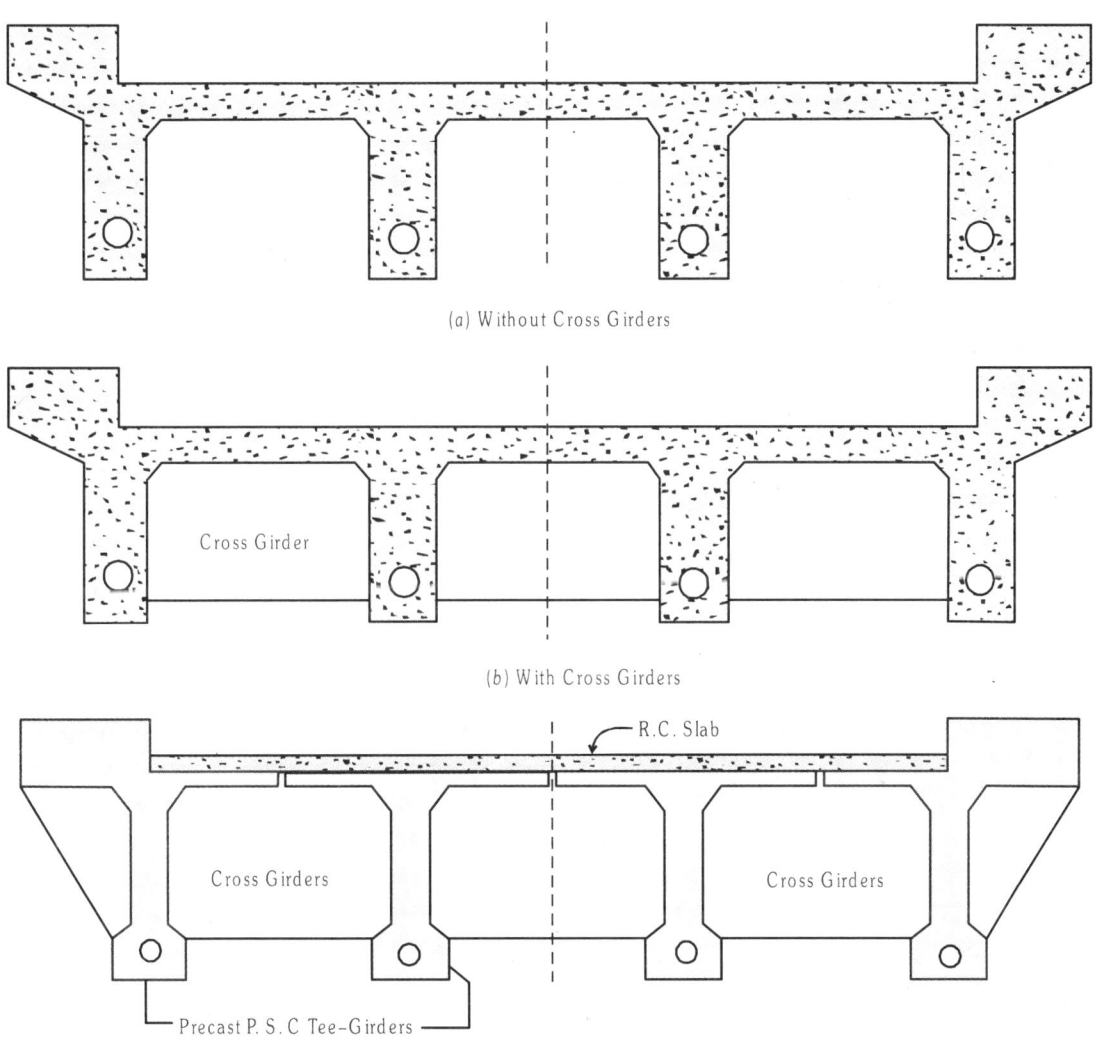

Fig. 6.1: Cross Sections of Tee Beam and Slab Bridge Decks

1. **Deck Slab:** The thickness of the continuous slab between the main girders depends upon the spacing of the longitudinal girders. For a two lane carriage way of 7.5 m, a minimum of three girders are required. However to reduce the thickness of the slab, four to five girders generally are used. The thickness of the structural concrete slab usually varies in the range of 200 to 300 mm and it is cast using concrete of Grade M-20 to M-30. The deck slab continuous on main and cross girders is designed as a two way slab using Pigeaud's curves outlined in section 5.2.

2. **Main or Longitudinal Beams:** The overall depth of the main girder is generally selected as a function of the span. Based on practical experience, an overall depth of 50 to 60 mm per metre of span length will satisfy the limit states of serviceability and collapse. The number of main girders is influenced by the width of the carriage way and normally the spacing between the main girders varies from 2 to 3 m. However a minimum of three girders are essential for the minimum carriage way of 7.5 m.

   The thickness of the flange of the tee beam is generally kept in the range of 200 to 250 mm and this will serve as the slab between the girders. The thickness of the web depends upon the diameter of the cable carrying the high tensile wires or strands. The cable housed in the web should have sufficient cover of not less than 75 mm as prescribed in IRC: 18-2000. The thickness of the web is also influenced by the service load shear force at supports. The bottom flange dimensions are governed by the requirements of housing the cables with suitable cover. The number and rows of cables in the horizontal and vertical planes, side and soffit cover requirements are also considered in fixing the dimensions of the bottom flange. 500 mm wide by 500 mm deep bottom flange can easily accommodate three rows of cables both in the horizontal and vertical directions satisfying the codal cover and spacing requirements.

3. **Cross Girders or Diaphragms:** Cross girders are generally provided at intervals of 4 to 5 m along the span to improve the flexural and torsional rigidity of the bridge deck. The thickness of cross girder should be not less than the web of the main girder and the depth should be not less than 75 per cent of the depth of the main girder. The width of the cross girder should be sufficient to house high tensile cables with due regard to cover requirements. Cross girders are prestressed using straight cables with nominal prestress to improve the flexural rigidity of the deck. In comparison with main girders, the flexural and shear stresses developed in cross girders are of smaller magnitude.

4. **Cantilever Portion:** The cantilever portion carries the Kerb, hand rails and foot path if any and also a part of the carriage way. The junction of the cantilever slab and the longitudinal girder forms the critical section to be designed to resist the hogging moments. For calculation of live load bending moments, the effective width of cantilever portion is assessed using the formula given in Clause 305.16.2 of IRC: 21-2000[3] Bridge Code which is presented in section 5.2.

5. **Foot Paths, Kerbs and Hand Rails:** Foot paths are generally designed to support a live load of 400 kg/m^2 and in bridges in the vicinity of towns and where large crowds are likely to occur, the foot way loading should be increased to 500 kg/m^2 as prescribed in Clause 209.1 of Indian Roads Congress Code IRC: 6-2000[4]. Kerbs having a width of 600 mm and depth of 225 mm with a slope of 1 in 8 for 200 mm height and having curved edge with a radius of 25 mm at the top are provided on either side of the road. If the kerb width is less than 600 mm, no live loads are to be applied in addition to the lateral load. The kerb is designed to resist a local lateral force of 750 kg/m applied horizontally at the top as outlined in Clause 209.2 of IRC: 6-2000[4].

   Hand rails are provided as a protective measure to keep the bridge users from falling to the depth over the sides of the bridge. Depending upon the situations, the hand rails can be either of the post and rail system or panel slab and post system. The R.C. posts are generally of size 150 by 100 mm spaced at intervals of less than 1.8 m and cast monolithic with the kerb. The rails are of rectangular shape and are precast having a size of 125 by 75 mm. Horizontal railing should be for a height of not less than 1 m, provided in two rows. The clear distance from the lower rail to the top of the kerb should be not more than 150 mm and the space between the bottom and top rails should be filled by means of closely spaced horizontal or inclined members.

6. **Parapets:** Parapets are provided for culverts with a minimum height of 450 mm above the road surface. Raina[5] has presented a detailed classification of highway bridge parapets together with detailed drawings to suit the different types of bridges. Parapets were widely used in 1950's for stone masonry arched bridges. Now-a-days hand rails made up of R.C.C. posts and precast concrete rails are generally preferred to parapets.

7. **Wearing Course:** A wearing course is generally provided over concrete bridge decks to protect the structural concrete from the direct wearing effects of moving vehicles and to provide the cross camber required for surface drainage. The wearing course is made up of either asphaltic concrete or cement concrete. Asphaltic concrete wearing course of 56 mm uniform thickness is preferred when the road pavement on the approach on either side of the bridge is of asphaltic concrete. This wearing course consists of a prime coat of mastic asphalt 6 mm thick over the deck slab followed by two layers each of 25 mm thick asphaltic concrete.

In the case of isolated bridges where use of asphaltic concrete is inconvenient, cement concrete wearing course of 75 mm thickness using M-30 grade concrete is adopted. This type of wearing course should be reinforced with mesh reinforcement comprising 6 mm diameter bars at 200 mm centres in both directions where the deck slab is in compression and with 6 mm diameter bars at 100 mm centres when the deck slab is in tension as in the case of continuous decks over the interior supports.

6.3 LOAD DISTRIBUTION METHODS FOR BEAM AND SLAB BRIDGE DECKS

The tee beam and slab bridge deck comprising longitudinal and cross girders with the deck slab may be considered as rigid grid structure for purposes of analysis under concentrated live loads. Concentrated wheel load on the deck is shared between the longitudinal girders depending upon the position of the load, the number of girders and their spacing. The problem of transverse distribution of live load among the longitudinal girders of the deck has been investigated by Morice and Little[6], Hendry and Jaeger[7], Cusens[8], Courbon[9] and Guyon-Massonet[10]. Bridge decks with three or more longitudinal girders are analysed for load distribution using any of the following rational methods.

1. Courbon's method
2. Morice and Little version of Guyon-Massonet method
3. Hendry-Jaeger method

The prominent features of these methods are summarized as follows:

1. **Courbon's Method:** Among the various load distribution methods, Courbon's method is the simplest and is applicable when the following conditions are satisfied.

(a) The ratio of span to width of deck is greater than 2 but less than 4.
(b) The longitudinal girders are interconnected by at least five symmetrically spaced cross girders.
(c) The cross girder extends to a depth of at least 0.75 times the depth of the longitudinal girders.

Courbon's method is popular due to the simplicity of computations as detailed below: In this method the wheel loads are placed on the deck in such way that the centroid of loads has the maximum eccentricity from the centre line of the road way. Figure 6.2 shows the typical position of two wheel loads (IRC Class AA tracked vehicle) arranged in such a way that the eccentricity of the loads measured from the centre of the road is maximum. Due to this eccentricity, the loads shared by each girder is increased or decreased depending upon the position of the girders. The load shared by any girder is computed by Courbon's theory in terms of a reaction factor expressed by the empirical relation as,

$$R_x = \left(\sum \frac{W}{n}\right)\left[1 + \left(\frac{\sum I}{\sum d_x^2 \cdot I}\right) d_x \cdot e\right]$$

where
R_x = Reaction factor for the girder under consideration
I = Second moment of area of each longitudinal girder
d_x = Distance of the girder under consideration from the centre line of road
W = Total concentrated live load
n = Number of longitudinal girders
e = Eccentricity of live load with respect to the axis of the bridge deck.

The live load bending moments and shear forces are computed for each of the girders. The maximum design moments and shear forces are obtained by adding the live load and dead load bending moments.

2. **Guyon-Massonet Method:** Guyon-Massonet method is based on the application of orthotropic plate theory to the bridge deck system. Morice and Little have successfully applied this theory to the analysis of bridge deck systems. The method has the advantage of using a single set of distribution coefficients for the two extreme cases of no torsion grillage and a full torsion slab thus enabling the determination of the load distribution behaviour of any type of bridge deck.

The longitudinal bending moments at various points along the cross-section are obtained by multiplying the mean longitudinal bending moment by the appropriate distribution coefficients for these points. The mean longitudinal bending moment is the bending moment developed by considering the total load on the span as uniformly spread over the whole width of the bridge. Hence the mean bending moment per girder can be expressed as

$$M_{mean} = (M/n)$$

where
M = Total mean longitudinal bending moment
n = Number of girders

The design bending moment is then computed as
Design B.M. = $(1.10 \times K \times M_{mean} \times I.F.)$
where
K = Distribution coefficient
I.F. = Impact factor

The factor 1.10 is used to compensate for the error involved in using only the first term of the Fourier series in finding the distribution coefficients as suggested by Rowe[1] based on experiments.

The distribution coefficient 'K' depends on the flexural and torsional parameters expressed as,

Flexural Parameter: $\theta = (b/2a)(i/j)^{0.25}$ ...(6.1)

Torsional parameter: $\alpha = [G(i_0 + j_0)/(2E\sqrt{ij})]$ ...(6.2)

where $2a$ = Span of the bridge
 $2b$ = Effective width of bridge
 i = Second moment of area per unit transverse width
 j = Second moment of area per unit longitudinal width
 $G \cdot i_0$ = Torsional stiffness per unit width
 $G \cdot j_0$ = Torsional stiffness per unit length

The values of distribution coefficient K_α is calculated from the interpolation formula expressed as,

$$K_\alpha = K_0 + (K_1 - K_0)\sqrt{\alpha} \qquad ...(6.3)$$

where K_0 and K_1 refers to the distribution coefficient corresponding to $\alpha = 0$ and $\alpha = 1$. Rowe has presented the values of K_0 and K_1 for five reference stations (0, $b/4$, $b/2$, $3b/4$ and b) and for various load positions and for values of θ from 0 to 3.0 in a graphical form. The values of K_0 and K_1 for range of θ between 0.2 and 0.8 have been presented in a tabular form for ready use in design office by Sarkar et al.[11]. These are compiled in Table 6.1. The reference stations and load positions for maximum distribution are shown in Fig. 6.3.

Table 6.1: Values of K_0 and K_1 for Various Values of θ

Ref. Point Load at	$\theta = 0.20$								
	$-b$	$-3b/4$	$-b/2$	$-b/4$	0	$b/4$	$b/2$	$3b/4$	b
					K_0				
0	0.94	0.99	0.97	1.02	1.06	1.02	0.97	0.99	0.94
$b/4$	0.25	0.42	0.63	0.84	1.02	1.19	1.35	1.56	1.73
$b/2$	−0.53	−0.15	0.25	0.63	0.97	1.35	1.72	2.10	2.40
$3b/4$	−0.66	−0.66	−0.15	0.42	0.99	1.56	2.10	2.70	3.27
b	−1.90	−1.20	−0.53	0.25	0.94	1.73	2.49	3.27	4.00
					K_1				
0	0.96	0.99	1.00	1.00	1.03	1.00	1.00	0.99	0.96
$b/4$	0.91	0.93	0.97	0.98	1.00	1.03	1.03	1.03	1.03
$b/2$	0.86	0.90	0.93	0.97	1.00	1.03	1.07	1.10	1.13
$3b/4$	0.80	0.85	0.90	0.93	0.99	1.03	1.10	1.16	1.23
b	0.75	0.80	0.86	0.91	0.96	1.03	1.13	1.23	1.35

				$\theta = 0.25$ K_0					
0	0.90	0.97	0.98	1.04	1.08	1.04	0.98	0.97	0.90
b/4	0.22	0.41	0.63	0.85	1.04	1.20	1.35	1.54	1.70
b/2	−0.53	−0.15	0.24	0.63	0.98	1.35	1.72	2.10	2.47
3b/4	−1.17	−0.64	−0.15	−0.41	−0.97	1.54	2.10	2.71	3.28
b	−1.85	−1.17	−0.53	−0.22	0.90	1.70	2.47	3.28	4.00
				K_1					
0	0.96	0.98	1.00	1.02	1.04	1.02	1.00	0.98	0.96
b/4	0.88	0.91	0.96	0.97	1.02	1.05	1.05	1.05	1.04
b/2	0.81	0.86	0.91	0.96	1.00	1.05	1.10	1.13	1.16
3b/4	0.75	0.80	0.86	0.91	0.98	1.05	1.13	1.22	1.30
b	0.69	0.75	0.81	0.88	0.96	1.04	1.16	1.30	1.46
				$\theta = 0.30$ K_0					
0	0.86	0.95	0.97	1.05	1.10	1.05	1.97	0.95	0.86
b/4	0.20	0.40	0.63	0.87	1.05	1.22	1.36	1.5	1.68
b/2	−0.54	−0.16	0.24	0.63	0.97	1.36	1.73	2.10	2.46
3b/4	−1.15	−0.63	−0.16	0.40	0.95	1.53	2.10	2.73	3.31
b	−1.79	−1.15	−0.54	0.20	0.86	1.68	2.46	3.31	4.10
				K_1					
0	0.94	0.97	1.00	1.02	1.05	1.02	1.00	0.97	0.94
b/4	0.85	0.89	0.94	0.97	1.02	1.06	1.06	1.05	1.06
b/2	0.77	0.82	0.89	0.94	1.00	1.06	1.13	1.17	1.21
3b/4	0.70	0.75	0.82	0.89	0.97	1.05	1.17	1.29	1.38
b	0.63	0.70	0.77	0.85	0.94	1.06	1.21	1.38	1.59
				$\theta = 0.35$ K_0					
0	0.80	0.93	0.98	1.08	1.15	1.08	0.9	0.93	0.80
b/4	0.17	0.39	0.63	0.89	1.08	1.25	1.38	1.50	1.62
b/2	−0.54	−0.17	0.24	0.63	0.98	1.38	1.75	2.10	2.43
3b/4	−1.11	−0.60	−0.17	0.39	0.93	1.50	2.10	2.75	3.34
b	−1.70	−1.11	−0.54	0.17	0.80	1.62	2.43	3.34	4.20
				K_1					
0	0.94	0.96	1.00	1.04	1.06	1.04	1.00	0.96	0.94
b/4	0.81	0.85	0.90	0.97	1.04	1.08	1.08	1.07	1.06
b/2	0.71	0.77	0.85	0.92	1.00	1.08	1.17	1.21	1.25
3b/4	0.65	0.70	0.77	0.85	0.96	1.07	1.21	1.35	1.46
b	0.56	0.65	0.71	0.81	0.94	1.06	1.25	1.46	1.72

Prestressed Concrete Tee Beam and Slab Bridge Deck

				$\theta = 0.40$ K_0					
0	0.71	0.90	0.99	1.11	1.20	1.11	0.99	0.90	0.71
$b/4$	0.12	0.36	0.64	0.91	1.11	1.29	1.40	1.47	1.56
$b/2$	−0.55	−0.17	0.23	0.63	0.99	1.37	1.76	2.10	2.40
$3b/4$	−1.07	−0.58	−0.17	0.36	0.90	1.47	2.10	2.77	3.38
b	−1.65	−1.07	−0.55	0.12	0.71	1.56	2.40	3.38	4.30
				K_1					
0	0.90	0.95	1.00	1.05	1.08	1.05	1.00	0.95	0.90
$b/4$	0.77	0.83	0.90	0.96	1.05	1.10	1.10	1.09	1.07
$b/2$	0.66	0.73	0.81	0.90	1.00	1.10	1.20	1.26	1.30
$3b/4$	0.58	0.65	0.73	0.83	0.95	1.09	1.26	1.41	1.55
b	0.50	0.58	0.66	0.77	0.90	1.07	1.30	1.55	1.88
				$\theta = 0.45$ K_0					
0	0.63	0.85	1.00	1.15	1.25	1.15	1.00	0.85	0.63
$b/4$	0.08	0.34	0.64	0.94	1.15	1.34	1.42	1.44	1.50
$b/2$	−0.54	−0.17	0.23	0.64	1.00	1.38	1.78	2.09	2.35
$3b/4$	−1.02	−0.56	−0.17	0.34	0.85	1.44	2.09	2.08	3.43
b	−1.55	−1.02	−0.54	0.08	0.63	1.50	2.35	3.43	4.50
				K_1					
0	0.88	0.95	1.00	1.06	1.10	1.06	1.00	0.95	0.88
$b/4$	0.73	0.79	0.87	0.96	1.06	1.14	1.14	1.10	1.09
$b/2$	0.60	0.67	0.76	0.87	1.00	1.14	1.25	1.30	1.35
$3b/4$	0.50	0.58	0.67	0.79	0.95	1.10	1.30	1.50	1.67
b	0.44	0.50	0.60	0.73	0.88	1.09	1.35	1.67	2.00
				$\theta = 0.50$ K_0					
0	0.55	0.79	1.00	1.21	1.32	1.21	1.00	0.79	0.55
$b/4$	0.00	0.30	0.63	0.96	1.21	1.40	1.44	1.40	1.40
$b/2$	−0.54	−0.17	0.22	0.63	1.00	1.40	1.80	2.08	2.30
$3b/4$	−0.96	−0.54	−0.10	0.30	0.79	1.40	2.08	2.84	3.50
b	−1.43	−0.96	−0.54	0.00	0.55	1.40	2.30	3.50	4.80
				K_1					
0	0.85	0.92	1.00	1.07	1.13	1.07	1.00	0.92	0.85
$b/4$	0.68	0.76	0.86	0.96	1.07	1.16	1.15	1.12	1.09
$b/2$	0.55	0.63	0.73	0.86	1.00	1.15	1.30	1.35	1.39
$3b/4$	0.45	0.53	0.63	0.76	0.92	1.12	1.35	1.58	1.76
b	0.38	0.45	0.55	0.68	0.85	1.09	1.39	1.76	2.15

				$\theta = 0.55$ K_0					
0	0.42	0.74	1.02	1.27	1.40	1.27	1.02	0.74	0.42
b/4	−0.10	0.25	0.63	0.98	1.27	1.45	1.46	1.35	1.26
b/2	−0.53	−0.18	0.21	0.63	1.02	1.43	1.84	2.07	2.25
3b/4	−0.89	−0.50	−0.18	0.25	0.74	1.35	2.07	2.87	3.70
b	−1.30	−0.89	−0.53	−0.10	0.42	1.26	2.25	3.70	5.10
				K_1					
0	0.82	0.90	1.00	1.09	1.15	1.09	1.00	0.90	0.81
b/4	0.65	0.71	0.84	0.96	1.09	1.18	1.17	1.14	1.09
b/2	0.50	0.58	0.69	0.84	1.00	1.17	1.35	1.40	1.44
3b/4	0.40	0.48	0.58	0.71	0.90	1.14	1.40	1.65	1.87
b	0.33	0.40	0.50	0.65	0.81	1.09	1.44	1.87	2.33
				$\theta = 0.60$ K_0					
0	0.31	0.66	1.02	1.35	1.50	1.35	1.02	0.66	0.31
b/4	−0.17	0.21	0.62	1.02	1.35	1.53	1.47	1.31	1.03
b/2	−0.52	−0.18	0.20	0.62	1.02	1.47	1.87	2.06	2.19
3b/4	−0.80	−0.47	−0.18	0.21	0.66	1.31	2.06	2.92	3.08
b	−1.05	−0.80	−0.52	−0.20	0.31	1.10	2.19	3.08	5.45
				K_1					
0	0.80	0.89	1.00	1.12	1.19	1.12	1.00	0.89	0.80
b/4	0.58	0.67	0.80	0.95	1.12	1.23	1.20	1.15	1.08
b/2	0.44	0.52	0.63	0.80	1.00	1.20	1.40	1.45	1.46
3b/4	0.34	0.41	0.52	0.67	0.89	1.15	1.45	1.75	1.96
b	0.28	0.34	0.44	0.58	0.80	1.08	1.46	1.96	2.50
				$\theta = 0.65$ K_0					
0	0.15	0.61	1.02	1.42	1.58	1.42	1.02	0.61	0.15
b/4	−0.26	0.15	0.61	1.04	1.42	1.60	1.51	1.26	0.90
b/2	−0.52	−0.18	0.20	0.61	1.02	1.51	1.91	2.06	2.13
3b/4	−0.71	−0.44	−0.18	−0.15	0.61	1.26	2.06	2.95	3.01
b	−0.80	−0.71	−0.52	−0.27	0.15	0.92	2.13	3.01	5.07
				K_1					
0	0.75	0.85	0.98	1.14	1.23	1.14	0.98	0.85	0.75
b/4	0.55	0.64	0.77	0.95	1.14	1.27	1.24	1.15	1.06
b/2	0.40	0.47	0.60	0.77	0.98	1.24	1.45	1.50	1.50
3b/4	0.30	0.36	0.47	0.64	0.85	1.15	1.50	1.84	2.06
b	0.24	0.30	0.40	0.53	0.75	1.06	1.50	2.06	2.65

Prestressed Concrete Tee Beam and Slab Bridge Deck

$\theta = 0.70$

K_0

0	−0.04	0.53	1.03	1.52	1.68	1.51	1.03	0.53	−0.04
b/4	−0.37	0.11	0.00	1.06	1.51	1.70	1.55	1.21	0.67
b/2	−0.50	−0.19	0.18	0.60	1.03	1.55	1.96	2.05	2.03
3b/4	−0.57	−0.40	−0.19	0.11	0.53	1.21	2.05	3.00	4.01
b	−0.48	−0.57	−0.50	−0.37	−0.04	0.73	2.03	4.01	6.03

K_1

0	0.71	0.83	0.98	1.17	1.28	1.17	0.98	0.83	0.71
b/4	0.50	0.59	0.74	0.94	1.17	1.33	1.27	1.15	1.04
b/2	0.33	0.43	0.55	0.74	0.98	1.27	1.51	1.55	1.52
3b/4	0.24	0.32	0.43	0.59	0.83	1.15	1.55	1.93	2.16
b	0.18	0.24	0.33	0.49	0.71	1.04	1.52	2.16	2.85

$\theta = 0.75$

K_0

0	−0.21	0.46	1.02	1.58	1.77	1.58	1.02	0.46	−0.21
b/4	−0.43	0.05	0.57	1.08	1.58	1.77	1.59	1.15	0.50
b/2	−0.49	−0.18	0.17	0.57	1.02	1.59	2.00	2.04	1.95
3b/4	−0.44	−0.35	−0.18	0.05	0.46	1.15	2.04	3.05	3.20
b	−0.30	−0.44	−0.49	−0.43	−0.21	0.56	1.95	3.20	6.70

K_1

0	0.66	0.80	0.98	1.20	1.33	1.20	0.98	0.80	0.66
b/4	0.45	0.55	0.72	0.94	1.20	1.37	1.30	1.15	1.02
b/2	0.30	0.39	0.51	0.72	0.98	1.30	1.57	1.60	1.55
3b/4	0.21	0.28	0.39	0.55	0.50	1.15	1.60	2.01	2.25
b	0.15	0.21	0.30	0.45	0.66	1.02	1.55	2.25	3.00

$\theta = 0.80$

K_0

0	−0.35	0.39	1.02	1.66	1.88	1.66	1.02	0.39	−0.35
b/4	−0.49	0.02	0.55	1.10	1.66	1.88	1.64	1.10	0.33
b/2	−0.48	−0.18	0.15	0.55	1.02	1.64	2.06	2.03	1.82
3b/4	−0.34	−0.30	−0.18	0.02	0.39	1.10	2.03	3.10	4.02
b	−0.16	−0.34	−0.48	−0.48	−0.35	0.39	1.82	4.02	7.02

K_1

0	0.63	0.78	0.98	1.22	1.38	1.22	0.98	0.78	0.63
b/4	0.40	0.51	0.68	0.93	1.22	1.43	1.34	1.14	1.00
b/2	0.25	0.34	0.47	0.68	0.98	1.34	1.63	1.64	1.55
3b/4	0.16	0.23	0.34	0.51	0.78	1.14	1.64	2.10	2.33
b	0.12	0.16	0.25	0.40	0.63	0.98	1.55	2.33	3.20

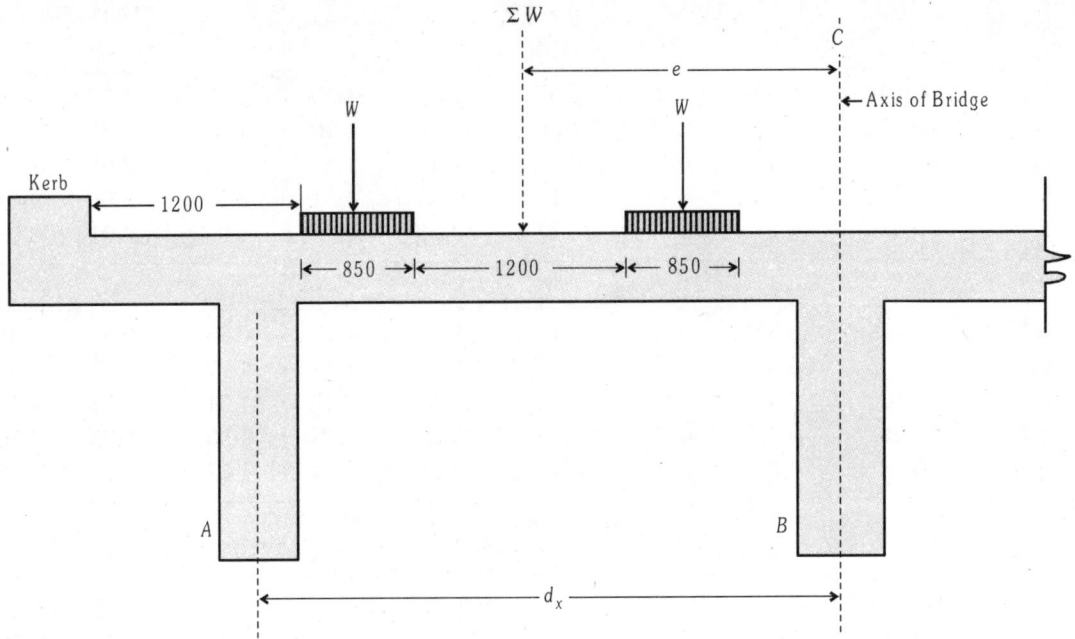

Fig. 6.2: Position of IRC Class AA Tracked Load for Maximum B.M. in Girder A

The maximum transverse moment occurs when an internal line of wheels coincides with the longitudinal centre line of the bridge, the maximum moment being at the centre of the bridge at the reference station O. The equation of transverse moment for a concentrated load 'W' at a distance 'u' from the left support is given by

$$M_y = (Wb/a)[\mu_\theta \sin(\pi u/2a) - \mu_{3\theta}\sin(3\pi u/2a) + \mu_{5\theta}\sin(5\pi u/2a) + \ldots] \quad \ldots(6.4a)$$

If there is a uniformly distributed load 'p' acting over a distance '2c' then,

$$M_y = \left(\frac{4pb}{\pi}\right)\left[\mu_\theta \sin\left(\frac{\pi c}{2a}\right) + \left(\frac{1}{3}\right)\mu_{3\theta}\sin\left(\frac{3\pi c}{2a}\right) + \left(\frac{1}{5}\right)\mu_{5\theta}\sin\left(\frac{5\pi c}{2a}\right) + \ldots\right]\ldots(6.4b)$$

where $\mu_\theta, \mu_{3\theta}, \mu_{5\theta}$ are the distribution coefficients corresponding to the flexural parameters respectively. Coefficient 'μ' is analogous to the distribution coefficient 'K' for longitudinal moments. 'μ_0' represents the distribution coefficient for $\alpha = 0$ and 'μ_1' for $\alpha = 1.0$. The value of μ corresponding to any other intermediate value of α can be evaluated using the interpolation relationship,

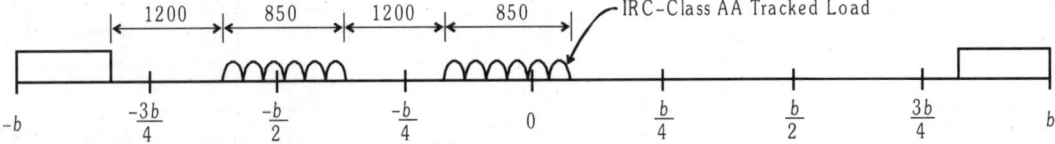

Fig. 6.3: Reference Stations and Position of IRC Class AA Loads for Maximum Distribution

$$\mu_\alpha = \mu_0 + (\mu_1 - \mu_0)\sqrt{\alpha}$$

The coefficients μ_0 and μ_1 are determined for values of θ, 3θ and 5θ from the charts shown in Figs. 6.4 and 6.5 for the reference station O, where the maximum transverse moment will occur for position of loads shown in Fig. 6.6. Graphs of these functions are plotted and values of 'μ' for actual load positions are determined. Then M_{y0} and M_{y1} are calculated for 'μ_0' and 'μ_1' respectively using the equations 6.4 (a) or 6.4 (b). The transverse moment M_y at the centre of the bridge is given by,

$$M_y = M_{y0} + (M_{y1} - M_{y0})\sqrt{\alpha} \qquad \ldots(6.5)$$

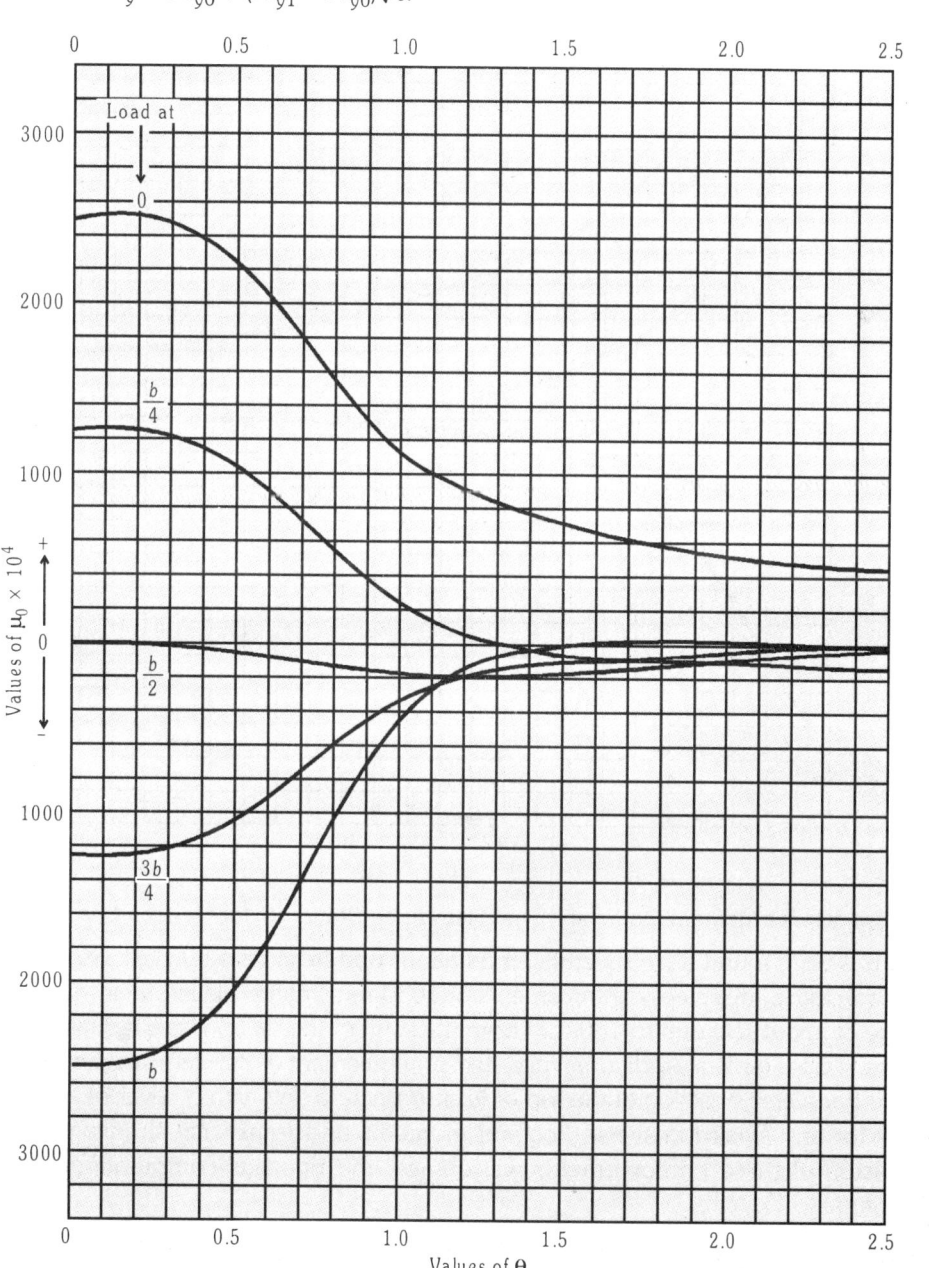

Fig. 6.4: Transverse Moment Coefficient μ_0 at Reference Station O for Various Load Eccentricities

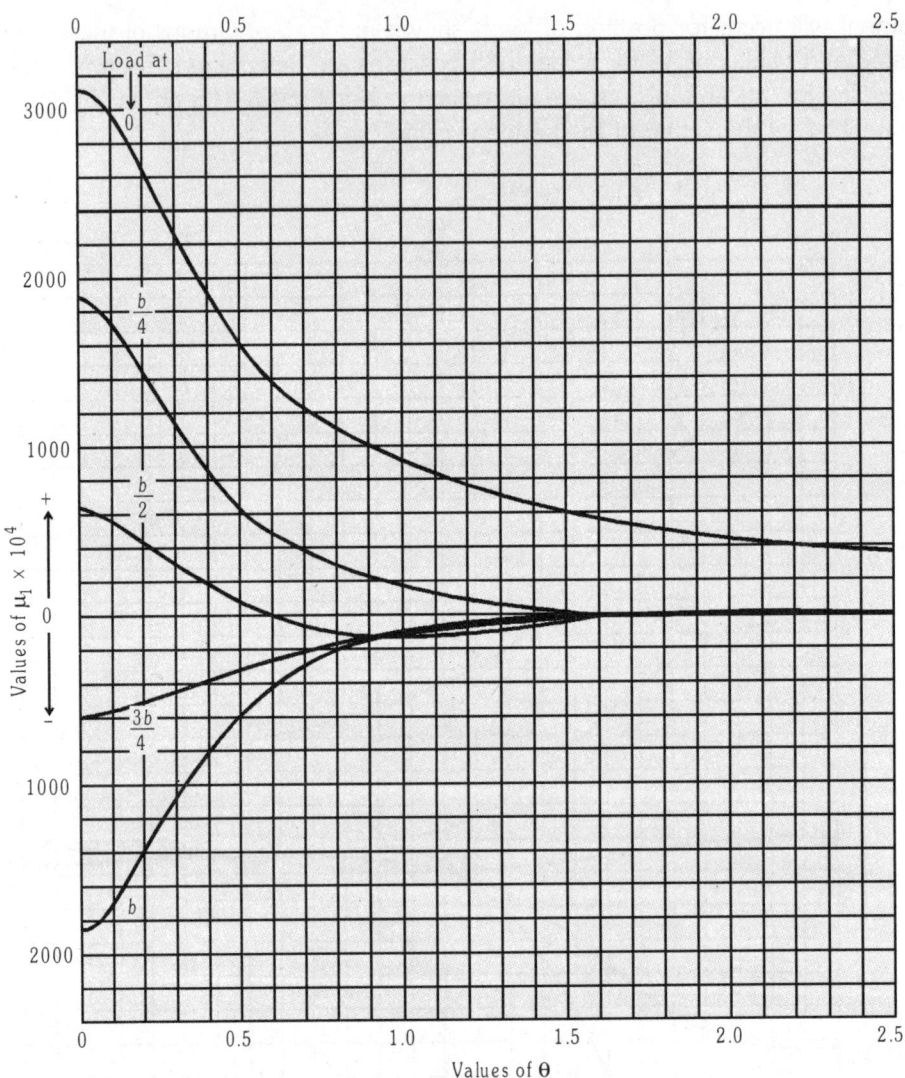

Fig. 6.5: Transverse Moment Cofficient μ_1 at Reference Station O for Various Load Eccentricities

It has been found from several computations and from tests by Best, Rowe and Gifford, that transverse mild steel reinforcements of 1030 mm^2/metre placed near the top of bottom flange or about 150 mm to 180 mm from the soffit of the sections is adequate for reinforced concrete cross girders. Additional mild steel reinforcement of 240 mm^2/m should be provided in the transverse direction at the top of *in situ* concrete slab with a cover of about 20 mm to cater for any small transverse hogging moments and transverse shrinkage stresses. The provision of these reinforcements will obviate the rigorous computations of transverse moments.

3. **Hendry-Jaegar Method:** In the analysis proposed by Hendry and Jaegar, the cross beams can be replaced by a uniform continuous transverse medium of equivalent stiffness. According

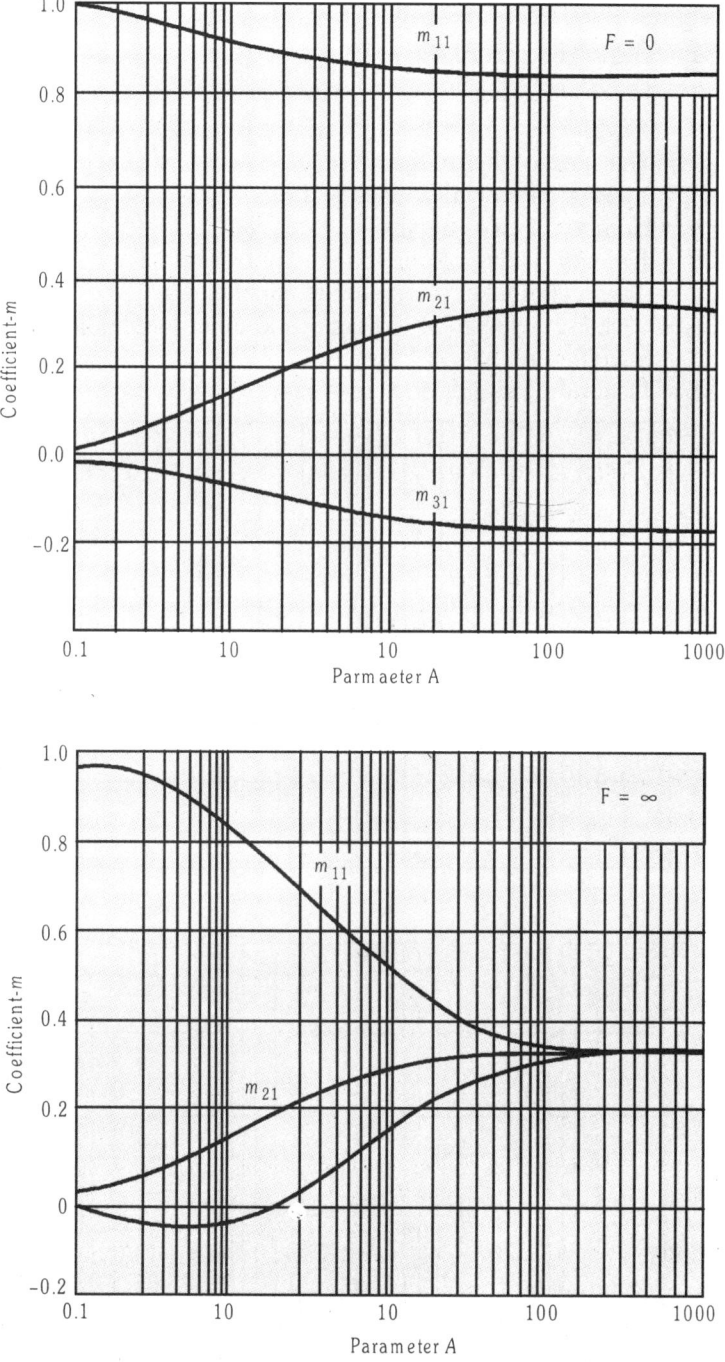

Fig. 6.6: Distribution Coefficient for three Girder Bridge with Load on Girder No. 1

to this method the load distribution in an interconnected bridge deck system depends upon three dimensionless parameters given by

$$A = (12/\pi^4)(L/h)^3 (nEI_r/EI)$$
$$F = (\pi^2/2n)(h/L)(CJ/EI_r)$$
$$C = (EI_1/EI_2)$$

...(6.6, 6.7 and 6.8)

where
L = The span of the bridge deck
H = Spacing of longitudinal girders
n = Number of cross beams
EI = Flexural rigidity of one longitudinal girder
CJ = Torsional rigidity of one longitudinal girder
EI_1, EI_2 = Flexural rigidities of the outer and inner longitudinal girders where these are different
EI_r = Flexural rigidity of one cross beam

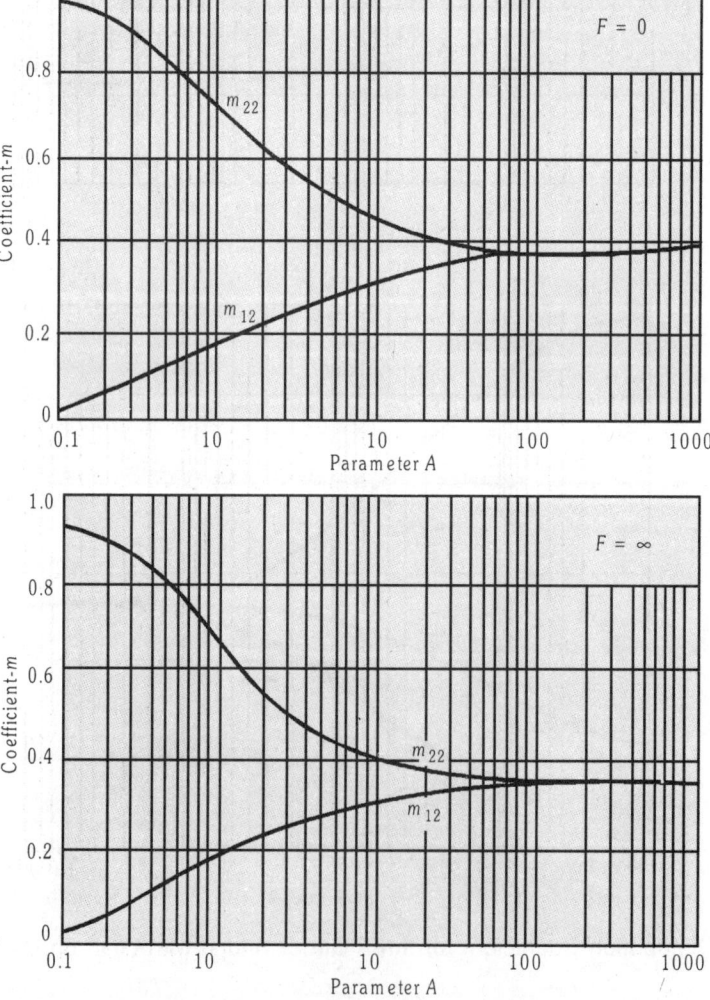

Fig. 6.7: Distribution Coefficient for three Girder Bridge with Load on Girder No. 2

Prestressed Concrete Tee Beam and Slab Bridge Deck

For slab bridges without cross beams, nEI_r, in equations 6.6 and 6.7 is to be replaced by LEI_r, which represents the total flexural rigidity of the slab deck. In general, the flexural rigidities of the outer and inner longitudinal girders of a tee beam deck are nearly equal in magnitude.

The first parameter 'A' represents a function of the ratio of span to the spacing of longitudinal girders and the ratio of transverse to longitudinal flexural rigidity. The second parameter 'F' is a measure of the ratio of torsional to flexural rigidity of longitudinal and cross girders respectively. The parameter is difficult to be evaluated due to the uncertainties in computations of torsional rigidity values for practical girder sections. In the case of tee beam bridges having three or four longitudinal girders with a number of cross beams. It is permissible to employ the distribution coefficients for $F = \infty$. The torsional rigidity of the transverse section is neglected in the analysis.

Hendry and Jaegar have presented graphs giving the values of the distribution coefficients (m) for different number of longitudinal girders (two or six) and for the two extreme values of $F = 0$ and $F = \infty$. Coefficients for intermediate values of F may be obtained by interpolation from the equation,

$$m_F = m_0 + (m_\infty - m_0)\sqrt{\frac{F\sqrt{A}}{(3 + F\sqrt{A})}} \qquad \ldots(6.9)$$

where m_F = required distribution coefficient
m_0 = coefficient for $F = 0$
m_∞ = coefficient for $F = \infty$

Typical graphs for distribution coefficients for a three girder system for $F = 0$ and $F = \infty$ are given in Figs. 6.6 and 6.7.

6.4 COMPARATIVE ANALYSIS OF VARIOUS LOAD DISTRIBUTION METHODS

Compare the maximum live load moments in the longitudinal girders of a tee beam and slab bridge deck having the following data, using Courbon, Guyon-Massonet and Hendry Jaegar methods of load distribution.

1. **Data:**
   Carraige width of road way = 7.5 m
   Span (centre to centre of bearings) = 16 m
   Live load: IRC Class AA tracked vehicle
   Average thickness of wearing coat = 80 mm
   Thickness of deck slab = 200 mm
   Three main girders are provided at 2.5 m centres
   Width of main girder = 300 mm
   Overall depth of main girder = 1600 mm
   Modulus of elasticity of concrete = E_c = 28 kN/mm^2
   Modulus of rigidity = $G = 0.4 E$
   Kerbs 600 mm wide by 300 mm deep are provided
   Cross girders are provided at 4 m intervals
   Breadth of cross girder = 300 mm

2. **Cross-section of the Bridge Deck:** The cross-section of the bridge deck and the plan showing the spacings of cross girders are shown in Fig. 6.8.

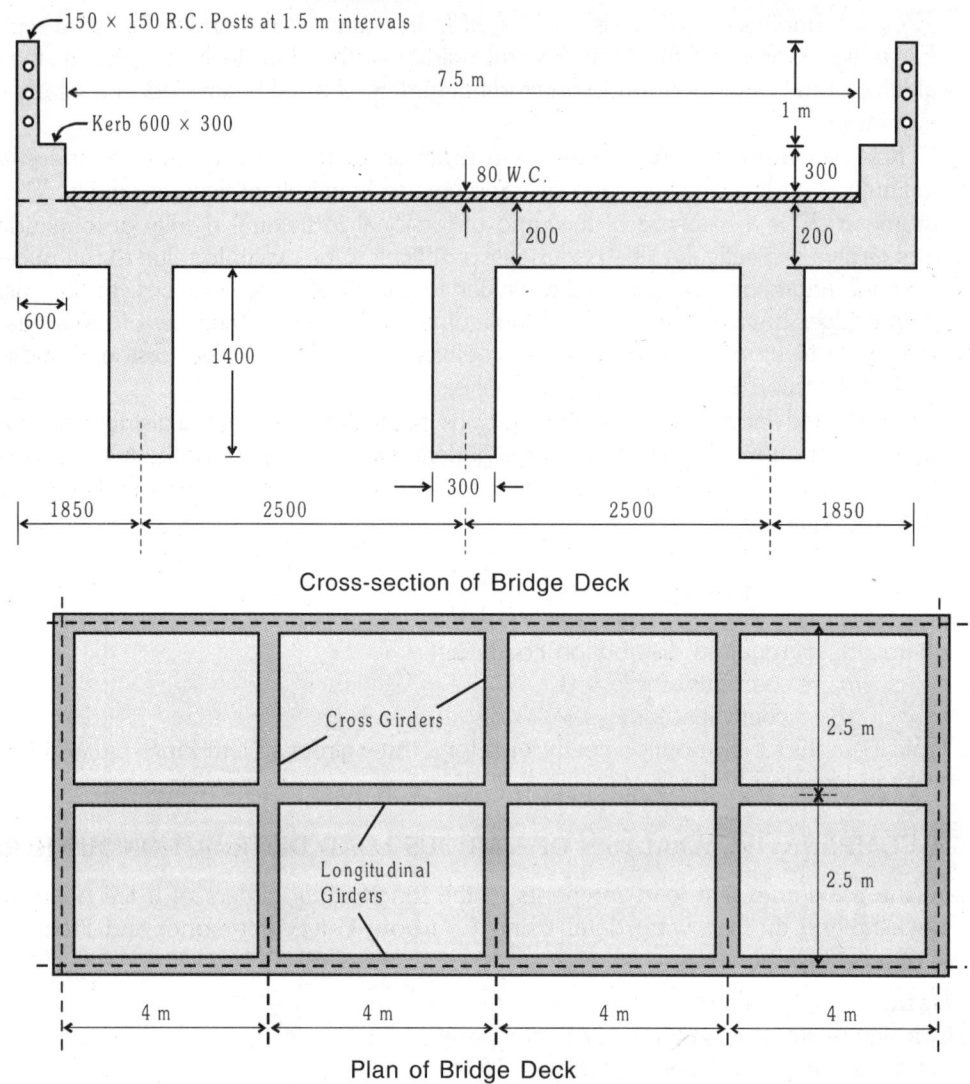

Fig. 6.8: Tee Beam and Slab Bridge Deck

3. Courbon's Method:

(a) **Reaction Factors:** Using Courbon's method, the IRC Class AA loads are arranged for maximum eccentricity on the deck as shown in Fig. 6.9.

Maximum eccentricity = $e = 1.1$ m

Load on each track = W_1

Reaction factor for any girder is computed by using the empirical relation

$$R_X = \left(\sum \frac{W}{n}\right)\left[1 + \left(\frac{\sum I}{\sum d_X^2 \cdot I}\right) d_x \cdot e\right]$$

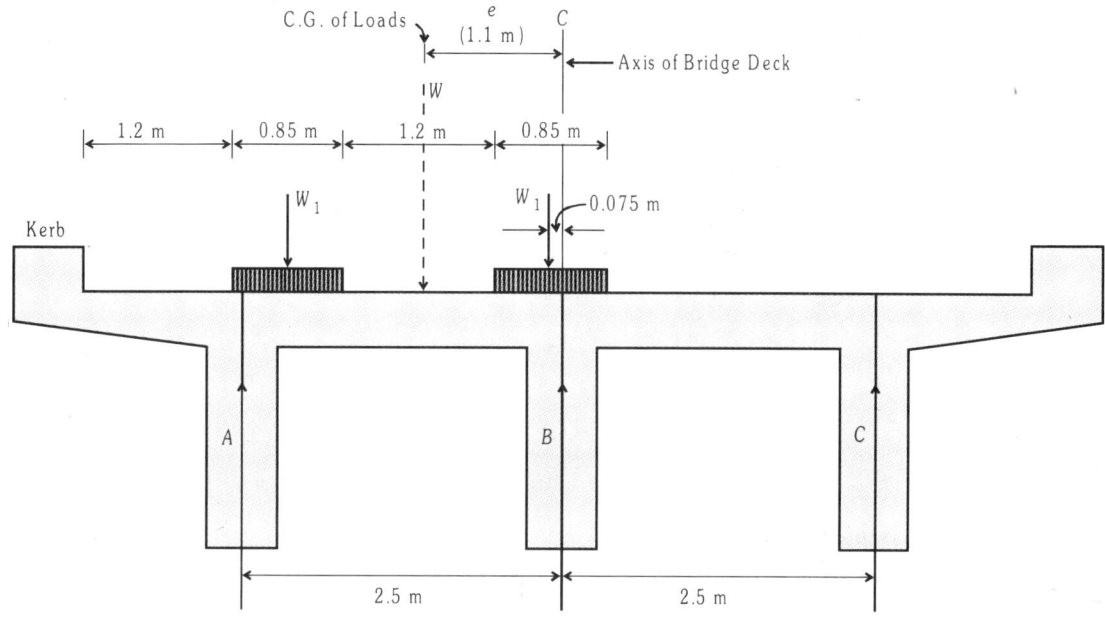

Fig. 6.9: Transverse Position of IRC Class AA Tracked Vehicle for Maximum Eccentricity

Substituting the values for loads, number of girders and eccentricity, we have the reaction factor for the outer girder (A) as

$$R_A = \left(\frac{2W_1}{3}\right)\left[1 + \left(\frac{3I}{2.5^2 \times 2I}\right)(2.5 \times 1.1)\right] = 1.107W_1$$

Reaction factor for inner girder (B) is computed as

$$R_B = \left(\frac{2W_1}{3}\right)[1 + 0] = 0.667W_1$$

If $\quad W$ = Axle load = 700 kN
$\quad W_1 = 0.5\,W$
∴ $\quad R_A = (1.107 \times 0.5\,W) = 0.5536\,W$
$\quad R_B = (0.667 \times 0.5\,W) = 0.3333\,W$

(b) Live Load Bending Moments

Span of the beam = 16 m
Impact factor (For IRC Class AA tracked vehicle) = 10%
The live load is placed centrally on the span as shown in Fig. 6.10 so as to result in maximum moment in the girder.
Using the influence line diagram for bending moment, bending moment at centre of span is computed as

$$M_{max} = 0.5(4 + 3.1)700 = 2485 \text{ kN·m}$$

Bending moment including impact and reaction factor for outer girder is computed as
M_{LL}(Exterior girder) = $(M_{max} \times \text{I.F.} \times \text{R.F.})$
$\quad = (2485 \times 1.1 \times 0.5536)$
$\quad = 1513$ kN·m

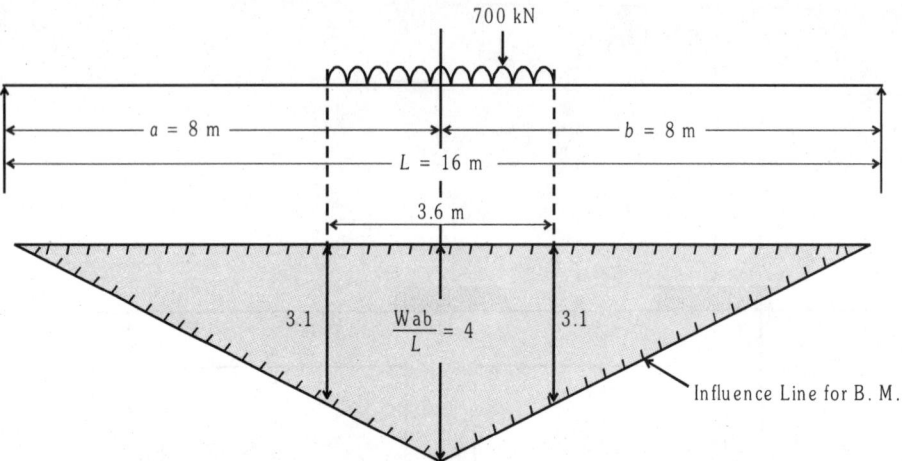

Fig. 6.10: Influence Line for Bending Moment

4. Guyon-Massonet Method

(a) **Cross-Sectional Properties of Girders:** The cross-section of the deck together with the cross-sections of the main and cross girders are shown in Fig. 6.11.

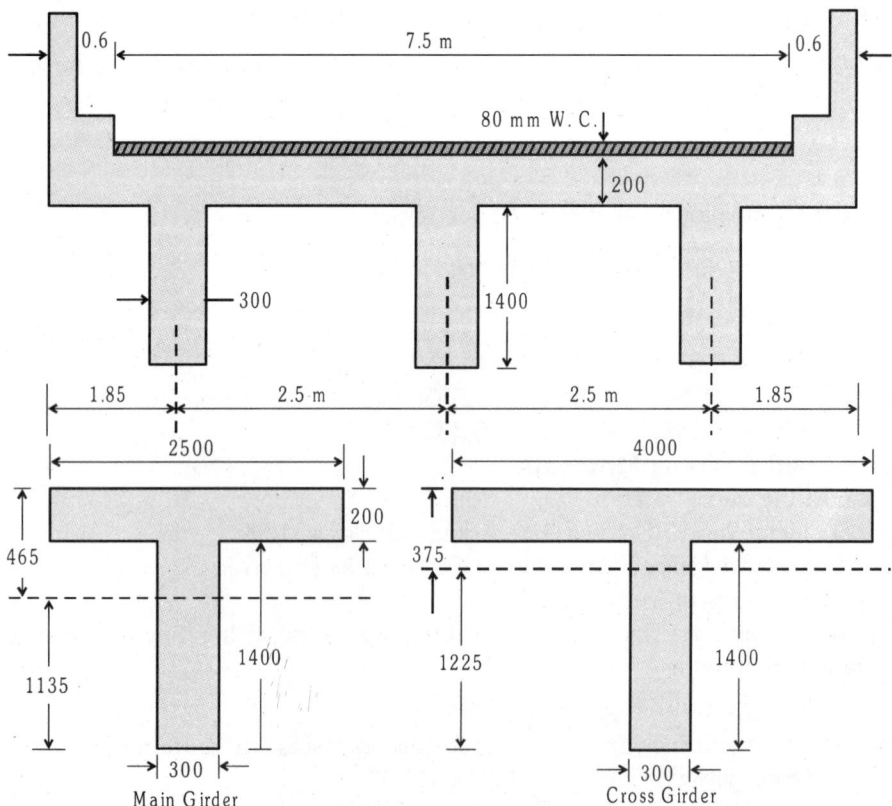

Fig. 6.11: Cross-section of Deck, Main and Cross Girders

Prestressed Concrete Tee Beam and Slab Bridge Deck

The cross-sectional properties are as follows:
Main Girder
$$I = (21.62 \times 10^{10}) \text{ mm}^4, \quad B = 2500 \text{ mm}, \quad y_t = 465 \text{ mm}, \quad y_b = 1135 \text{ mm}$$
$$i = (I/B) = [(21.62 \times 10^{10})/2500] = (0.864 \; 10^8) \text{ mm}^4/\text{mm}$$
$$Z_t = (I/y_t) = [(21.62 \times 10^{10})/465] = (4.64 \times 10^8) \text{ mm}^3$$
$$Z_b = (I/y_b) = [(21.62 \times 10^{10})/1135] = 1.90 \times 10^8 \text{ mm}^3$$
Cross Girder
$$J = (24.74 \times 10^{10}) \text{ mm}^4, \quad B = 4000 \text{ mm}, \quad y_t = 375 \text{ mm}, \quad y_b = 1225 \text{ mm}$$
$$j = (J/B) = (24.74 \times 10^{10})/4000 = (0.618 \times 10^8) \text{ mm}^4/\text{mm}$$
$$Z_t = (J/y_t) = (24.74 \times 10^{10})/375 = (6.59 \times 10^8) \text{ mm}^3$$
$$Z_b = (J/y_b) = (24.74 \times 10^{10})/1225 = (2.01 \times 10^8) \text{ mm}^3$$

(b) Torsional Inertia of Girders

$$I_0 \text{ or } J_0 = R \cdot a^3 \cdot b$$

where 'a' and 'b' are shorter and larger sides of a rectangle and 'R' is the torsion coefficient given in Table 6.2.

Table 6.2: Values of Torsion Coefficient 'R'

(b/a)	R	(b/a)	R
1.0	0.141	3.0	0.263
1.2	0.166	4.0	0.281
1.5	0.196	5.0	0.291
2.0	0.229	10.0	0.312
2.5	0.249	Above 10	0.333

Main Girder
$$(b/a) = (2500/200) = 12.5 \qquad \therefore R = 0.333$$
and $(b/a) = (1400/300) = 4.66 \qquad \therefore R = 0.287$
$$I_0 = (0.333 \times 200^3 \times 2500) + (0.287 \; 300^3 \times 1400) = (1.75 \times 10^{10}) \text{ mm}^4$$
$$\therefore \quad i_0 = (I_0/R) = (1.75 \times 10^{10})/2500 = (0.07 \times 10^8) \text{ mm}^4/\text{mm}$$

Cross Girder
$$(b/a) = (4000/200) = 20 \qquad \therefore R = 0.333$$
$$(b/a) = (1400/300) = 4.66 \qquad \therefore R = 0.287$$
$$J_0 = (0.333 \times 200^3 \times 4000) + (0.287 \; 300^3 \times 1400) = (1.15 \times 10^{10}) \text{ mm}^4$$
$$\therefore \quad j_0 = (J_0/B) = (1.15 \times 10^{10})/4000 = (0.028 \times 10^8) \text{ mm}^4/\text{mm}$$

(c) Distribution Coefficients for Main Girders

$$\theta = (b/2a)(i/j)^{0.25}$$

where $2b$ = effective width of bridge = 8.7 m
$2a$ = span of bridge = 16 m

∴ $\theta = (4.35/16)[(0.864 \times 10^8)/(0.618 \times 10^8)]^{0.25} = 0.3$

$\alpha = (G/2E)(i_0 + j_0)/\sqrt{ij}$

Since $G = 0.4E$

$$\alpha = \left\{ \frac{0.2\left[(0.07 \times 10^8) + (0.028 \times 10^8)\right]}{\sqrt{(0.864 \times 10^8)(0.618 \times 10^8)}} \right\} = 0.026$$

$\sqrt{\alpha} = 0.161$

The values of K_0 and K_1 for $\theta = 0.3$ are compiled in Table 6.3

(d) Distribution Coefficients for IRC Class AA Loading: The distribution coefficients for IRC Class AA tracked vehicle loading are compiled in Table 6.4.

Table 6.3: Values of K_0 and K_1 for $\theta = 0.3$

Ref. Point Load at	$-b$	$-3b/4$	$-b/2$	$-b/4$	0	$b/4$	$b/2$	$3b/4$	b
				K_0					
0	0.86	0.95	0.97	1.05	1.1	01.05	0.97	0.95	0.86
$b/4$	0.20	0.40	0.63	0.87	1.05	1.22	1.36	1.53	1.68
$b/2$	−0.54	−0.16	0.24	0.63	0.97	1.36	1.73	2.10	2.46
$3b/4$	−1.15	−0.63	−0.16	0.40	0.95	1.53	2.10	2.73	3.31
b	−1.79	−1.15	−0.54	0.20	0.86	1.68	2.46	3.31	4.10
				K_1					
0	0.94	0.97	1.00	1.02	1.05	1.02	1.00	0.97	0.94
$b/4$	0.85	0.89	0.94	0.97	1.02	1.06	1.06	1.05	1.06
$b/2$	0.77	0.82	0.89	0.94	1.00	1.06	1.13	1.17	1.21
$3b/4$	0.70	0.75	0.82	0.89	0.97	1.05	1.17	1.29	1.38
b	0.63	0.70	0.77	0.85	0.94	1.06	1.21	1.38	1.59

The values of distribution coefficients for different load positions with weighting factors is shown in Fig. 6.12 are computed as λK_0 and λK_1 and the distribution coefficient for Class AA loading is computed as shown in Table 6.4. The maximum distribution coefficient occurs at the centre of the end beam.

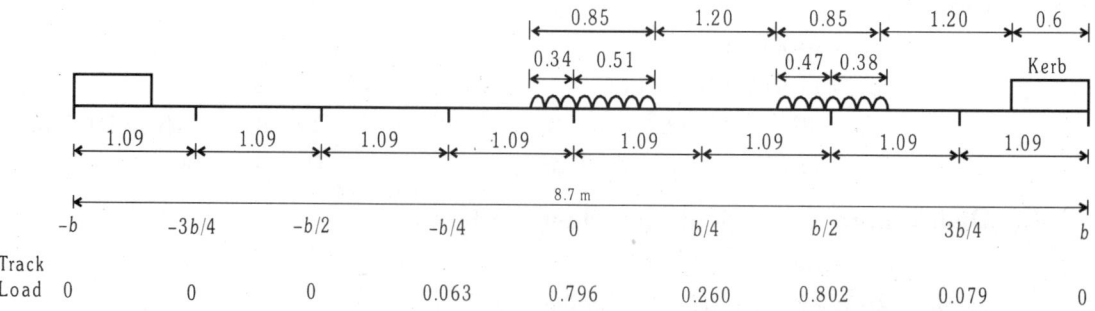

Fig. 6.12: Weighting Factors for IRC Class AA Tracked Loading

Table 6.4: Distribution Coefficients for Class AA Loading

Load at	Weighting factor	Values of λK_0								
		$-b$	$-3b/4$	$-b/2$	$-b/4$	0	$b/4$	$b/2$	$3b/4$	b
0	0.796	0.684	0.756	0.772	0.835	0.875	0.835	0.772	0.756	0.684
b/4	0.260	0.052	0.104	0.163	0.226	0.273	0.317	0.353	0.397	0.436
b/2	0.802	−0.433	−0.128	0.192	0.505	0.777	1.090	1.387	1.684	1.972
3b/4	0.079	−0.090	−0.049	−0.012	0.031	0.075	0.120	0.165	0.215	0.261
−b/4	0.063	0.105	0.096	0.085	0.076	0.066	0.054	0.039	0.025	0.012
$\Sigma \lambda K_0$ =		0.318	0.779	1.200	1.673	2.066	2.416	2.716	3.077	3.365
$\Sigma \lambda K_0/2$ =		0.159	0.389	0.600	0.837	1.033	1.208	1.358	1.539	1.683

Load at	Weighting factor	Values of λK_1								
		$-b$	$-3b/4$	$-b/2$	$-b/4$	0	$b/4$	$b/2$	$3b/4$	b
0	0.796	0.748	0.772	0.796	0.811	0.835	0.811	0.796	0.772	0.748
b/4	0.260	0.221	0.231	0.244	0.252	0.265	0.275	0.275	0.273	0.275
b/2	0.802	0.617	0.657	0.713	0.753	0.802	0.850	0.906	0.938	0.970
3b/4	0.079	0.055	0.059	0.064	0.070	0.076	0.082	0.092	0.101	0.109
−b	0.063	0.066	0.066	0.066	0.066	0.064	0.061	0.059	0.056	0.053
$\Sigma \lambda K_1$ =		1.707	1.785	1.883	1.952	2.042	2.079	2.129	2.140	2.155
$\Sigma \lambda K_1/2$ =		0.854	0.893	0.942	0.976	1.021	1.039	1.064	1.070	1.078
$[\Sigma \lambda K_1/2 - \Sigma \lambda K_0/2]\sqrt{\alpha}$ =		0.111	0.081	0.055	0.022	−0.002	0.027	−0.046	−0.075	−0.097
$\Sigma \lambda K_0/2 + [\Sigma \lambda K_1/2 - \Sigma \lambda K_0/2]\sqrt{\alpha}$ =		0.270	0.470	0.655	0.859	1.031	1.181	1.312	1.464	1.586

The maximum distrubution coefficient occurs at the centre of end beam and is given by

$$DK_w = 1.312 + [(1.464 - 1.312)/1.09]\,0.33 = 1.358$$

(e) Maximum Live Load Bending Moment in Longitudinal Girders: The maximum moments in longitudinal girders are computed for the position of live loads shown in Fig. 6.10. Maximum live load bending moment in the end girder occurs at the centre of span and is computed as

$$M_{mean} = [(350 \times 8) - (350 \times 0.9)] = 2485 \text{ kN} \cdot \text{m}$$

The mean bending moment is shared by 3 longitudinal girders. Applying the impact factor and distribution coefficient to this moment, the maximum moment is computed as

$$M_{max} = [(\text{I.F.} \times DKw \times M_{mean})/3]$$
$$= [(1.1 \times 1.358 \times 2485)/3]$$
$$= 1237 \text{ kN} \cdot \text{m}$$

5. Hendry-Jaegar Method

(a) **Design Parameters**: The main design parameters A and F are computed as

$$A = \left(\frac{12}{\pi^4}\right)\left(\frac{L}{h}\right)^3 \left(\frac{nEI_r}{EI}\right)$$

$$= \left(\frac{12}{\pi^4}\right)\left(\frac{16}{2.5}\right)^3 \left(\frac{5 \times 28 \times 24.74 \times 10^{10}}{28 \times 21.62 \times 10^{10}}\right)$$

$$= 184.84$$

$$F = \left(\frac{\pi^2}{2n}\right)\left(\frac{h}{L}\right)\left(\frac{GJ}{EI_r}\right)$$

$$= \left(\frac{\pi^2}{2 \times 5}\right)\left(\frac{2.5}{16}\right)\left(\frac{0.4E \times 1.75 \times 10^{10}}{E \times 24.74 \times 10^{10}}\right)$$

$$= 0.004$$

$$m_F = m_0 + (m_\infty - m_0)\sqrt{\frac{F\sqrt{A}}{(3 + F\sqrt{A})}}$$

From Fig. 6.6 the following values of m_0 and m_∞ are obtained for the three girder system.
For exterior girder, $m_0 = 0.82$ and $m_\infty = 0.35$
Since F is very nearly equal to zero, $m_F = m_0$
For exterior girder, $m_F = 0.82$
Live load moment in exterior girder $= [M_{mean} \times \text{I.F.} \times m_F]$
$$= [2485 \times 1.1 \times 0.82]$$
$$= 2241 \text{ kN} \cdot \text{m}$$

The live load design moment obtained in the exterior girder by various methods are compared in Table 6.5.

Table 6.5 Comparison of Live Load Moments in Outer Girder by Different Methods

Sl. No.	Design Method	Live Load Moment (kN·m)
1	Courbon	1513
2	Guyon-Massonet	1237
3	Hendry-Jaegar	2241

The comparative analysis indicates that the live load bending moment is the least by the Guyon-Massonet method since it considers a large number of influencing parameters in comparison with the other methods. Hendry-Jaegar's method yields conservatively higher values of moment while Courbon's method results in moments nearer to the Guyon-Massonet method. Hence it is recommended that preliminary analysis may be made by Courbon's method followed by detailed design calculations by the more rigorous Guyon-Massonet method.

6.5 DESIGN OF POST TENSIONED PRESTRESSED CONCRETE TEE BEAM AND SLAB BRIDGE DECK

Design a post tensioned prestressed concrete tee beam and slab bridge deck for a national high way intersection. The design should conform to the specifications of IRC: 6-2000 and IRC: 18-2000 codes and the following data:

1. **Data**
   Effective span = 30 m
   Width of road way = 7.5 m
   Kerbs: 600 mm wide on each side
   Foot path: 1.5 m wide on each side
   Thickness of wearing coat = 80 mm
   Live load: IRC Class AA tracked vehicle
   Grade of concrete: M-50
   Compressive strength of concrete at transfer = 40 N/mm^2
   Loss ratio = 0.85
   Spacings of cross girders = 5 m
   Adopt Fe-415 Grade HYSD bars and 15.2 mm—7 ply high tensile steel strands conforming to IRC: 6006-1983 for post tensioning the girders
   Type of structure: Class 1 type member (No tensile stresses at any stage)

2. **Permissible Stresses:** The permissible stresses in concrete at transfer and service loads as recommended in IRC: 21-2000 and IRC: 18-2000 are as follows:
   $\sigma_{cb} = 11.5$ N/mm^2, $\sigma_{st} = 200$ N/mm^2, $j = 0.879$
   $Q = 1.844$ and $m = 10$
   For reinforced concrete slab using M-50 Grade concrete and Fe-415 HYSD bars,
   For prestressed concrete beams, $f_{ck} = 50$ N/mm^2 and $f_{ci} = 40$ N/mm^2
   $f_{ct} = 18$ N/mm^2 < $0.5 f_{ci} = (0.5 \times 40) = 20$ N/mm^2
   Loss ratio = $\eta = 0.80$
   $f_{cw} = 16$ N/mm^2 < $0.33 f_{ck} = (0.33 \times 50) = 16.5$ N/mm^2
   $f_{tt} = f_{tw} = 0$ (Class 1 type member)
   Modulus of elasticity of concrete in beams is computed as
   $$E_c = 5000\sqrt{f_{ck}} = 5000\sqrt{50} = 35 \text{ kN/mm}^2$$

3. **Cross-Section of Deck**
   Four main girders are provided spaced at intervals of 2.5 m
   Thickness of deck slab assumed as 200 mm

Wearing coat thickness = 80 mm
Foot paths 1.5 m wide on either side and kerbs 600 mm wide by 300 mm deep are provided.
Spacing of cross girders = 5 m
Spacing of main girders = 2.5 m
The bridge deck is cast *in situ* and the girders are post tensioned
The cross-section of the deck is shown in Fig. 6.13.

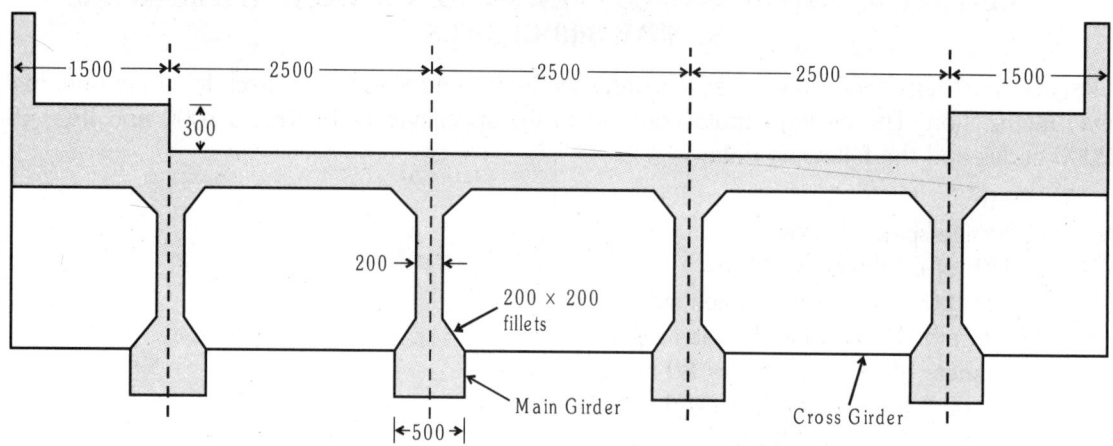

Fig. 6.13: Cross Section of Tee Beam Bridge Deck

4. Design of Interior Slab Panel

(a) Dead Load Bending Moments and Shear Forces

Dead weight of slab = $(1 \times 1 \times 0.20 \times 240) = 4.80$ kN/m²

Dead weight of wearing coat = $(0.08 \times 22) = 1.76$

Total dead load ... = 6.56 kN/m²

Considering an interior slab panel 5 m by 2.5 m enclosed between main and cross girders,

total load on panel = $(5 \times 2.5 \times 6.56) = 82$ kN

$(u/B) = 1$ and $(v/L) = 1$ since panel is loaded with uniformly distributed load

$K = (B/L) = (2.5/5.0) = 0.5$ and $(1/K) = 2.0$

From Pigeaud's curves (Refer Fig. 6.14) read out the values of the moment coefficients,

$m_1 = 0.047$ and $m_2 = 0.01$

$M_B = W(m_1 + \mu m_2) = 82(0.047 + 0.15 \times 0.01) = 3.98$ kN·m

$M_L = W(m_2 + \mu m_1) = 82(0.01 + 0.15 \times 0.047) = 1.40$ kN·m

Design B.M. including continuity factor is

$M_B = (0.8 \times 3.98) = 3.18$ kN·m

$M_L = (0.8 \times 1.40) = 1.12$ kN·m

Dead load shear force = $(0.5 \times 6.56 \times 2.3) = 7.544$ kN

Prestressed Concrete Tee Beam and Slab Bridge Deck

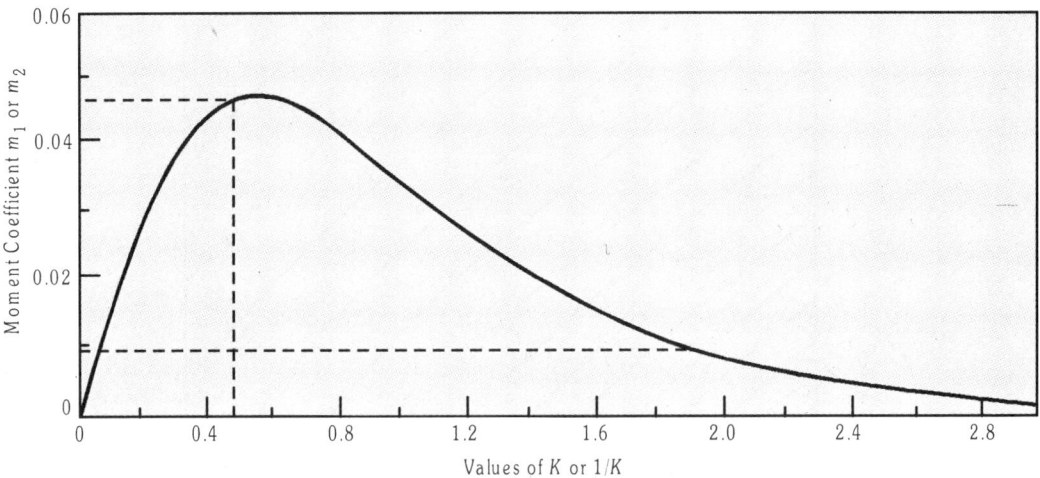

Fig. 6.14: Moment Coefficients for Slabs Completely Loaded with Uniformly Distributed Load (Coefficient is m_1 for K and m_2 for $1/K$)

(b) Live load Bending Moments and Shear Forces: Live load is made up of IRC Class AA tracked vehicle. One wheel is placed at the centre of panel as shown in Fig. 6.15.

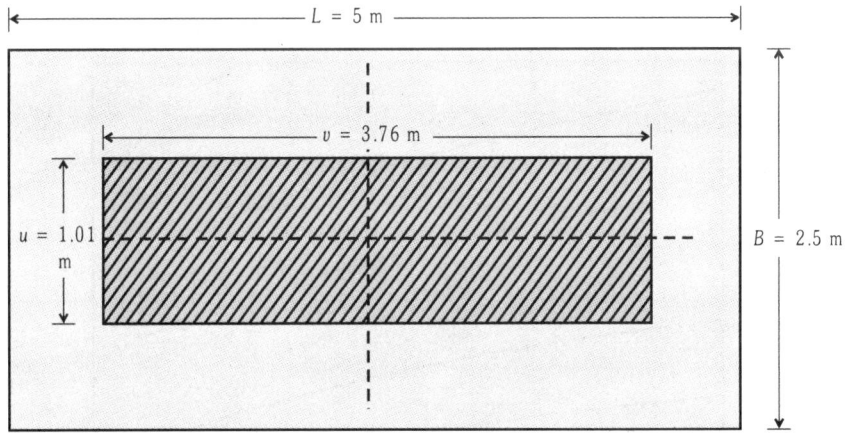

Fig. 6.15: Position of IRC Class AA Tracked Vehicle for Maximum Bending Moment

$$u = (0.85 + 2 \times 0.08) = 1.01 \text{ m}$$
$$v = (3.60 + 2 \times 0.08) = 3.76 \text{ m}$$
Ratio $(u/B) = (1.01/2.5) = 0.404$
$(v/L) = (3.76/5.0) = 0.752$
$K = (B/L) = (2.5/5.0) = 0.5$

Referring to Pigeaud's curves (Fig. 6.16) read out the values of moment coefficients,
$m_1 = 0.098$ and $m_2 = 0.02$
$M_B = W(m_1 + 0.15 m_2) = 350(0.098 + 0.15 \times 0.02) = 35.35 \text{ kN·m}$

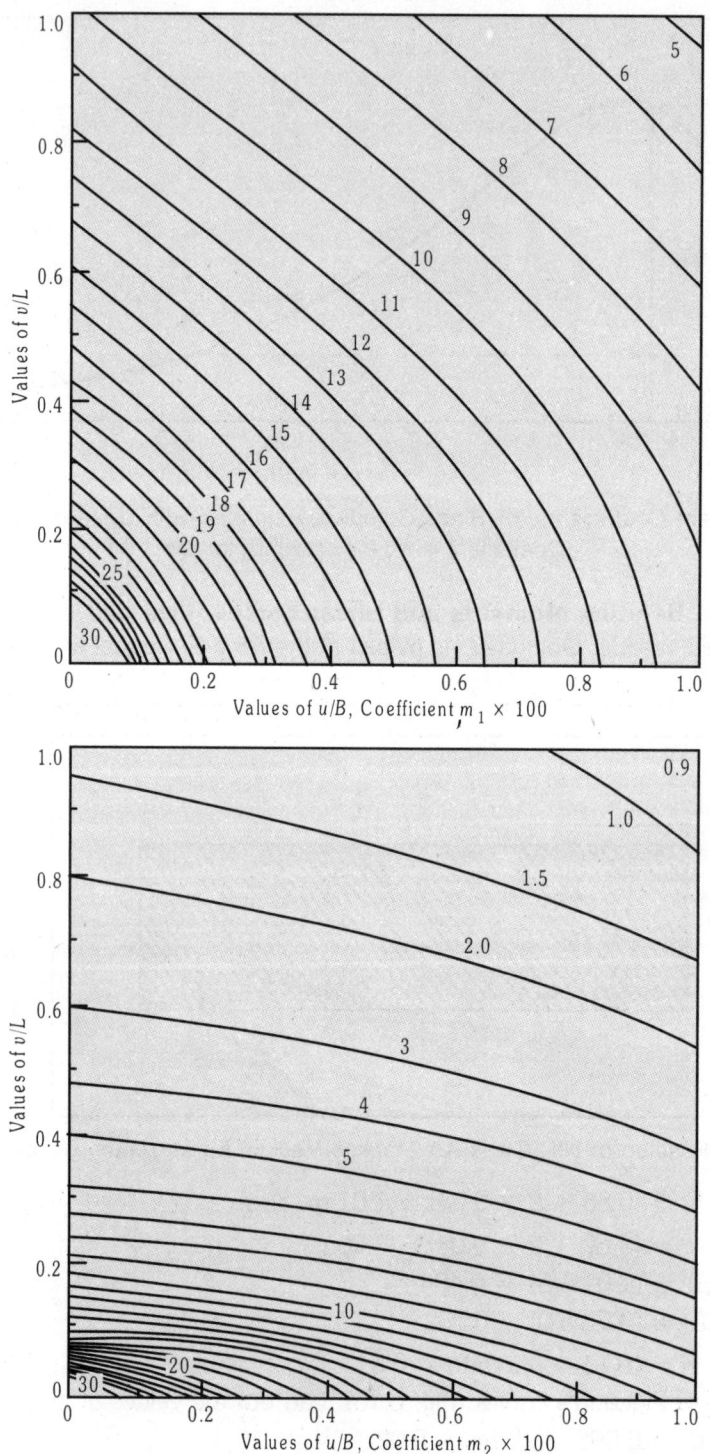

Fig. 6.16: Moment Coefficients m_1 and m_2 for $K = 0.5$ (Pigeaud's Curves)

As the slab is continuous, design B.M. = $0.8 M_B$
Hence design B.M. including impact and continuity factors is given by
M_B (short span) = $(1.25 \times 0.8 \times 35.35)$ = 35.35 kN·m
Similarly $M_L = W(m_2 + 0.15 m_1) = 350 (0.02 + 0.15 \times 0.098) = 12.14$ kN·m
M_L (long span) = $(1.25 \times 0.8 \times 12.14)$ = 12.14 kN·m
Live Load Shear Force is computed using the following method:
Dispersion in the direction of span = $[0.85 + 2(0.08 + 0.20)] = 1.41$ m
For maximum shear, load is kept such that the whole dispersion is in span. The wheel load is kept at $(1.41/2) = 0.075$ m from the edge of the beam as shown in Fig. 6.17.
Effective width of slab = $Kx[1 - (x/L)] + b_w$
Breadth of cross girder = 200 mm
Clear length of panel = $(5 - 0.2) = 4.8$ m

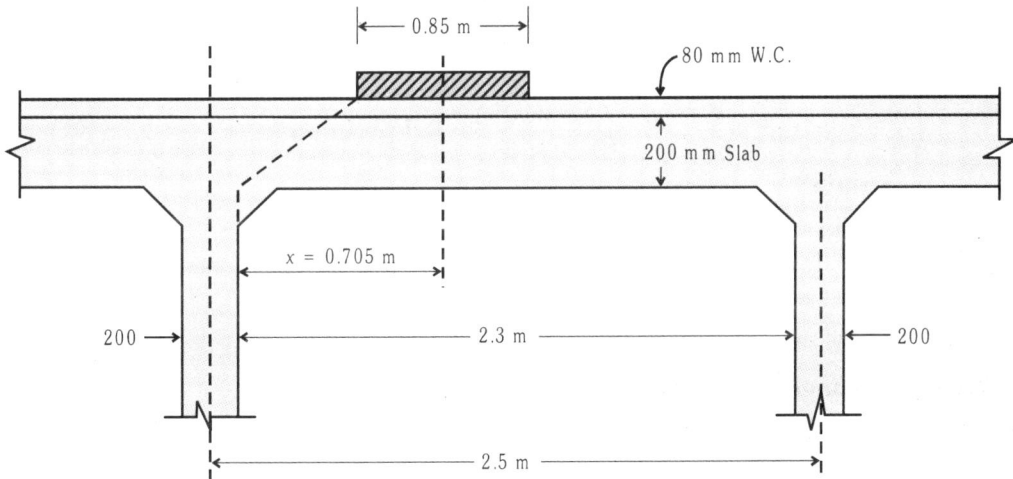

Fig. 6.17: Position of Wheel Loads for Maximum Shear Force

∴ $(B/L) = (4.8/2.3) = 2.08$
From Table 5.1, read out the value of constant K for continuous slab as 2.60
Effective width of slab = $(2.6 \times 0.705)[1 - (0.705/2.3)] + 3.6 + (2 \times 0.08) = 5.032$ m
Load per metre width = $(350/5.032) = 69.55$ kN
Shear force per metre width = $69.55 (2.3 - 0.705)/2.3 = 48.23$ kN
Shear force with impact = $(1.25 \times 48.23) = 60.28$ kN.

(c) **Design Moments and Shear Forces**
Total $M_B = (35.35 + 3.18) = 38.53$ kN·m
Total $M_L = (12.14 + 1.12) = 13.26$ kN·m
Design Shear force = $V = (7.544 + 60.28) = 67.82$ kN

(d) **Design of Slab Section and Reinforcements**

Effective depth = $d = \sqrt{\dfrac{M}{Qb}} = \sqrt{\dfrac{38.53 \times 10^6}{1.844 \times 10^3}} = 145$ mm

Minimum clear cover = 40 mm
Adopt effective depth = d = 150 mm and overall depth = D = 200 mm
Short span reinforcement is computed as

$$A_{st} = \left[\frac{M}{\sigma_{st}\,jd}\right] = \left[\frac{38.53 \times 10^6}{200 \times 0.879 \times 150}\right] = 1461 \text{ mm}^2$$

Provide 16 mm diameter bars at 130 mm centres (A_{st} = 1547 mm²)
Effective depth for long span using 12 mm diameter bars is = (150 – 8 – 6) = 136 mm

$$A_{st} = \left[\frac{13.26 \times 10^6}{200 \times 0.879 \times 136}\right] = 550 \text{ mm}^2 > 0.12 \text{ per cent of cross-section}$$

Provide 10 mm diameter bars at 150 mm centres (A_{st} = 754 mm²) for crack width control as per the specifications of IRC: 21-2000.

(e) Check for Shear Stresses (As per IRC: 21-2000)
Nominal shear stress

$$= \tau_v = \left(\frac{V}{bd}\right) = \left(\frac{67.82 \times 10^3}{10^3 \times 150}\right) = 0.452 \text{ N/mm}^2 < \tau_{max} = 1.25 \text{ N/mm}^2$$

Ratio $\left(\dfrac{100 A_{st}}{bd}\right) = \left(\dfrac{100 \times 1547}{1000 \times 150}\right) = 1.03$

Referring to Table 4.2, read out the permissible shear stress in concrete as
$K\tau_c = (1.20 \times 0.42) = 0.504 \text{ N/mm}^2 > 0.452 \text{ N/mm}^2$

Hence shear stresses are within safe permissible limits.

5. Design of Longitudinal Girders

(a) **Reaction Factors:** Using Courbon's theory, the IRC Class AA loads are arranged for maximum eccentricity as shown in Fig. 6.18.

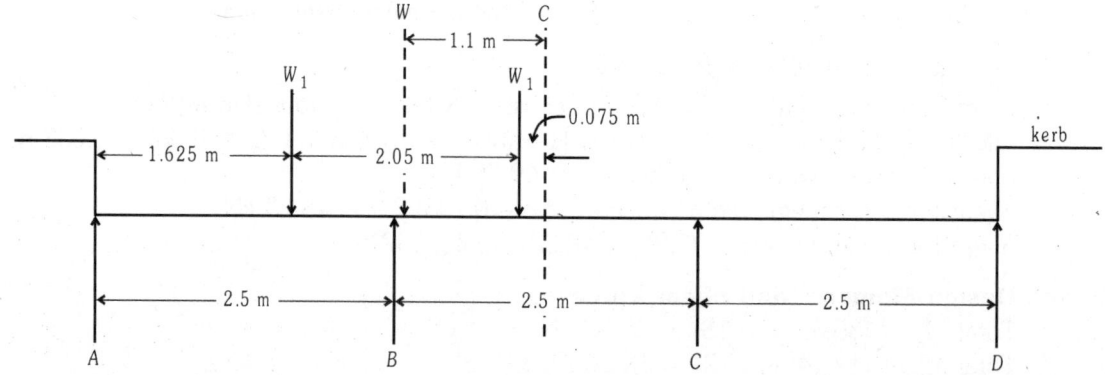

Fig. 6.18: Transverse Dispostion of IRC Class AA Tracked Vehicle on Deck

Reaction factor for outer girder A is expressed as

$$R_A = \frac{2W_1}{4}\left[1 + \frac{4I \times 3.75 \times 1.1}{(2I \times 3.75^2) + (2I \times 1.25^2)}\right] = 0.764 W_1$$

Reaction factor for inner girder is

$$R_B = \frac{2W_1}{4}\left[1 + \frac{4I \times 1.25 \times 1.1}{(2I \times 3.75^2) + (2I \times 1.25^2)}\right] = 0.588W_1$$

If W = Axle load = 700 kN
$W_1 = 0.5W$
$R_A = (0.764 \times 0.5\,W) = 0.382\,W$
$R_B = (0.588 \times 0.5\,W) = 0.294\,W$

(b) Dead Load from Slab per Girder: The dead load of deck slab is calculated with reference to Fig. 6.19.

Loads due to slab, foot path and parapet railing is computed as
1. Deck slab = $(0.20 \times 1.5 \times 24)$ .... = 7.2 kN/m
2. Foot path and kerb = $(0.3 \times 1.5 \times 24)$ .... = 10.8
3. Parapet railing (Lumpsum) .... = 2.0
   Total dead load .... = 20.0 kN/m
   Total dead load of deck = $[(2 \times 20) + (6.56 \times 7.5)]$ ...... = 89.2 kN/m

It is assumed that deck load is shared equally by all the four girders
Hence dead load per girder = $(89.2/4) = 22.3$ kN/m.

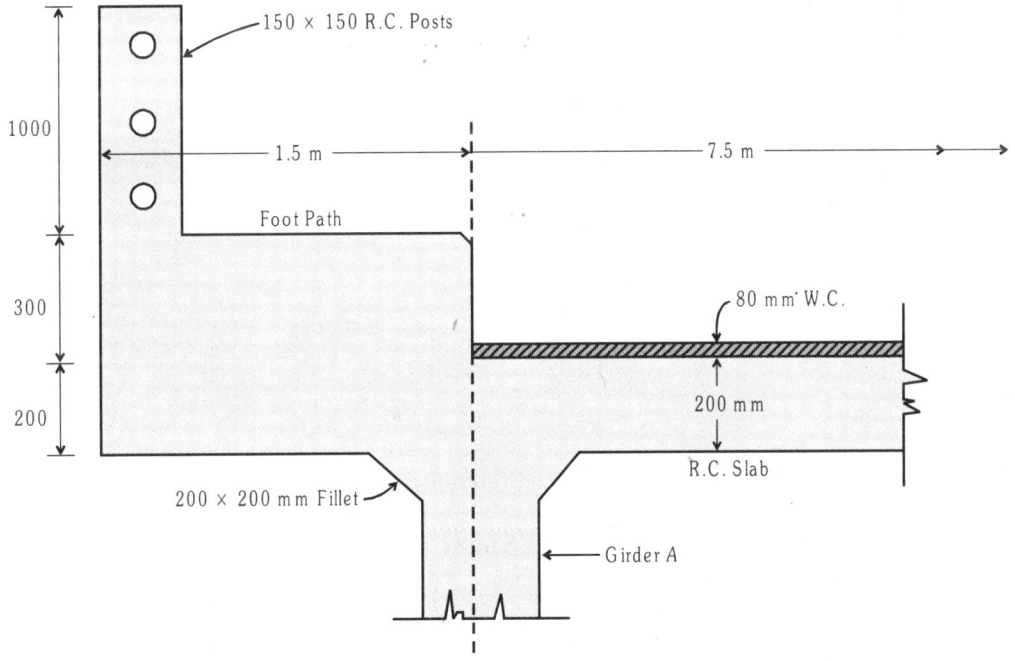

Fig. 6.19: Details of Foot Path, Kerb, Parapet and Deck Slab

(c) Dead Load of Main Girder: The overall depth of the main girder is assumed as 60 mm for every metre of span.
Span of the girder = 30 m
Overall depth of the girder = $(60 \times 30) = 1800$ mm

The size of the bottom flange is estimated to accommodate 4 to 6 cables in two rows with suitable spacing between cables and cover to cables of not less than 50 mm or the diameter of the cable, whichever is higher. The cross-section of the main girder selected is shown in Fig. 6.20.

Dead weight of rib = (1.20 × 0.2 × 24) .... = 5.76 kN/m
Dead weight of bottom flange = (0.5 × 0.4 × 24) .... = 4.80
Total self weight of main girder .... = 10.56 kN/m
Weight of cross girder = (0.2 × 1.30 × 24) .... = 6.24 kN/m

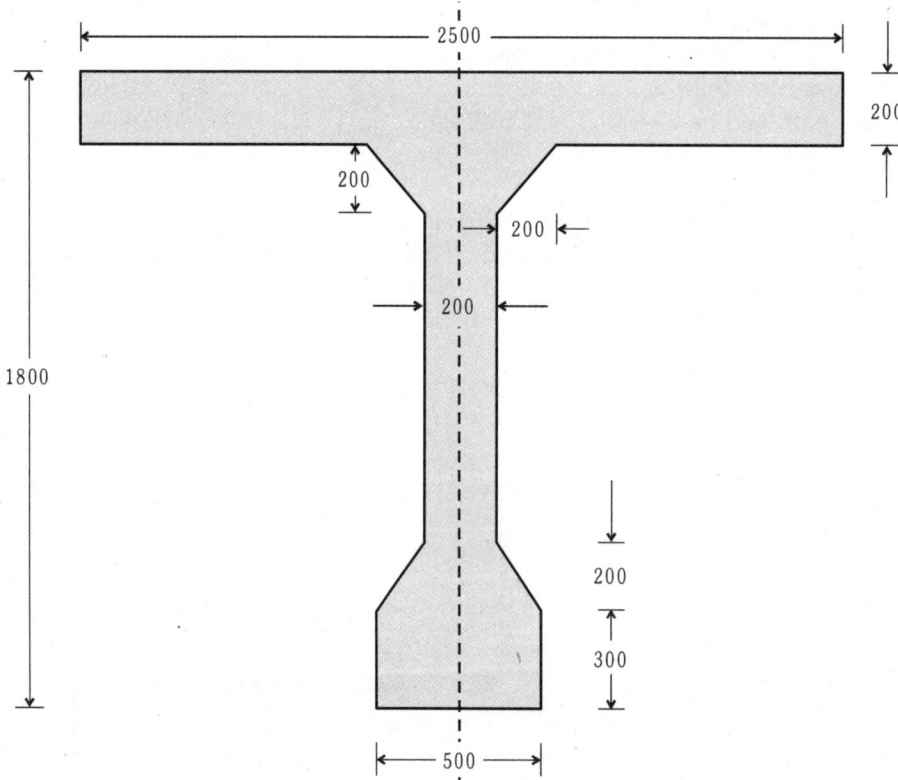

Fig. 6.20: Cross-section of Prestressed Concrete Girder

(d) Dead Load Bending Moments and Shear Forces in Main Girder

Reaction from deck slab on each girder .... = 22.30 kN/m
Weight of cross girder .... = 6.24 kN/m
Reaction on main girder = (6.24 × 2.5) .... = 15.60 kN/m
Extra load due to fillets (Lumpsum) .... = 0.40 kN/m
Total load of main girder .... = 16.00 kN/m
Self weight of main girder .... = 10.56 kN/m
Dead load on each girder = (22.30 + 10.56) .... = 32.86 kN/m
Self weight of finishes, etc. (Lumpsum) .... = 1.14 kN/m
Total dead load on each girder .... = 34.00 kN/m

The maximum dead load bending moment and shear forces are computed using the loads on the girder shown in Fig. 6.21.

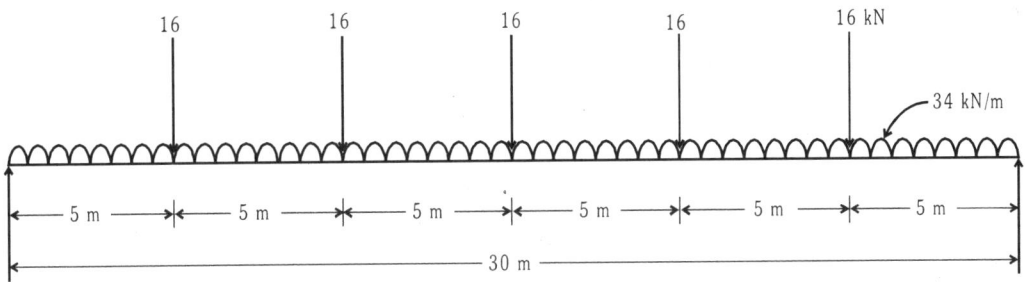

Fig. 6.21: Dead Load on Main Girder

$M_{max} = [(0.125 \times 34 \times 30^2) + (0.25 \times 16 \times 30) + (16 \times 10) + (16 \times 5)] = 4185$ kN·m
$V_{max} = [(0.5 \times 34 \times 30) + (0.5 \times 16 \times 5)] = 550$ kN

(e) Live Load Bending Moments in Main Girder
Span of the girder = 30 m
Impact Factor (IRC Class AA tracked vehicle) = 10%
The live load is placed centrally on the span as shown in Fig. 6.22
Bending moment at centre of span = 0.5 (6.6 + 7.5)700 = 4935 kN·m
B.M. including Impact and Reaction Factors for outer girder is
Live load B.M. = (4935 × 1.1 × 0.382) = 2074 kN·m
B.M. in inner girder = (4935 × 1.1 × 0.294) = 1596 kN·m

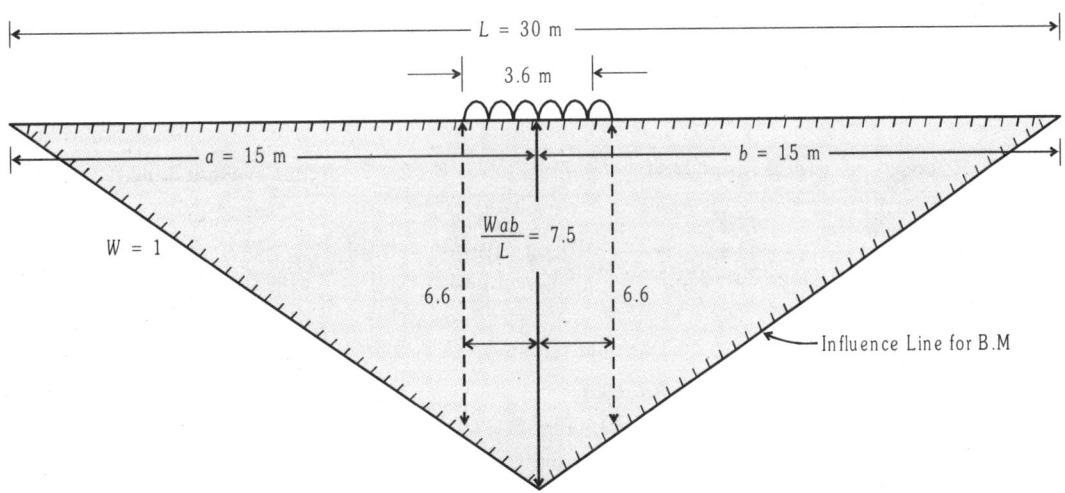

Fig. 6.22: Influence Line for Bending Moment in Girder

(f) Live Load Shear Force in Girders: The live load shear force will be maximum when the IRC Class AA loads occupy the position as shown in Fig. 6.23.

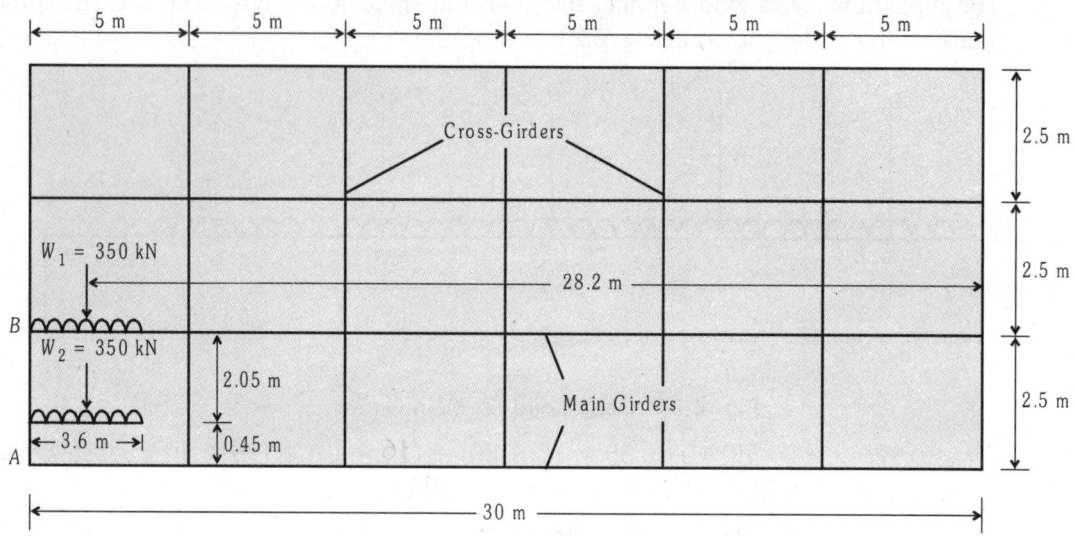

Fig. 6. 23: Position of IRC Class AA Loads for Maximum Shear Force in Inner Girder

Reaction of W_2 on girder $B = [(350 \times 0.45)/2.5] = 63$ kN
Reaction of W_2 on girder $A = [(350 \times 2.05)/2.5] = 287$ kN
Total load on girder $B = (350 + 63) = 413$ kN
Maximum reaction in girder $B = [(413 \times 28.2)/30] = 388$ kN
Maximum reaction in girder $A = [(287 \times 28.2)/30] = 270$ kN
Maximum live load shear with impact factor in inner girder $= (388 \times 1.1) = 427$ kN
Shear force in outer girder $= (270 \times 1.1) = 297$ kN.

(g) Design Bending Moments and Shear Forces: The design bending moments and shear forces are compiled in Table 6.6

Table 6.6: Abstract of Design Bending Moments and Shear Forces in Main Girder

Bending Moment	Dead Load B.M.	Live Load B.M.	Total Design B.M.	Unit
Outer Girder	4185	2074	6259	kN.m
Inner Girder	4185	1596	5781	kN.m
Shear Force	**Dead Load S.F.**	**Live Load S.F.**	**Total Design S.F.**	**Unit**
Outer Girder	550	297	847	kN
Inner Girder	550	427	977	kN

(h) Properties of Main Girder Section

The main girder section is as shown in Fig. 6.24 for computational purposes. The properties of the section are as follows:

$A = (94 \times 10^4)$ mm^4
$y_t = 640$ mm, $y_b = 1160$ mm, $I = (3694 \times 10^8)$ mm^4
$Z_t = (I/y_t) = [(3694 \times 10^8)/640] = (5.77 \times 10^8)$ mm^3
$Z_b = (I/y_b) = [(3694 \times 10^8)/1160] = (3.18 \times 10^8)$ mm^3

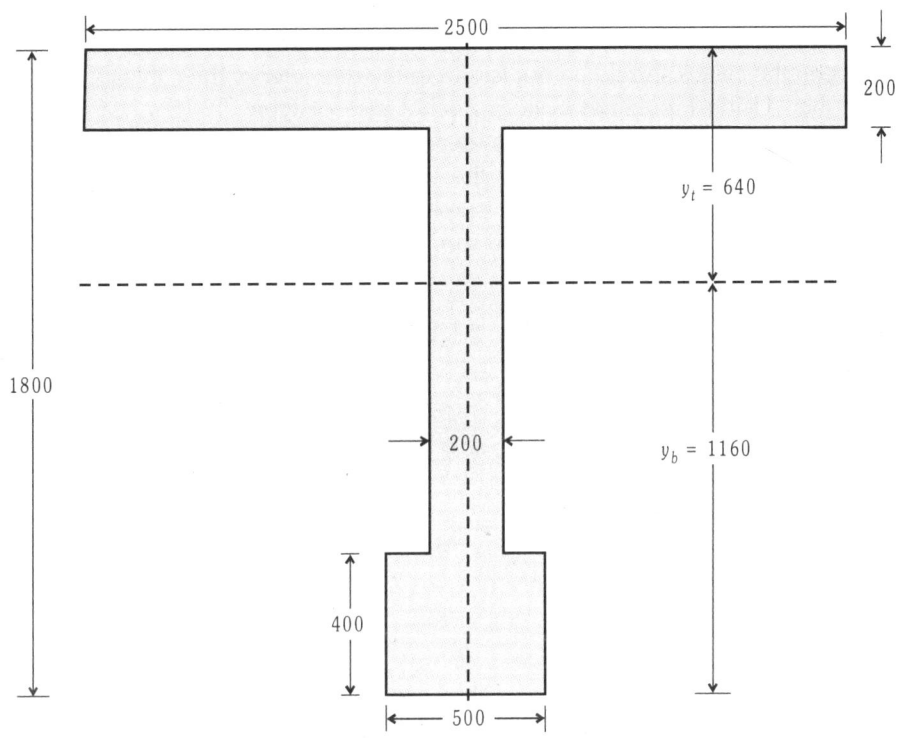

Fig. 6.24: Cross-section of Main Girder

(i) Check for Minimum Section Modulus

$f_{ck} = 50$ N/mm² $\eta = 0.85$
$f_{ct} = 18$ N/mm² $M_g = 4185$ kN.m
$f_{ci} = 40$ N/mm² $M_q = 2074$ kN.m
$f_{cw} = 16$ N/mm² $M_d = (M_g + M_q) = 6259$ kN.m
$f_{tt} = f_{tw} = 0$
$f_{br} = (\eta f_{ct} - f_{tw}) = [(0.85 \times 18) - 0] = 15.3$ N/mm²
$f_{tr} = (f_{cw} - \eta f_{tt}) = 16$ N/mm²

$$f_b = \left[\frac{f_{tw}}{\eta} + \frac{M_d}{\eta Z_b}\right] = \left[0 + \frac{6259 \times 10^6}{0.85 \times 3.18 \times 10^8}\right] = 23.15 \text{ N/mm}^2$$

$$Z_b = \left[\frac{M_q + (1-\eta)M_g}{f_{br}}\right] = \left[\frac{(2074 \times 10^6) + (1-0.85)4185 \times 10^6}{15.3}\right]$$

$$= (1.76 \times 10^8) \text{ mm}^3 < (3.18 \times 10^8) \text{ mm}^3$$

Hence the section provided is adequate.

(j) Prestressing Force:
In this case, $M_g > M_q$ and hence the prestressing force is computed by providing the maximum possible eccentricity with the required cover to the cables

and due provision being made for the number of rows (at least 2) in the bottom flange. Minimum clear cover to the cables in the row nearest to soffit = 75 mm and spacing between the cables should be not less than the diameter of the cable or 50 mm whichever is higher. Using Freyssinet system anchorages of type 7K-15 (7 strands of 15.2 mm diameter) housed in 65 mm diameter cable ducts (IS: 6006–1983) and allowing for sufficient cover to cables such that the centroid of cables is located at a distance of 200 mm from the soffit, the maximum possible eccentricity is computed as

$$e = (1160 - 200) = 960 \text{ mm}$$

Prestressing force is computed as,

$$P = \left[\frac{Af_b Z_b}{Z_b + Ae}\right] = \left[\frac{(94 \times 10^4) \times 23.15 \times 3.18 \times 10^8}{(3.18 \times 10^8) + (94 \times 10^4 \times 960)}\right] = (5670 \times 10^3) \text{ N} = 5670 \text{ kN}$$

Force in each cable = $(7 \times 0.8 \times 260.7) = 1460$ kN
Number of cables = $(5670/1460) = 4$
Area of each strand = 140 mm^2
Area of 7 strands in each cable = $(7 \times 140) = 980$ mm^2
Area of 7 strands in 4 cables = $(4 \times 980) = 3920$ mm^2
The cables are arranged at centre of span as shown in Fig. 6.25.

(k) Permissible Tendon Zone

At support section, $e \leq [(Z_b f_{ct}/P) - (Z_b/A)]$
$\leq [(3.18 \times 10^8 \times 18)/(5670 \times 10^3) - (3.18 \times 10^8)/(94 \times 10^4)]$
≤ 671 mm

And $e \geq (Z_b f_{tw}/\eta P) - (Z_b/A)$
$\geq [0 - (3.18 \times 10^8)/(94 \times 10^4)]$
≥ 338 mm

The four cables are arranged to follow a parabolic profile with the resultant force having an eccentricity of 310 mm towards the soffit at the support section. The position of cables at support section is shown in Fig. 6.26.

6. Check for Stresses

For the centre of span section, we have the parameters,

$P = 5670$ kN $\qquad A = (94 \times 10^4)$ mm^4
$e = 960$ mm $\qquad Z_t = (5.77 \times 10^8)$ mm^3
$M_g = 4185$ kN·m $\qquad Z_b = (3.18 \times 10^8)$ mm^3
$M_q = 2074$ kN·m $\qquad \eta = 0.85$

$(P/A) = [(5670 \times 10^3)/(94 \times 10^4)] = 6.03$ N/mm^2
$(Pe/Z_t) = [(5670 \times 10^3 \times 960)/(5.77 \times 10^8)] = 9.43$ N/mm^2
$(Pe/Z_b) = [(5670 \times 10^3 \times 960)/(3.18 \times 10^8)] = 17.10$ N/mm^2
$(M_g/Z_t) = [(4185 \times 10^6)/(5.77 \times 10^8)] = 7.25$ N/mm^2
$(M_g/Z_b) = [(4185 \times 10^6)/(3.18 \times 10^8)] = 13.16$ N/mm^2
$(M_q/Z_t) = [(2074 \times 10^6)/(5.77 \times 10^8)] = 3.59$ N/mm^2

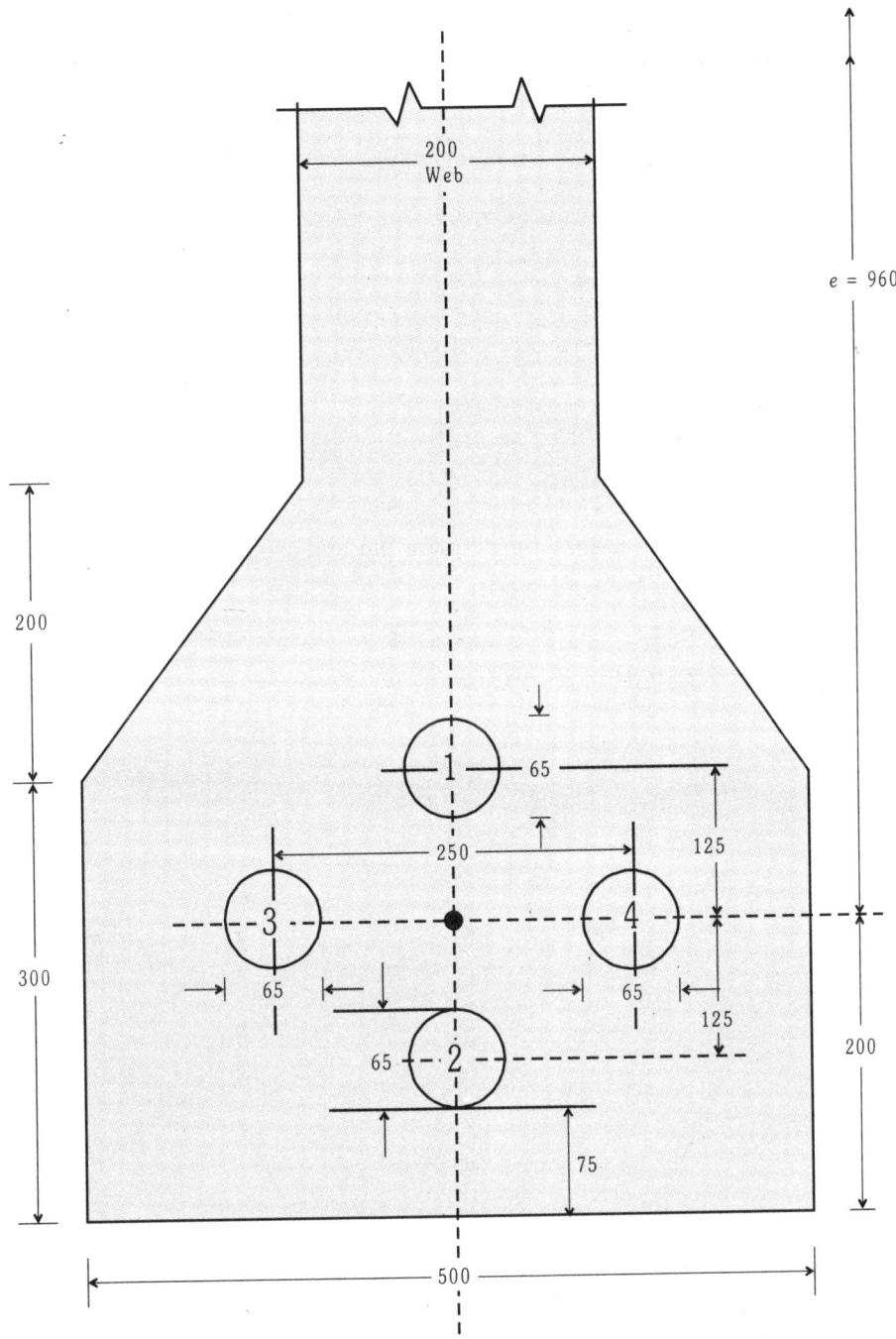

Fig. 6.25: Arrangement of Cables at Centre of Span Section

$(M_q/Z_b) = [(2074 \times 10^6)/(3.18 \times 10^8)] = 6.52 \text{ N/mm}^2$

At transfer stage, the stresses developed are computed as,

$$\sigma_t = \left[\frac{P}{A} - \frac{Pe}{Z_t} + \frac{M_g}{Z_t}\right] = [6.03 - 9.43 + 7.25] = 3.85 \text{ N/mm}^2$$

$$\sigma_b = \left[\frac{P}{A} + \frac{Pe}{Z_b} - \frac{M_g}{Z_b}\right] = [6.03 + 17.10 - 13.16] = 9.97 \text{ N/mm}^2$$

At service load stage, the stresses developed are computed as,

$$\sigma_t = \left[\frac{\eta P}{A} - \frac{\eta Pe}{Z_t} + \frac{M_g}{Z_t} + \frac{M_q}{Z_t}\right]$$

$$= [0.85(6.03 - 9.43) + 7.25 + 3.59] = 7.95 \text{ N/mm}^2 \text{ (Compression)}$$

$$\sigma_b = \left[\frac{\eta P}{A} + \frac{\eta Pe}{Z_b} - \frac{M_g}{Z_b} - \frac{M_q}{Z_b}\right]$$

$$= [0.85(6.03 + 17.10) - 13.16 - 6.52] = -0.0195 \text{ N/mm}^2 \text{ (Tension)}$$

All the stresses at top and bottom fibres at transfer and service loads are well within the safe permissible limits.

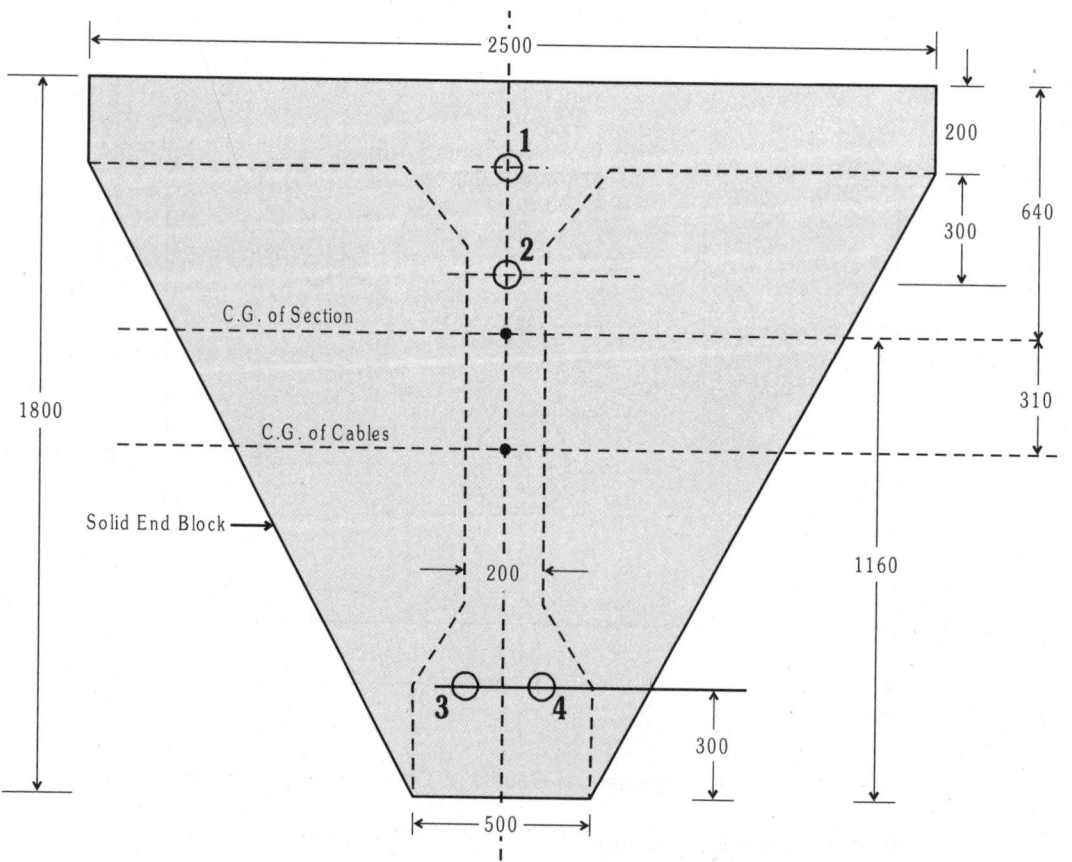

Fig. 6.26: Arrangement of Cables at Support Section

7. **Check for Ultimate Flexural Strength:** The centre of span section is checked for flexural strength according to the specifications of IRC: 18-2000.

$A_p = 3920$ mm^2 $\qquad M_g = 4185$ kN·m
$b = 2500$ mm $\qquad M_q = 2074$ kN·m
$d = 1600$ mm $\qquad b_w = 200$ mm
$f_{ck} = 50$ N/mm^2 $\qquad f_p = 1723$ N/mm^2
$D_f = 200$ mm

M_u (required) $= [1.5 M_g + 2.5 M_q] = [(1.5 \times 4185) + (2.5 \times 2074)] = 11463$ kN·m

   1. **Failure by Yielding of Steel**

$M_{us} = [0.9\ d\ A_p\ f_p] = [(0.9 \times 1600 \times 3920 \times 1723)/10^6] = 9725$ kN·m

Balance moment $= M_{us} = [11463 - 9725] = 1738$ kN·m
Additional area of high tensile steel A_p required for balance moment placed at an effective depth of 1600 mm is computed as,

$$A_p = \left[\frac{1738 \times 10^6}{0.9 \times 1600 \times 1723}\right] = 700 \text{ mm}^2$$

Effective area of untensioned steel ($f_y = 415$ N/mm^2) is calculated as

$$A_s = \left(\frac{A_p f_p}{f_y}\right) = \left(\frac{700 \times 1723}{415}\right) = 2906 \text{ mm}^2$$

Provide 6 bars of 25 mm diameter distributed in the bottom flange ($A_s = 2945$ mm^2)

   2. **Failure by Crushing of Concrete**

$$M_{uc} = 0.176 b_w d^2 f_{ck} + \frac{2}{3} \times 0.8(b - b_w)\left[d - \frac{D_f}{2}\right] D_f f_{ck}$$

$= (0.176 \times 200 \times 1600^2 \times 50) + 0.67 \times 0.8(2500 - 200)$
$\qquad\qquad\qquad\qquad\qquad\qquad [1600 - 100](200 \times 50)$
$= (22997 \times 10^6)$ N·mm
$= 22997$ kN·m > 11463 kN·m (Hence safe)

8. **Check for Ultimate Shear Strength**

Ultimate shear strength required $= [1.5 V_g + 2.5 V_q]$
$\qquad\qquad\qquad\qquad\qquad = [(1.5 \times 550) + (2.5 \times 427]$
$\qquad\qquad\qquad\qquad\qquad = 1892$ kN

According to IRC: 18-2000 the ultimate shear resistance of support section uncracked in flexure is given by

$$V_{cw} = 0.76 b_w h \sqrt{f_t^2 + 0.8 f_{cp} f_t} + \eta P \sin\theta$$

where b_w = width of the web = 200 mm
$\qquad\quad h$ = overall depth of girder
$\qquad\quad f_t$ = maximum principal tensile stress at centroidal axis
$\qquad\quad\quad = 0.24\ \sqrt{f_{ck}} = 0.24\sqrt{50} = 1.7$ N/mm^2

f_{cp} = compressive prestress at centroidal axis

$$= \left(\frac{\eta P}{A}\right) = \left(\frac{0.85 \times 5670 \times 10^3}{94 \times 10^4}\right) = 5.12 \text{ N/mm}^2$$

Eccentricity of cables at centre of span = 960 mm
Eccentricity of cables at support = 310 mm
Net eccentricity = e = (960 – 310) = 650 mm
Slope of cable = θ = $(4e/L)$ = [(4 × 650)/(30 × 10000] = 0.0866

$\therefore \quad V_{cw} = (0.67 \times 200 \times 1800)\sqrt{(1.7^2 + 0.8 \times 5.12 \times 1.7)}$
$\qquad\qquad + (0.85 \times 5670 \times 10^3 \times 0.0866)$

$\qquad = 1174367$ N

$\qquad = 1174$ kN

Shear resistance required = 1892 kN
Shear capacity of section = 1174 kN
Balance shear force = V = 718 kN
Using 10 mm diameter two legged stirrups of FE-415 HYSD bars, the spacing S_v at supports is given by

$$S_v = \left[\frac{0.87 f_y A_{sv} d}{V}\right] = \left[\frac{0.87 \times 415 \times 2 \times 79 \times 1750}{718 \times 1000}\right] = 139 \text{ mm}$$

Provide 10 mm diameter stirrups at 130 mm centres at supports gradually increasing the spacing to 300 mm towards the centre of span.

9. **Supplementary Reinforcement:** Minimum longitudinal reinforcements of not less than 0.18 per cent of gross cross-section are to be provided to safeguard against shrinkage cracking.

$$A_{st} = [0.0018 \times 94 \times 10^4] = 1692 \text{ mm}^2$$

Provide 8 bars of 20 mm diameter distributed in the compression flange as shown in Fig. 6.27.

10. **Design of End Block:** Solid end blocks are provided at the end supports over a length equal to the depth of the girder which is 1.8 m. Typical equivalent prisms on which the anchorage forces are considered to be effective are detailed in Fig. 6.28.
The bursting tension is computed using the data given in Table 4.3.
In the horizontal plane we have the data,

$\qquad P_k = (2 \times 1460)$ kN, $2y_{po} = 225$ mm and $2y_o = 800$ mm
$\therefore \quad$ The ratio $(y_{po}/y_o) = (450/800) = 0.50$
Bursting tension $F_{bsv} = (0.17 \times 2 \times 1460) = 496$ kN
Area of steel to resist this tension is obtained as

$$A_{st} = \left[\frac{496 \times 1000}{0.87 \times 415}\right] = 1375 \text{ mm}^2$$

Provide 12 mm diameter bars at 100 mm centres in the horizontal direction. In the vertical plane the ratio of (y_{po}/y_o) being higher, the magnitude of bursting tension is smaller. However the same reinforcements are provided in the form of a mesh in the vertical and horizontal directions in front of the anchorages as shown in Fig. 6.28.

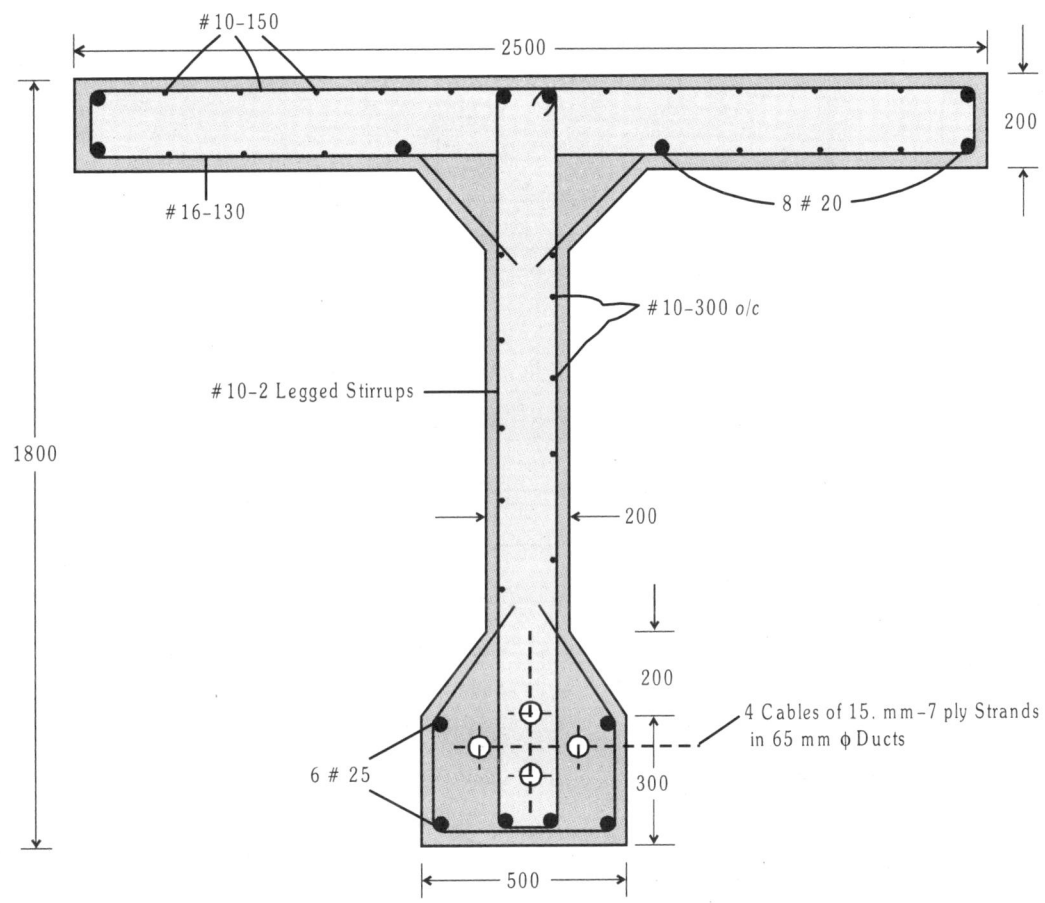

Fig. 6.27: Reinforcement Details at Centre of Span Section

11. **Cross Girders:** Analysis of bending moments and shear forces in cross girders under IRC Class AA loads reported by the author elsewhere[11, 12] are of very low magnitude and the depth of the cross girders being nearly 75 per cent of the main girder, minimum steel reinforcements in the girders will be sufficient to resist these small moments and shear forces. Cross girders of 200 mm width and 1500 mm depth are provided with nominal reinforcements comprising 12 mm diameter bars distributed in the section. Two bars each at top, mid depth and near the soffit are provided. Nominal stirrups of 10 mm diameter two legged links are provided at 200 mm centres. Two cables comprising 12 numbers of 7 mm diameter high tensile wires are provided at middle third points along the depth of the cross girder.

12. **Reinforcement Details:** The details of reinforcements in the mid span section of the main girder is shown in Fig. 6.27 and the anchorage zone reinforcements in the end block is detailed in Fig. 6.28.

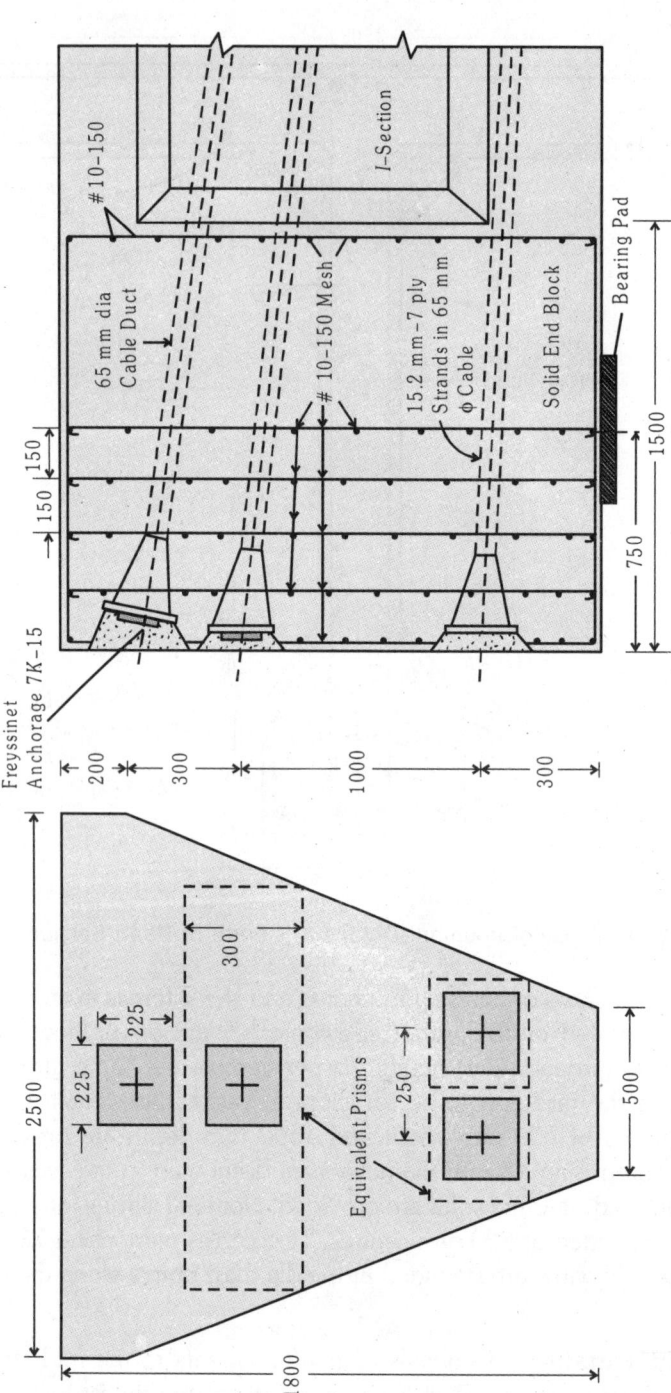

Fig. 6.28: Equivalent Prisms and End Block Reinforcement

Prestressed Concrete Tee Beam and Slab Bridge Deck

REFERENCES

1. ROWE, R. E., *Concrete Bridge Design*. C. R. Books Ltd, London, 1962, First Edition, p. 336.
2. VICTOR, D. J. and LAKSHMANAN, N., Reactions in three girder bridge decks. *Journal of the Indian Roads Congress*, Vol. 36-1, Oct. 1975, Paper No. 299, pp. 41-68.
3. IRC: 21-2000, *Standard Specifications and Code of Practice for Road Bridges*, Section-III, *Cement Concrete* (Plain and Reinforced), Third Revision, Indian Roads Congress, 2000, pp. 52-55.
4. IRC:6-2000, *Standard Specifications and Code of Practice for Road Bridges* (Section II), *Loads and Stresses* (Fourth Revision), Indian Roads Congress, New Delhi, 2000, pp. 16-17.
5. RAINA, V. K., *Concrete Bridge Practice, Analysis, Design and Economics*. Tata McGraw Hill Publishers, New Delhi, 1991, pp. 474-483.
6. MORICE, P. B. and LITTLE, G., The analysis of right bridge decks subjected to abnormal loading. Cement and Concrete Association, Db 11, Publication No. 46.002 London, July 1956, p. 43.
7. HENDRY, A. W. and JAEGAR, L.G., *The Analysis of Grid Frame Works and Related Structures*, Chatto and Windus, London, 1958, p. 308.
8. Cussens, A. R., Load Distribution in Concrete Bridge Decks. Construction Industry Research and Information Association Report 53, London, December, 1974, p. 38.
9. COURBON, J., Application de la Resistance des Materiaux au Calculdes Ponts, Dunod, Paris, 1950.
10. MASSONET, C., *Method of Calculation for Bridges with Several Longitudinal Beams taking into Consideration their Torsional Resistance*. International Association for Bridges and Structural Engineering (IABSE), Vol. 10, 1950, pp. 147-182.
11. SARKAR. S., *et al.*, *Hand Book for Prestressed Concrete Bridges*. Structural Engineering Research Centre, Roorkee, 1969, pp. 1-238.
12. KRISHNA RAJU, N., *Prestressed Concrete*, (Fourth Edition). Tata McGraw Hill Publising Co, New Delhi, 2007, p. 794.

EXERCISES

1. Design a prestressed concrete tee beam and slab bridge deck for a National Highway crossing to suit the following data:
   Clear width of road way = 15 m
   Width of kerbs = 600 mm
   Effective span = 25 m
   Thickness of wearing coat = 100 mm
   Spacings of main girders = 2.5 m
   Spacings of cross girders = 5 m
   Foot paths: one metre on either side
   Live load : IRC Class A or AA whichever is critical
   Grade of concrete: M-45
   Type of steel reinforcements: Fe-415 HYSD bars and H. T. Steel strands conforming to IS: 6006.
   Design the deck slab, main girder and cross girder using Courbon's method.
   The designs should conform to the specifications of relevant IRC Codes.
   Sketch the details of reinforcements in the various structural elements at critical cross-sections.

2. A National Highway crossing requires a post tensioned P.S.C. tee beam and slab bridge deck to cross a river. The following data is available:
   Effective span of the main girders = 30 m
   Clear width of road way = 15 m

Width of kerbs = 600 mm
Number of main girders = 7
Foot paths: one metre on either side
Live load : IRC Class AA or A whichever produces the worst effect
Thickness of the wearing coat = 80 mm
Spacings of cross girders = 45 m
Spacings of main girders = 2.5 m
M-50 Grade concrete and Fe-415 Grade HYSD bars along with 7K-15 H.T. Strands are available for use
Design the deck slab, main girders and cross girders using Guyon-Massonet method. Sketch the details of reinforcements in the slab and beams at critical sections.

3. Design a suitable prestressed concrete tee beam and slab bridge deck to suit the following data:
Effective span of the main girders = 24 m
Width of road way = 15 m
Kerbs: 600 mm on either side with 1 metre foot paths
Spacings of cross girders = 6 m
Spacings of main girders = 2.5 m
Thickness of wearing coat = 80 mm
Concrete: M-45 Grade
Steel: Fe-415 HYSD bars and H.T. steel strands conforming to IS: 6006.
Live load: IRC Class AA or A whichever gives the worst effect
Design one of the interior slab panel, main girder and cross girder and sketch the details of reinforcements at critical sections.

7
PRESTRESSED CONCRETE CONTINUOUS SPAN BRIDGE DECKS

7.1 ADVANTAGES OF CONTINUOUS SPAN BRIDGE DECKS

Prestressed concrete continuous girder bridge decks exhibit many advantages in comparison with simply supported bridge decks. Continuous bridge decks being statically indeterminate exhibit the following benefits:

1. In continuous members, the bending moments are more evenly distributed between the centre of span and the support sections.
2. The ultimate load carrying capacity is higher than in statically determinate structures due to the phenomenon of redistribution of moments.
3. The cross-sections of continuous members are comparatively smaller than that of simply supported members resulting in lighter structures.
4. Continuity in framed structures leads to increased stability.
5. In continuous post tensioned girders, the curved cable can be suitably positioned to resist the span and support moments.
6. Precast segmental construction can be adopted by connecting them by high tensile cables.
7. The depth of the girders can be suitably varied depending upon the magnitude of moments at various sections along the span.
8. Continuous girders require less number of anchorages in comparison with a series of simply supported beams. Only one pair of post tensioning anchorages and a single stressing operation can serve in combining the several precast elements.
9. Continuity is often preferred for long span bridges for economical reasons.
10. In continuous prestressed bridge decks, the deflections under dead and super imposed loads are comparatively smaller in comparison with simply supported members.

Some of the disadvantages of continuous bridge decks are as follows:

1. Secondary stresses may develop due to settlement of supports, differential temperature, shrinkage and creep effects.
2. Prestressed concrete continuous beams must either be designed with concordant cables or else the effect of lack of concordance should be considered in the design.

3. Continuous bridges require greater care and skill in construction. Continuous bridge decks are often built using statically determinate beam elements and full continuity established by using cap cables near supports.

7.2 METHODS OF PRESTRESSING CONTINUOUS BRIDGE DECKS

Continuity in prestressed concrete construction is achieved by adopting any of the following methods:

1. By using continuous curved cables to suit the bending moments along the spans as shown in Fig. 7.1 (a).
2. By using a straight cable in conjunction with variable section of girders so that the desirable eccentricity of the cable is ensured at centre of span and supports as shown in Fig. 7.1 (b).
3. Simply supported spans with normal cables cast and subsequently made continuous by using cap cables at the junction of the beams [Fig. 7.1 (c)].
4. Simply supported spans with normal cables subsequently made continuous by using straight cables near the top at the junction of the beam [Fig. 7.1(d)].

Several methods of developing continuity in prestressed concrete structures have been critically examined by Lin[1] and Visvesvaraya et al[2]. Cantilever method of construction developed by Finisterwalder[3] has revolutionalised the construction of long span prestressed concrete bridges using precast units. Most of the flyovers in urban areas are made up of continuous prestressed concrete bridges assembled by prestressing modular precast segmental units of variable cross-section. The reader may refer to a separate monograph[4] by author for exhaustive information regarding continuous prestressed concrete bridges.

7.3 CROSS-SECTIONS OF PRESTRESSED CONCRETE CONTINUOUS BRIDGE DECKS

Several novel types of cross-sections have been developed over the years for the construction of prestressed concrete continuous bridge decks.

The factors influencing the selection of cross-sections of continuous bridge decks are

1. Individual span lengths between supports
2. Total length of the bridge
3. Number of lanes of traffic
4. Type of loading expected on bridge deck
5. Availability of suitable construction techniques at site
6. Type of crossing such as river, viaduct, road, etc.
7. Speed of construction.

Figure 7.2 shows different types of bridge deck configurations used for prestressed concrete continuous bridge decks.

Voided slabs shown in Fig. 7.2 (a) have been used in the span range of 10 to 20 m on wall type piers both as precast or cast *in situ* elements. The span/depth ratios being generally 30 and as high as 40 ± have been used. This type of deck having high torsional resistance is ideally suited for curved alignment especially on single columns.

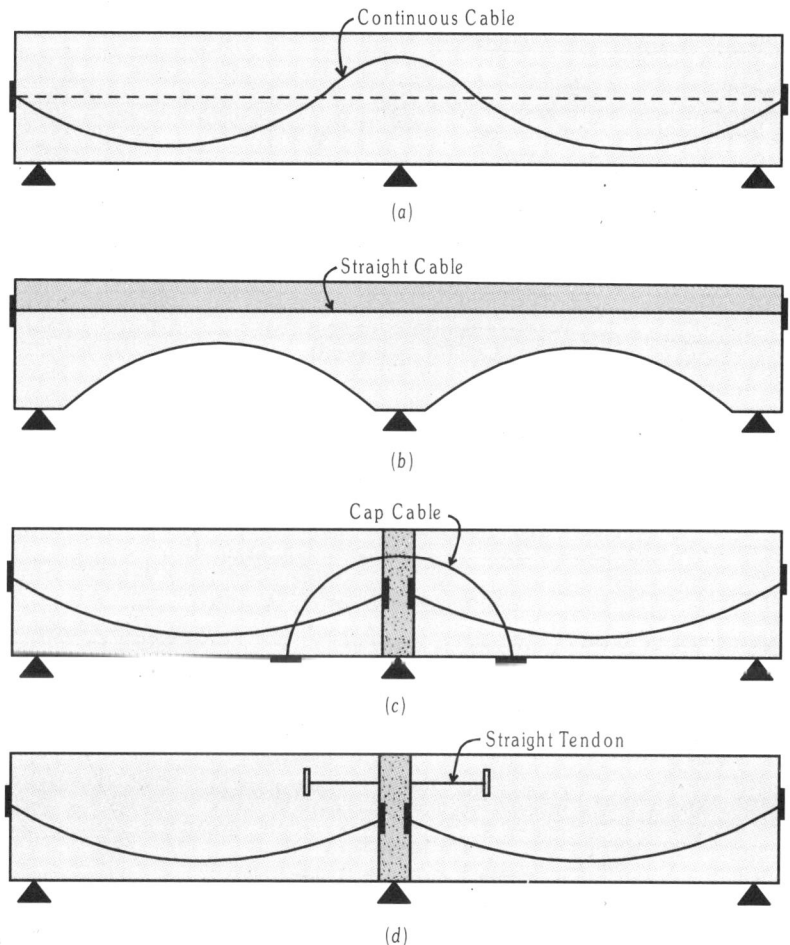

Fig. 7.1: Methods of Prestressing Continuous Bridge Decks

Single box precast post tensioned girders [Fig. 7.2 (b)] have been widely used in the span range of 25 to 200 m for two or three lane traffic. For multilane traffic wider decks are required and in such cases two or three celled box girders are adopted as shown in Fig. 7.2 (c) and (d).

Raina[5] has reported the details of cast *in situ* prestressed concrete box decks used for two, three and four continuous spans. The spans varied in the range of 30, 45 and 60 m and the span/depth ratios varied from 20 to 33. The quantity of prestressing steel used in kg/m^2 of deck plan area increased with the span/depth ratio from a minimum of 10.9 to a maximum of 33.4. Cantilever construction is ideally suited for continuous bridge decks proceeding from either side of the piers with moveable form work. Modular precast box girders can be lifted by cranes to the desired alignment and the bridge spans may be constructed by successive prestressing of the units.

7.4 DESIGN OF TWO SPAN CONTINUOUS PRESTRESSED CONCRETE BRIDGE DECK

A two span post tensioned prestressed concrete beam and slab bridge deck is to be designed for a National Highway crossing. Design the continuous P.S.C. bridge deck using the following data:

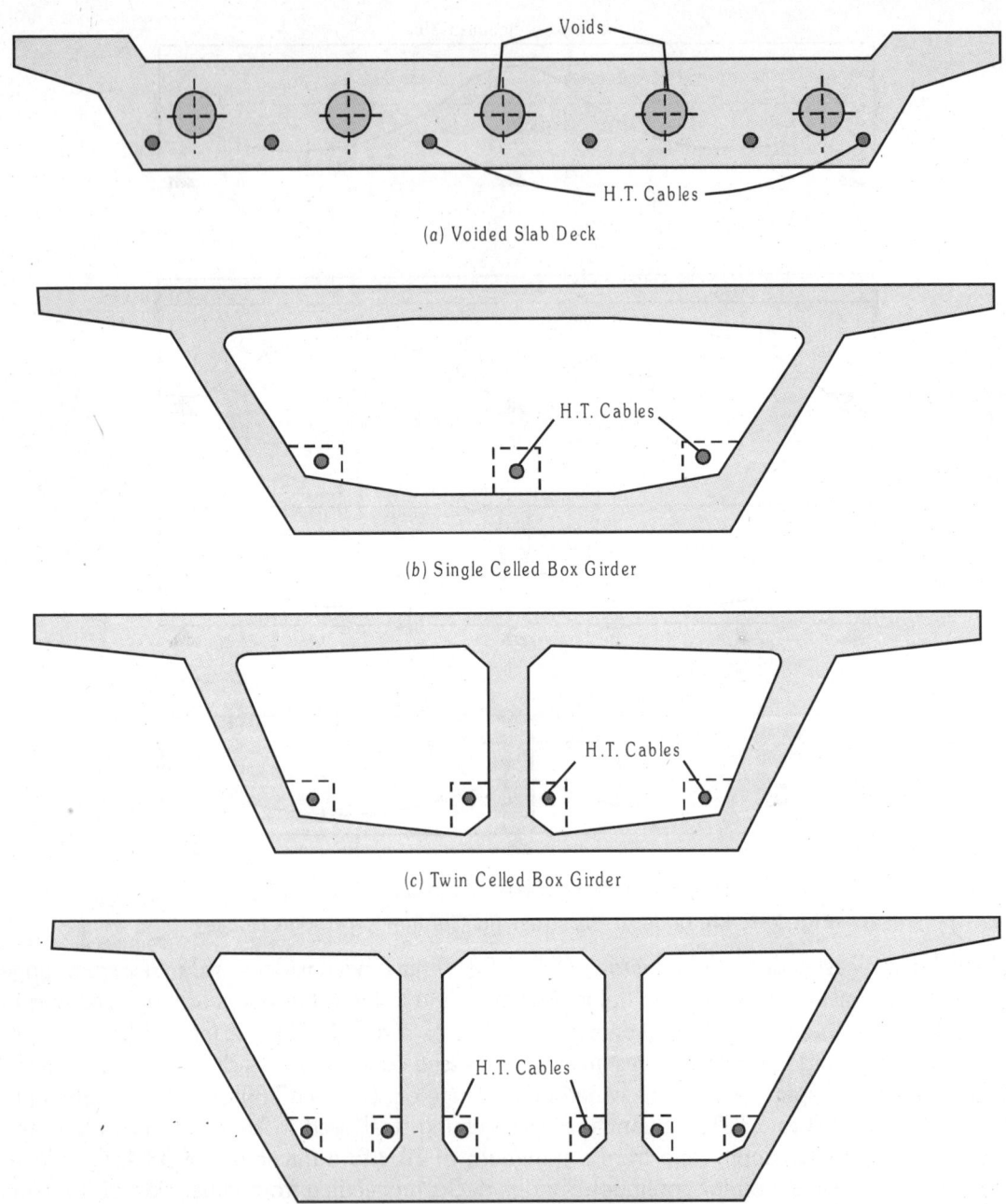

Fig. 7.2: Prestressed Concrete Continuous Bridge Deck Configurations

1. **Data**

   Two continuous spans each of 40 m between the bearings.
   Width of road way = 7.5 m
   Kerbs: 600 mm wide on each side

Prestressed Concrete Continuous Span Bridge Decks

Thickness of wearing coat = 80 mm
Live load: IRC Class AA tracked vehicle
For cast *in situ* deck slab adopt M-25 Grade concrete and Fe-415 Hysd bars
For prestressed concrete girders adopt M-60 Grade concrete with cube strength of concrete at transfer as 40 N/mm^2
Loss ratio = 0.8
High tensile steel strands of 15.2 mm diameter conforming to IS: 6006-1983 are available for use.
Design the salient structural components of the bridge deck as Class 1 type structure conforming to the codes IRC: 6-2000, IRC: 18-2000 and IRC: 21-2000. Sketch the typical details of reinforcements and cables in the beam and slab deck.

2. **Permissible Stresses**: The permissible stresses for M-25 grade concrete and Fe-415 HYSD bars according to IRC: 21-2000 are compiled below. For design coefficients refer to the manograph[6] by the author.
For M-25 Grade concrete and Fe-415 HYSD bars, we have the design coefficients as

σ_{cb} = 8.3 N/mm^2 Q = 1.10
σ_{st} = 200 N/mm^2 j = 0.90
m = 10 η = 0.8

For M-60 Grade concrete used in P.S.C. Girder (IRC: 21-2000)

f_{ck} = 60 N/mm^2 f_{ci} = 45 N/mm^2
f_{ct} = 0.5 f_{ci} = (0.5 × 45) = 22.5 N/mm^2 limited to 20 N/mm^2
f_{cw} = 0.33 f_{ck} = (0.33 × 60) = 20 N/mm^2
$f_{tt} = f_{tw}$ = 0 (Class 1 type member—No tensile stresses)
E_c = 5000$\sqrt{f_{ck}}$ = 5000$\sqrt{60}$ = 38730 N/mm^2 = 38.73 kN/mm^2

3. **Cross-section of Bridge Deck**
Spacings of main girders = 2.5 m
Spacings of cross girders = 5.0 m
Thickness of deck slab = 250 mm
Thickness of wearing coat = 80 mm
Kerbs 600 mm wide by 300 mm deep are provided at each end
Overall depth of main girder is assumed at 50 mm per metre of span.
∴ Overall depth of girder = h = (50 × 40) = 2000 mm
Width of top and bottom flange = (0.4 h) = (0.4 × 2000) = 800 mm
Thickness of web to house a cable of 100 mm diameter with side covers of 75 mm on either side = (100 + 75 + 75) = 250 mm
The section properties of the main girder are as follows:
Cross-sectional area = A = 0.94 m^2
Second moment of area = I = 0.45 m^4
$y_b = y_t$ = 1 m
Section modulus = $Z_t = Z_b = Z$ = 0.45 m^3
The cross-sectional details of the bridge deck and main girders are shown in Figs. 7.3 and 7.4 respectively.

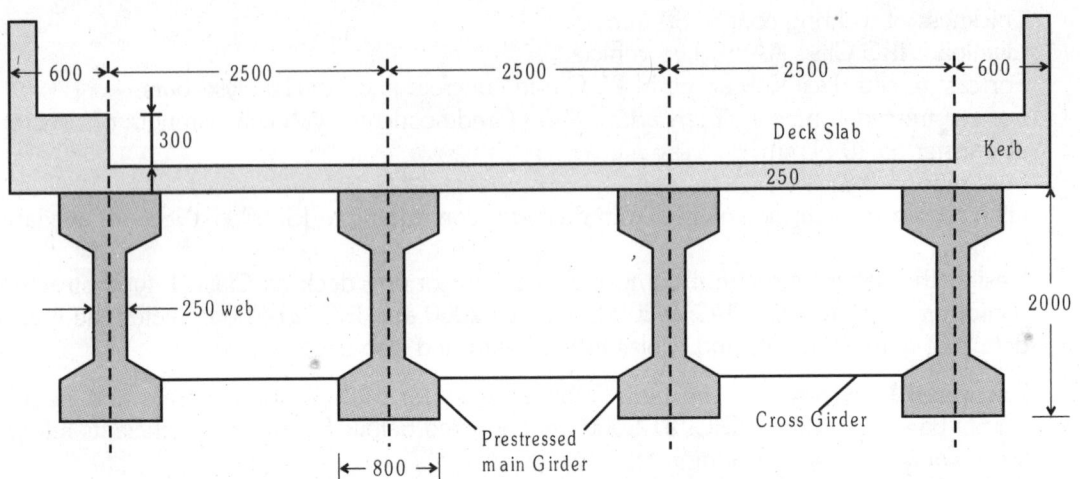

Fig. 7.3: Cross-section of Prestressed Concrete Continuous Bridge Deck

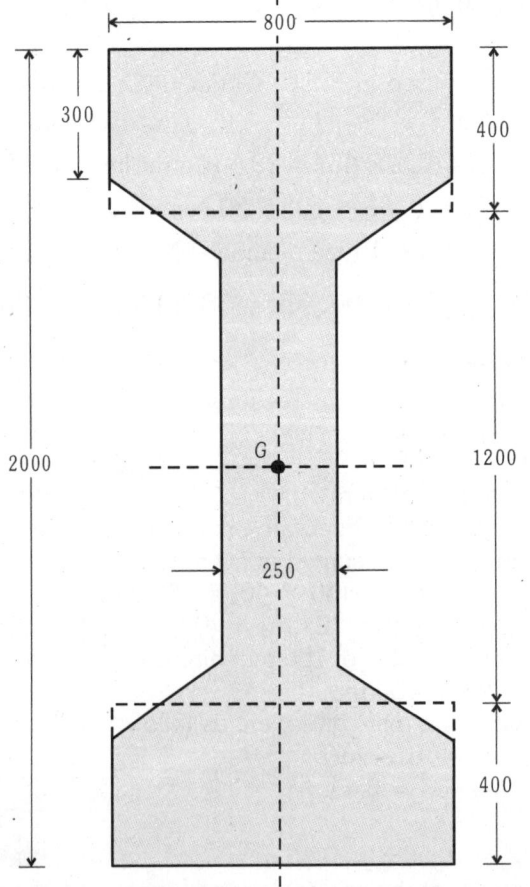

Fig. 7.4: Cross-section of Continuous Prestressed Concrete Girder

4. Design of Interior Slab Panel
(a) Bending Moments

Size of slab panel = 5 m by 2.5 m Hence $L = 5$ m and $B = 2.5$ m
Thickness of slab = 250 mm
Dead weight of slab = $(1 \times 1 \times 0.25 \times 25)$ = 6.25 kN/m²
Dead weight of wearing coat = (0.08×22) = 1.76 kN/m²
Total dead load = g ... = 8.00 kN/m²
Live load is IRC Class AA tracked vehicle. One wheel is placed at the centre of slab panel as shown in Fig. 7.5.

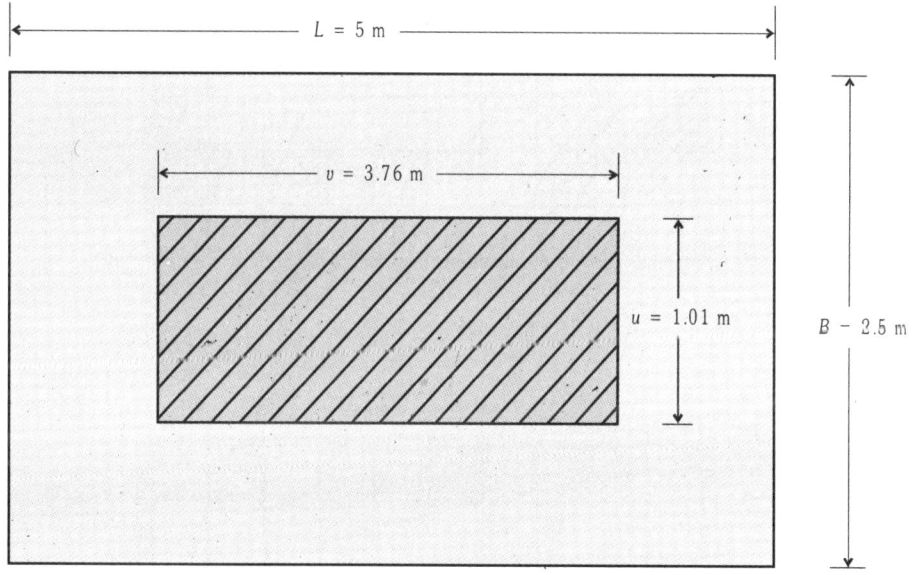

Fig. 7.5: Position of IRC Class AA Wheel Load for Maximum Moment

$$u = (0.85 + 2 \times 0.08) = 1.01 \text{ m}$$
$$v = (3.60 + 2 \times 0.08) = 3.76 \text{ m}$$
$$(u/B) = (1.01/2.5) = 0.404$$
$$(v/L) = (3.76/5.0) = 0.752$$
$$K = (B/L) = (2.5/5.0) = 0.5$$

Referring to Pigeaud's curves (Fig. 7.6), read out the values of coefficients as
$$m_1 = 0.098 \quad \text{and} \quad m_2 = 0.02$$
$$M_B = W[m_1 + \mu m_2]$$
$$= 350[0.098 + (0.15 \times 0.02)] = 35.35 \text{ kN·m}$$

As the slab is continuous, design B.M. = $0.8 M_B$
Design bending moment including the impact and continuity factors is given by
M_B (short span) = $(1.25 \times 0.8 \times 35.35) = 35.35$ kN·m
Similarly, M_L (long span) = $W[m_2 + \mu m_1]$
$$= 350[0.02 + 0.15 \times 0.098] = 12.14 \text{ kN·m}$$
Design bending moment, $M_L = (1.25 \times 0.8 \times 12.14) = 12.14$ kN·m

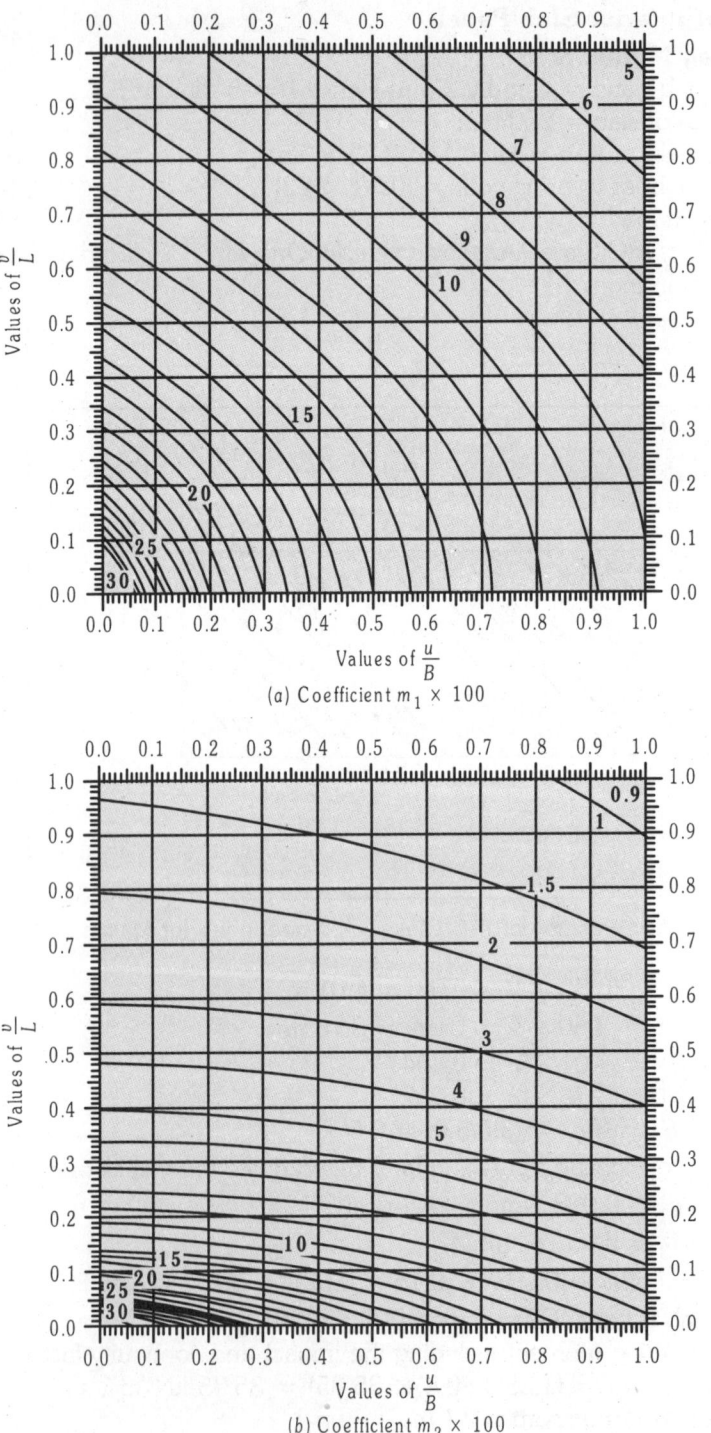

Fig. 7.6: Moment Coefficients m_1 and m_2 for $K = 0.5$ (Pigeaud's Curves)

(b) Shear Forces

Dispersion in the direction of span = [0.85 + 2(0.08 + 0.25)] = 1.51 m
For maximum shear force, the load is kept such that the whole dispersion is in the span. The load is kept at (1.51/2) = 0.755 m from the edge of the beam as shown in Fig. 7.7.

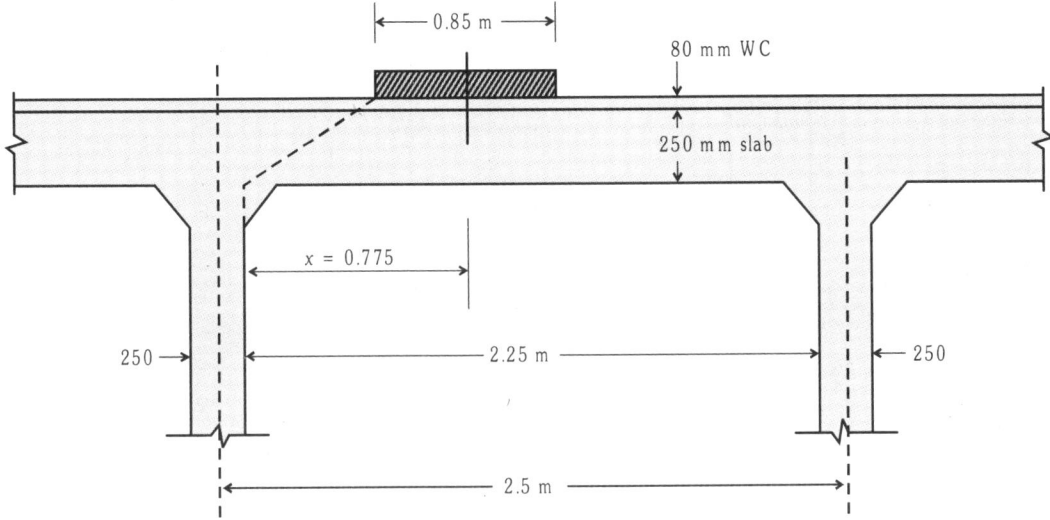

Fig. 7.7: Position of Wheel Load for Maximum Shear

Effective width of the slab = $Kx[1- (x/L)] + b_w$
Breadth of the cross girder = 200 mm
Clear length of the panel = (5 – 0.2) = 4.8 m
∴ (B/L) = (4.8/2.3) = 2.08
From Table 5.1, the value of 'K' for continuous slab is read out as 2.60
Effective width of the slab = (2.6 × 0.755)[1 – (0.755/2.3) + 3.6 + (2 × 0.08)]
= 5.079 m
Load per metre width = (350/5.079) = 70 kN
Shear force/metre width = [70 (2.3 – 0.755)/2.3] = 47 kN
Shear force with impact = (1.25 × 47) = 58.75 kN

(c) Dead Load Bending Moment and Shear Forces

Dead Load = 8 kN/m²
Total load on panel = (5 × 2.5 × 8) = 100 kN
   (u/B) =1 and (v/L) = 1
as the panel is loaded with a uniformly distributed load
   K = (B/L) = (2.5/5.0) = 0.5 and (1/K) = 2.0
From Pigeaud's curves (Fig. 7.8), read out the moment coefficients
   m_1 = 0.047 and m_2 = 0.01
   M_B = 100 (0.047 + 0.15 × 0.01) = 4.85 kN·m
   M_L = 100 (0.01 + 0.15 × 0.047) = 1.70 kN·m

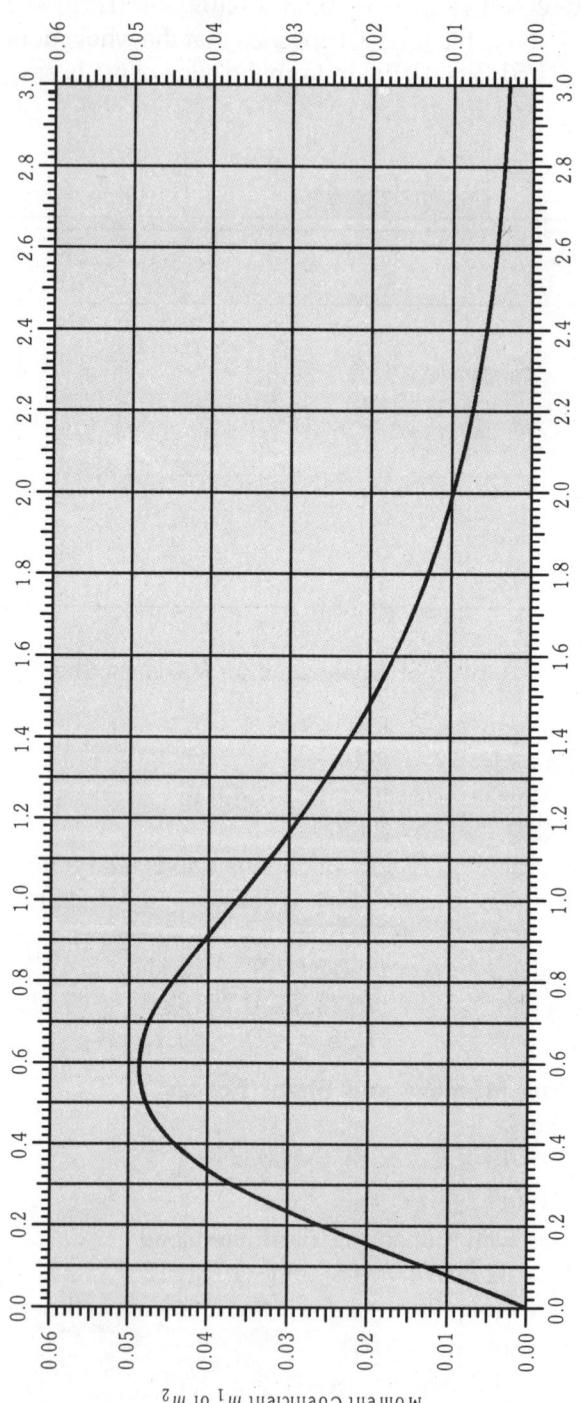

Fig. 7.8: Moment Coefficients for Slab Completely Loaded with Uniformly Distributed Load (Coefficient m_1 for K and m_2 for $1/K$)

Design B.M. including the continuity factor
$$M_B = (0.8 \times 4.85) = 3.88 \text{ kN·m}$$
$$M_L = (0.8 \times 170) = 1.36 \text{ kN·m}$$
Dead Load Shear Force = $(0.5 \times 8 \times 2.3) = 9.2$ kN

(d) **Design Moments and Shear Forces**
Total $M_B = (35.35 + 3.88) = 39.23$ kN·m
Total $M_L = (12.14 + 1.36) = 13.50$ kN·m
Total $V = (58.75 + 9.20) = 67.95$ kN

(e) **Design of Slab Section and Reinforcements**
$$\text{Effective depth} = d = \sqrt{\frac{M}{Qb}} = \sqrt{\frac{39.23 \times 10^6}{1.10 \times 10^3}} = 189 \text{ mm}$$

According to IRC: 21-2000, the minimum clear cover to steel reinforcement should be not less than 40 mm
Hence adopt effective depth = d = 200 mm
$$A_{st} = \left[\frac{M}{\sigma_{st}\, jd}\right] = \left[\frac{39.23 \times 10^6}{200 \times 0.90 \times 200}\right] = 1090 \text{ mm}^2$$

Use 16 mm diameter bars at 180 mm centres ($A_{st} = 1117$ mm²)
Effective depth for long span using 10 mm diameter bars = $(200 - 8 - 50) = 187$ mm
$$A_{st} = \left(\frac{13.50 \times 10^6}{200 \times 0.90 \times 187}\right) = 401 \text{ mm}^2$$

But minimum reinforcement according to IRC: 18-2000 is 0.15 per cent of the cross-sectional area. Hence area of steel in the long span direction is computed as
A_{st} (Long span) = $(0.0015 \times 1000 \times 250) = 375$ mm²
Adopt 10 mm diameter bars at 180 mm centres ($A_{st} = 436$ mm²)

(f) **Check for Shear Strength (As per IRC: 21-2000)**
$$\text{Nominal shear stress} = \tau_v = \left(\frac{V}{bd}\right) = \left(\frac{67.95 \times 1000}{1000 \times 200}\right) = 0.337 \text{ N/mm}^2$$

$$\text{Ratio} \left(\frac{100 A_{st}}{bd}\right) = \left(\frac{100 \times 1117}{1000 \times 200}\right) = 0.558$$

From Table 12B of IRC: 21-2000, read out the permissible shear stress in M-25 Grade concrete as 0.32 N/mm².
For solid slabs, permissible shear stress is 'K' where $K = 1.10$ for 250 mm thick slab from Table 12 (C) of IRC: 21-2000.
Permissible shear stress = $(1.1 \times 0.32) = 0.352$ N/mm² $> \tau_v$ Hence safe.

5. Design of Continuous Longitudinal Girder

(a) **Reaction Factors:** Using Courbon's theory, the IRC Class AA tracked loads are arranged for maximum eccentricity as shown in Fig. 7.9. The reaction factor maximum for any of the longitudinal girder is computed using the general relation,

$$R_x = \frac{\Sigma W}{n}\left[1+\left\{\frac{\Sigma I}{\Sigma d_x^2 I}\right\}d_x \cdot e\right]$$

where R_x = Reaction factor for the girder under consideration
 I = Second moment of area of each longitudinal girder
 d_x = Distance of the girder under consideration from the centroidal axis of the bridge
 ΣW = Total concentrated live load
 n = Number of longitudinal girders
 e = Eccentricity of live load with respect to the axis of the bridge

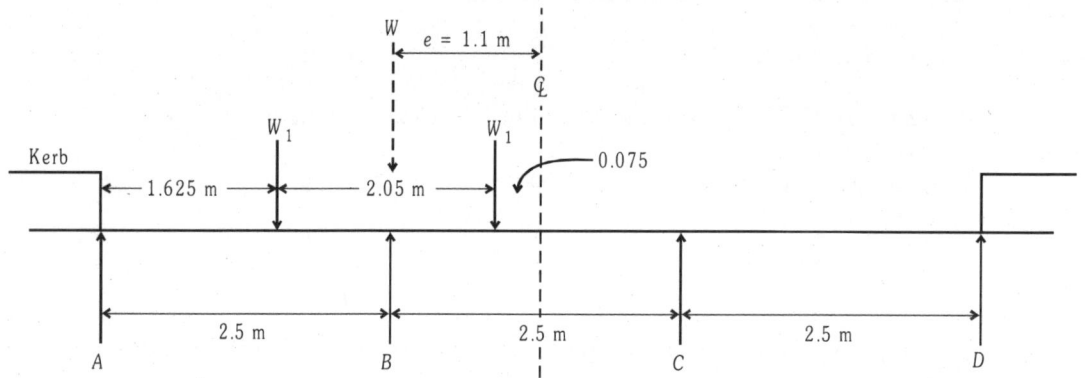

Fig. 7.9: Transverse Disposition of IRC Class AA Tracked Vehicle Loads

In the present case the corresponding values for the exterior girder A are as follows:
 $\Sigma W = 2 W_1$
 $d_x = 3.75$ m
 $e = 1.1$ m

$$R_A = \frac{2W_1}{4}\left[1+\frac{4I \times 3.75 \times 1.1}{(2I \times 3.75^2)+(2I \times 1.25^2)}\right] = 0.764 W_1$$

If W = axle load = 700 kN
 $W_1 = 0.5 W$
$\therefore$ $R_A = (0.764 \times 0.5W) = 0.382 W$

(b) Loads Acting on the Main Girder
Self weight of slab and wearing coat = 8 kN/m^2
Load from slab and W.C. on girder = (8×25) = 20 kN/m
Self weight of main girder = (0.94×25) = 23.5 kN/m
Weight of cross girders assumed to act as uniformly distributed load = 3.5 kN/m
$\therefore$ Total dead load of main girder = g = $(20 + 23.5 + 3.5)$ = 47 kN/m

(c) Dead Load Moments and Shear Forces
Referring to bending moment and shear force coefficients given in Appendix 1 of the monograph by the author[7], dead load moment at mid support section is computed as,
$$M_{gB} = 0.125\ gL^2 = (0.125 \times 47 \times 40^2) = 9400 \text{ kN} \cdot \text{m}$$

Dead load moment at mid span section is calculated as
$$M_{gD} = 0.071\ gL^2 = (0.071 \times 47 \times 40^2) = 5340\ \text{kN·m}$$
Dead load shear is maximum near the mid support section and is computed as
$$V_g = 0.62\ gL = (0.62 \times 47 \times 40) = 1166\ \text{kN}$$

(d) Live Load Moments in Girder: Referring to the influence line diagram for bending moment (Influence line ordinates obtained from Reference 8) shown in Fig. 7.10, the live load is placed near the maximum ordinate so as to yield the maximum positive live load moment which is computed as
$$M_D = \left(\frac{7.3 + 8.12}{2}\right) 700 = 5397\ \text{kN·m}$$

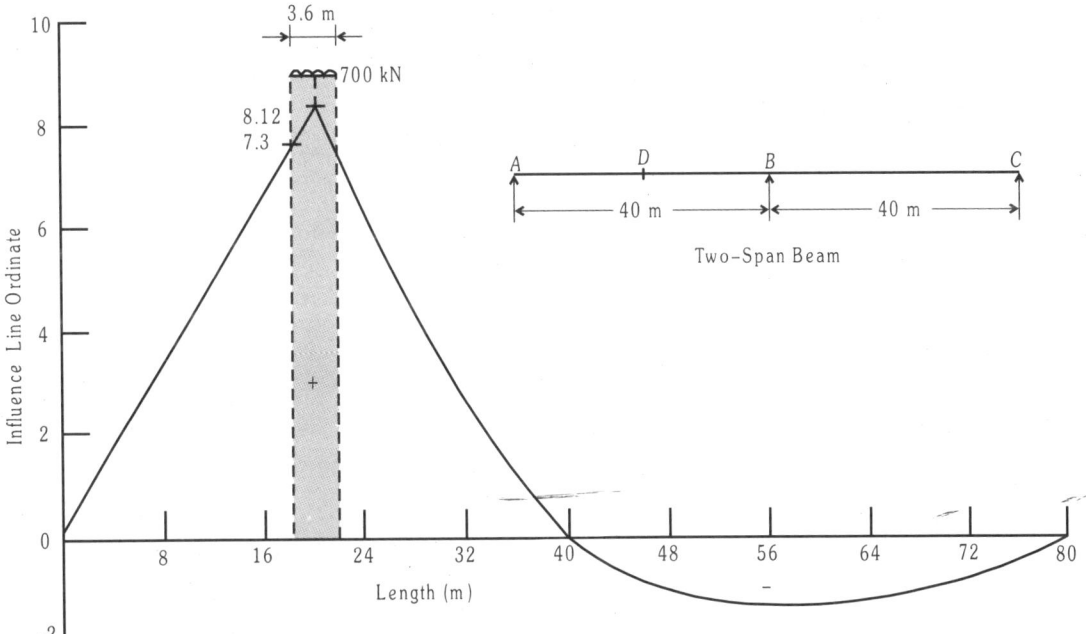

Fig. 7.10: Influence Line for Bending Moment at Mid Span Section

The maximum negative moment due to live loads developing at mid support B is obtained from the influence line diagram shown in Fig. 7.11. The maximum negative bending moment at B is computed as
$$M_B = (3.76 \times 700) = 2632\ \text{kN·m}$$
The live load bending moments including the reaction and impact factors for the exterior girder are
$$M_{qD} = (5397 \times 1.1 \times 0.382) = 2268\ \text{kN·m}$$
$$M_{qB} = (2632 \times 1.1 \times 0.382) = 1106\ \text{kN·m}$$

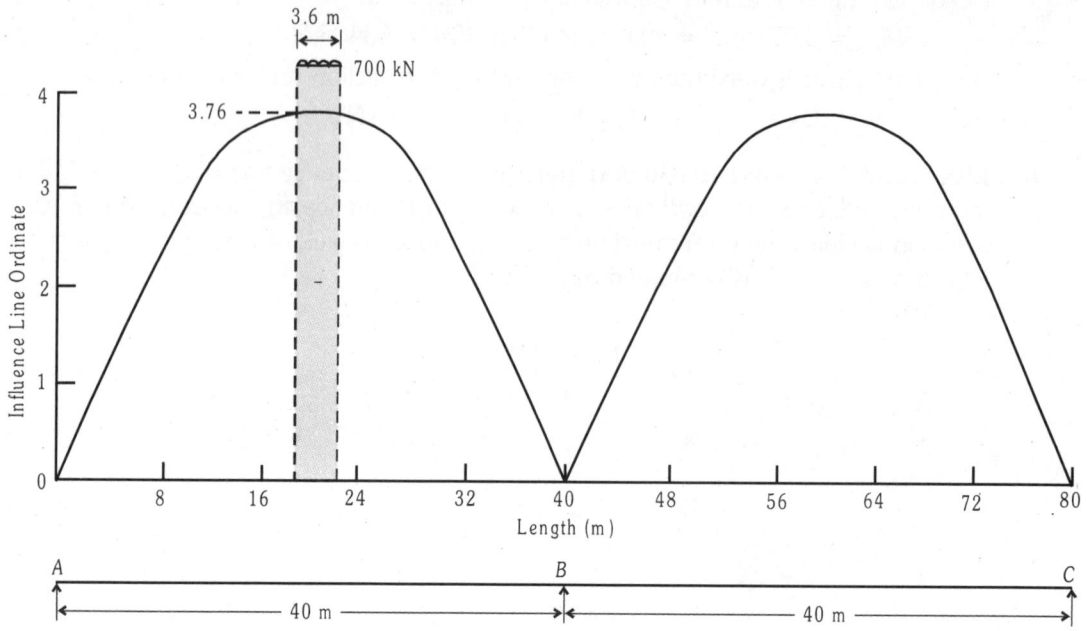

Fig. 7.11: Influence Line for Bending Moment at Mid Support B

(e) Live Load Shear Forces in Girder: The maximum live load shear develops in the interior girder when the IRC Class AA loads are placed near the mid support B as shown in Fig. 7.12.

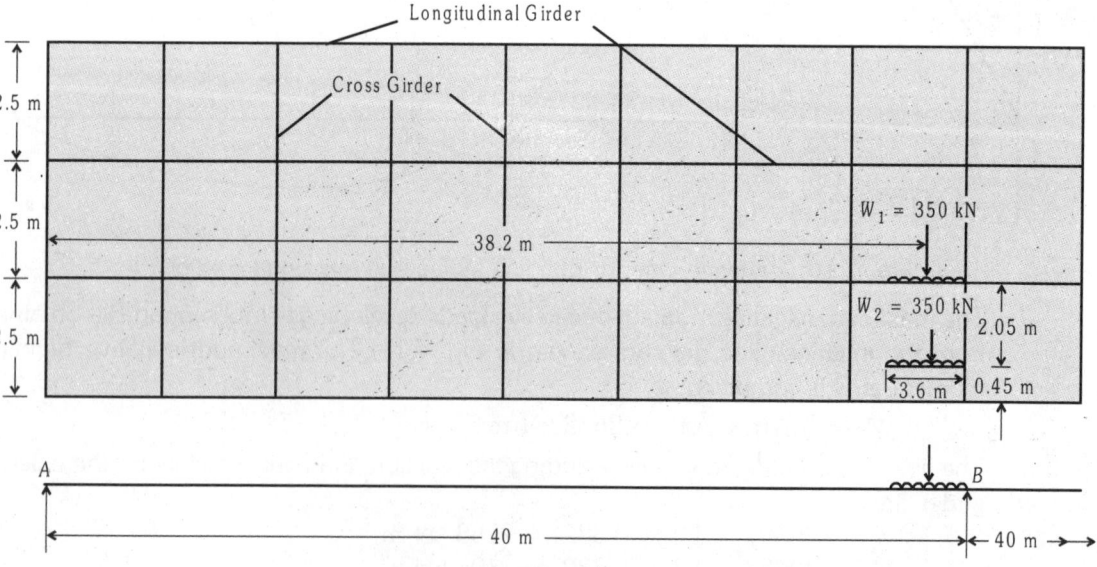

Fig. 7.12: Position of IRC Class AA Loads for Maximum Shear in Inner Girder

Prestressed Concrete Continuous Span Bridge Decks

Reaction of W_2 on girder $B = \left(\dfrac{350 \times 0.45}{2.5}\right) = 63$ kN

Reaction of W_2 on girder $A = \left(\dfrac{350 \times 2.05}{2.5}\right) = 287$ kN

Total load on girder $B = (350 + 63) = 413$ kN

Maximum reaction in the inner girder $B = \left(\dfrac{413 \times 38.2}{40}\right) = 394$ kN

Maximum live load shear with impact factor in girder $B = (394 \times 1.1) = 434$ kN

(f) **Design Bending Moments and Shear Forces:** The design bending moments and shear forces at service loads and at the limit state of collapse as stipulated in IRC: 18-2000 are compiled in Table 7.1.

Table 7.1: Working and Ultimate Load Moments and Shear Forces in Longitudinal Girders

	Dead Load B.M.	Live Load B.M.	Total Working Load B.M.	Required M_u	Units
(a) Bending Moments (Outer Girder)					
Mid Span Section (D)	(M_g) 5340	(M_q) 2268	$(M_g + M_q)$ 7608	$(1.5M_g + 2.5M_q)$ 13680	kN·m
Mid support Section (B)	(M_g) 9400	(M_q) 1106	$(M_g + M_q)$ 10506	$(1.5M_g + 2.5M_q)$ 16865	kN·m
	Dead Load S.F.	Live Load S.F.	Total Working Load S.F.	Required V_u	Units
(b) Shear Forces (Inner Girder)					
Near Mid Support Section	(V_g) 1166	(V_q) 434	$(V_g + V_q)$ 1600	$(1.5V_g + 2.5V_q)$ 2834	kN

(g) **Check for Minimum Section Modulus:** Maximum design moments develop at mid support section B

$M_g = 9400$ kN·m

$M_q = 1106$ kN·m

∴ $M_d = (M_g + M_q) = 10506$ kN·m

$f_{br} = (\eta f_{ct} - f_{tw}) = [(0.8 \times 20) - 0] = 16$ N/mm²

$$f_b = \left[\dfrac{f_{tw}}{\eta} + \dfrac{M_d}{\eta Z_b}\right] = \left[0 + \dfrac{(10506 \times 10^6)}{(0.8 \times 0.45 \times 10^9)}\right] = 29.2 \text{ N/mm}^2$$

$$Z_b \geq \left[\dfrac{M_q + (1-\eta)M_g}{f_{br}}\right] \geq \left[\dfrac{(1106 \times 10^6) + (1 - 0.8) 9400 \times 10^6}{16}\right] \geq (0.186 \times 10^9) \text{ mm}^3$$

which is less than (0.45×10^9) mm^3
Hence the section provided is adequate.

(h) Prestressing Force: In the present case, $M_g > M_q$ and consequently the minimum prestressing force computed will lie outside the section. Hence the maximum possible eccentricity with due respect to cover is provided and the enhanced prestressing force is calculated using the Eq. 4.6.

Maximum possible eccentricity at support section B is determined by assuming three Freyssinet cables of type 19K-15 (19 strands of 15.2 mm diameter) in 95 mm diameter cable ducts with a clear spacing not less than the diameter of the cable and with a clear cover of not less than 75 mm according to section 16 of IRC: 18-2000. Hence maximum possible eccentricity at support B is computed as,

$$e = [1000 - 75 - 95 - 100 - 50] = 680 \text{ mm}$$

Modified prestressing force is evaluated from the relation

$$P = \left[\frac{Af_b Z_b}{Z_b + Ae}\right]\left[\frac{(0.94 \times 10^6 \times 29.20 \times 0.45 \times 10^9)}{(0.45 \times 10^9) + (0.94 \times 10^6 \times 680)}\right] = 11338597 \text{ N} = 11338 \text{ kN}$$

Force in each cable = $(19 \times 0.8 \times 265) = 4028$ kN
Provide 3 cables carrying an initial prestressing force of
$$P = (3 \times 4000) = 12000 \text{ kN}$$
Area of each strand of 15.2 mm diameter = 140 mm^2
Area of 19 strands in each cable = $(19 \times 140) = 2660$ mm^2
Total area of high tensile steel in 3 cables = $A_p = (3 \times 2660) = 7980$ mm^2

(i) Concordant Cable Profile: For the two span continuous beam, a concordant parabolic cable profile is adopted with a maximum eccentricity 'e' at mid support section B and 'e/2' at mid span sections. Hence the cables are arranged in a parabolic concordant profile with an effective eccentricity of 680 mm towards the top fibre at mid support section B and an eccentricity of 340 mm towards the soffit at mid span section D as shown in Fig. 7.13. The centroid of the cables is concentric at the end supports A and C. The selected parabolic profiles of the three cables along the span with cross-sections at critical sections are shown in Fig. 7.14.

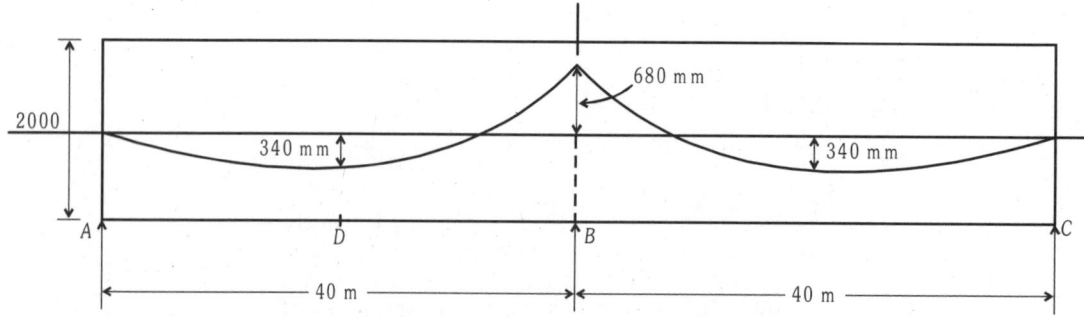

Fig. 7.13: Concordant Cable Profile

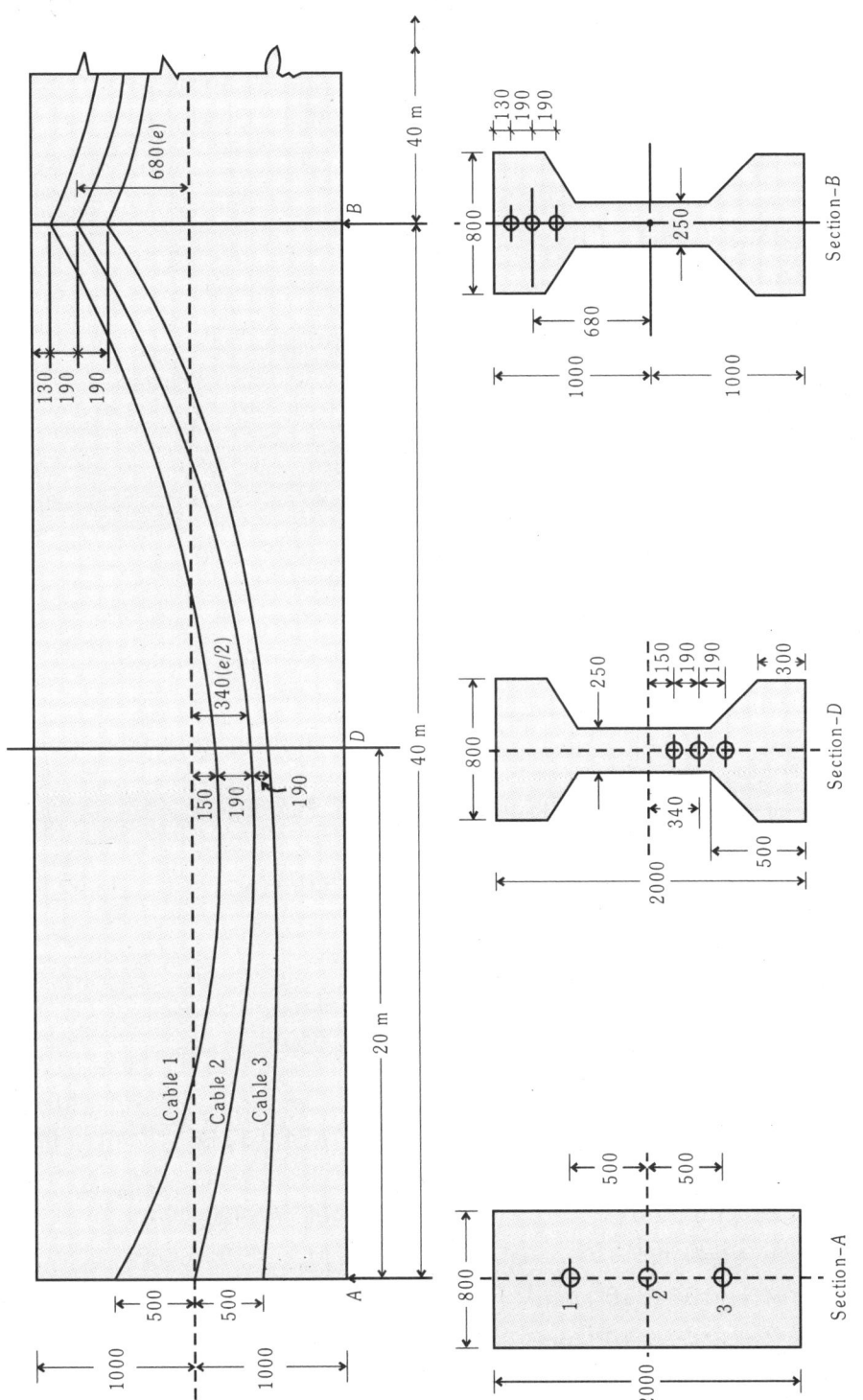

Fig. 7.14: Cable Layout in Two Span Prestressed Concrete Girder

6. Check for Stresses at Service Loads
(a) Centre of Span Section

$P = 12000$ kN $\qquad M_g = 5340$ kN·m
$e = 340$ mm $\qquad M_q = 2268$ kN·m
$A = (0.94 \times 10^6)$ mm² $\qquad \eta = 0.80$
$Z = (0.45 \times 10^9)$ mm³

$$\left(\frac{P}{A}\right) = \left(\frac{12000 \times 10^3}{0.94 \times 10^6}\right) = 12.76 \text{ N/mm}^2$$

$$\left(\frac{Pe}{Z}\right) = \left(\frac{12000 \times 10^3 \times 340}{0.45 \times 10^9}\right) = 9.06 \text{ N/mm}^2$$

$$\left[\frac{M_g}{Z}\right] = \left[\frac{5340 \times 10^6}{0.45 \times 10^9}\right] = 11.86 \text{ N/mm}^2$$

$$\left[\frac{M_q}{Z}\right] = \left[\frac{2268 \times 10^6}{0.45 \times 10^9}\right] = 5.04 \text{ N/mm}^2$$

At the transfer stage, the extreme fibre stresses are

$$\sigma_t = \left[\left(\frac{P}{A}\right) - \left(\frac{Pe}{Z}\right) + \left(\frac{M_g}{Z}\right)\right] = [12.76 - 9.06 + 11.86] = 15.56 \text{ N/mm}^2$$

$$\sigma_b = \left[\left(\frac{P}{A}\right) + \left(\frac{Pe}{Z}\right) - \left(\frac{M_g}{Z}\right)\right] = [12.76 + 9.06 - 11.86] = 9.96 \text{ N/mm}^2$$

At the limit state of service loads, the stresses are

$$\sigma_t = \left[\eta\left(\frac{P}{A}\right) - \eta\left(\frac{Pe}{Z}\right) + \left(\frac{M_g}{Z}\right) + \left(\frac{M_q}{Z}\right)\right] = [10.21 - 7.24 + 11.86 + 5.04]$$
$$= 19.87 \text{ N/mm}^2$$
$$< 20 \text{ N/mm}^2$$

$$\sigma_b = \left[\eta\left(\frac{P}{A}\right) + \eta\left(\frac{Pe}{Z}\right) - \left(\frac{M_g}{Z}\right) - \left(\frac{M_q}{Z}\right)\right]$$
$$= [10.21 + 7.24 - 11.86 - 5.04] = 0.55 \text{ N/mm}^2$$

(b) Mid Support Section

$P = 12000$ kN $\qquad M_g = 9400$ kN·m
$e = 680$ mm $\qquad M_q = 1106$ kN·m
$A = (0.94 \times 10^6)$ mm² $\qquad d = 0.80$
$Z = (0.45 \times 10^9)$ mm³

$$\left(\frac{P}{A}\right) = 12.76 \text{ N/mm}^2$$

Prestressed Concrete Continuous Span Bridge Decks

$$\left(\frac{Pe}{Z}\right) = \left(\frac{12000 \times 10^3 \times 680}{0.45 \times 10^9}\right) = 18.12 \text{ N/mm}^2$$

$$\left(\frac{M_g}{Z}\right) = \left(\frac{9400 \times 10^6}{0.45 \times 10^9}\right) = 20.88 \text{ N/mm}^2$$

$$\left(\frac{M_q}{Z}\right) = \left(\frac{1106 \times 10^6}{0.45 \times 10^9}\right) = 2.45 \text{ N/mm}^2$$

At the stage of transfer:

$\sigma_t = [12.76 + 18.12 - 20.88] = 10.00$ N/mm^2

$\sigma_b = [12.76 - 18.12 + 20.88] = 15.52$ N/mm^2

At the limit state of service loads:

$\sigma_t = [0.8 (12.76 + 18.12) - 20.88 - 2.45] = 1.37$ N/mm^2

$\sigma_b = [0.8 (12.76 - 18.12) + 20.88 + 2.45] = 19.04$ N/mm^2

The stresses at various critical stages are within the maximum permissible limits of 20 N/mm^2.

7. Check for Ultimate Flexural Strength

(a) Centre of Span Section

$A_p = 7980$ mm^2 $f_{ck} = 60$ N/mm^2

$b = 800$ mm $f_p = 1862$ N/mm^2

$b_w = 250$ mm $d = 1340$ mm

$D_f = 400$ mm M_u (Required) $= 13680$ kN·m

According to IRC: 18-2000 specifications,

(i) Failure by Yielding of Steel

$M_u = (0.9 \, d A_p f_p)$

$= (0.9 \times 1340 \times 7980 \times 1862)$

$= (17920 \times 10^6)$ N·mm

$= 17920$ kN·m > 13680 kN·m

(ii) Failure by Crushing of Concrete

$M_u = [0.176 \, b_w \, d^2 \, f_{ck}] + [(0.67 \times 0.8(b - b_w)(d - 0.5D_f)D_f f_{ck}]$

$= [(0.176 \times 250 \times 1340^2 \times 60)] + [(0.67 \times 0.8)(800 - 250)$

$(1340 - 0.5 \times 400)(400 \times 60)]$

$= (12806 \times 10^6)$ N·mm

$= 12806$ kN·m < 13680 kN·m

Since the actual $M_u < M_u$ (required), additional untensioned reinforcements are designed to resist the balance moment.

Balance moment $= M_{bal} = [13680 - 12806] = 874$ kN·m

Using Fe-415 Grade HYSD bars at an effective depth of 1900 mm,

$$A_{us} = \left[\frac{M_{bal}}{0.87f_y(d - 0.5D_f)}\right] = \left[\frac{874 \times 10^6}{0.87 \times 415(1900 - 0.5 \times 400)}\right] = 1424 \text{ mm}^2$$

Provide 6 bars of 20 mm diameter (A_{st} = 1885 mm²)

(b) Mid Support Section

A_p = 7980 mm² $\qquad f_{ck}$ = 60 N/mm²
b = 800 mm $\qquad f_p$ = 1862 N/mm²
b_w = 250 mm $\qquad d$ = 1680 mm
D_f = 400 mm $\qquad M_u$(Required) = 16865 kN·m

$M_u = [0.176 b_w d^2 f_{ck}] + [(0.67 \times 0.8(b - b_w)(d - 0.5D_f)D_f f_{ck}]$
$\quad = [(0.176 \times 250 \times 1680^2 \times 60)] + [(0.67 \times 0.8)(800 - 250)$
$\qquad (1680 - 0.5 \times 400)(400 \times 60)]$
$\quad = (17922 \times 10^6)$ N·mm
$\quad = 17922$ kN·m > 16865 kN·m

The ultimate flexural strength of mid support section is greater than the required design ultimate strength.

8. Check for Ultimate Shear Strength

Design shear force = V_u = 2834 kN
According to IRC: 18-2000 Code, the ultimate shear resistance of the support section uncracked in flexure is given by

$$V_{cw} = 0.67 b_w h \sqrt{f_t^2 + 0.8 f_{cp} f_t} + \eta P \sin\theta$$

where $\quad b_w$ = 250 mm
$\qquad h$ = 2000 mm

$$f_t = 0.24 \sqrt{f_{ck}} = 0.24\sqrt{60} = 1.85 \text{ N/mm}^2$$

$$f_{cp} = \left(\frac{\eta P}{A}\right) = \left(\frac{0.8 \times 12000 \times 1000}{0.94 \times 10^6}\right) = 10.2 \text{ N/mm}^2$$

$$\theta = \left(\frac{4e}{L}\right) = \left(\frac{4 \times 680}{40 \times 1000}\right) = 0.068$$

$V_{cw} = [(0.67 \times 250 \times 2000 \times \sqrt{1.85^2 + 0.8 \times 10.2 \times 1.85}$
$\qquad\qquad + (0.8 \times 12000 \times 10^3 \times 0.068)]$
$\quad = (2093 \times 10^3)$ N
$\quad = 2093$ kN < 2834 kN

Balance shear force = (2834 − 2093) = 741 kN
Using 12 mm diameter two legged stirrups

$$\text{Spacing } S_V = \left[\frac{0.87 \times 415 \times 2 \times 113 \times 1900}{741 \times 1000}\right] = 209 \text{ mm}$$

Prestressed Concrete Continuous Span Bridge Decks

Provide 12 mm diameter two legged stirrups at a spacing of 200 mm centres near the supports which is gradually increased to 300 mm centres towards the centre of span.

9. **Supplementary Reinforcements**: According to IRC: 18-2000, longitudinal reinforcements of not less than 0.18% (for f_{ck} > M-45) of the gross cross sectional area are to be provided to safeguard against shrinkage cracking.

$$A_{st} = \left[\frac{0.18 \times 0.94 \times 10^6}{100}\right] = 1692 \text{ mm}^2$$

Provide 15 bars of 12 mm diameter distributed in the top and bottom flanges and web. The reinforcements provided at mid span and mid support sections are shown in Fig. 7.15.

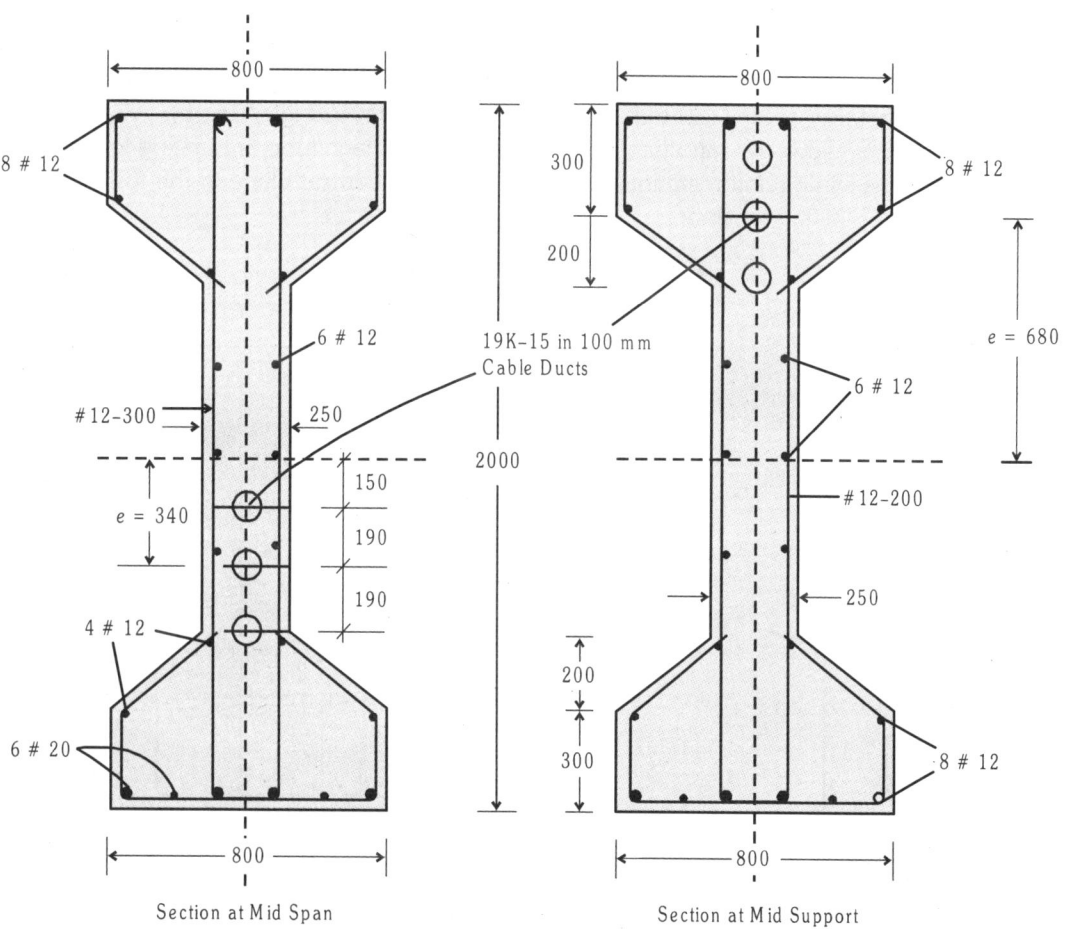

Section at Mid Span   Section at Mid Support

Fig. 7.15: Reinforcement and Cable Details at Mid Span and Mid Support Sections

10. **Design of End Blocks**: At the end supports A and C, solid end blocks, 800 mm wide by 2000 mm deep are provided for a length of 2 m from each end face of the girder. The equivalent prisms on which the anchorage forces are considered to be effective are shown

in Fig. 7.16 (a). The bursting tension in the end block is evaluated using the data compiled in Table 4.3.

In the horizontal plane, we have

$$2y_{po} = 340 \text{ mm}$$
$$2y_o = 800 \text{ mm}$$
$$P_k = 4000 \text{ kN}$$

$$\text{Ratio}\left(\frac{y_{po}}{y_o}\right) = \left(\frac{340}{800}\right) = 0.425$$

Bursting Tension, $F_{bst} = (0.20 \times 4000) = 800 \text{ kN}$
Using Fe-415 HYSD bars

$$A_{st} = \left(\frac{800 \times 1000}{0.87 \times 415}\right) = 2215 \text{ mm}^2$$

Provide 16 mm diameter bars at 150 mm centres in the horizontal plane distributed in the region from $0.2y_o$ to $2y_o$ (80 mm to 800 mm) as shown in Fig. 7.16 (b).

In the vertical plane, the ratio (y_{po}/y_o) being larger, the magnitude of bursting tension is less. However the same reinforcements are provided in the vertical plane in the form of a mesh to resist the bursting tension.

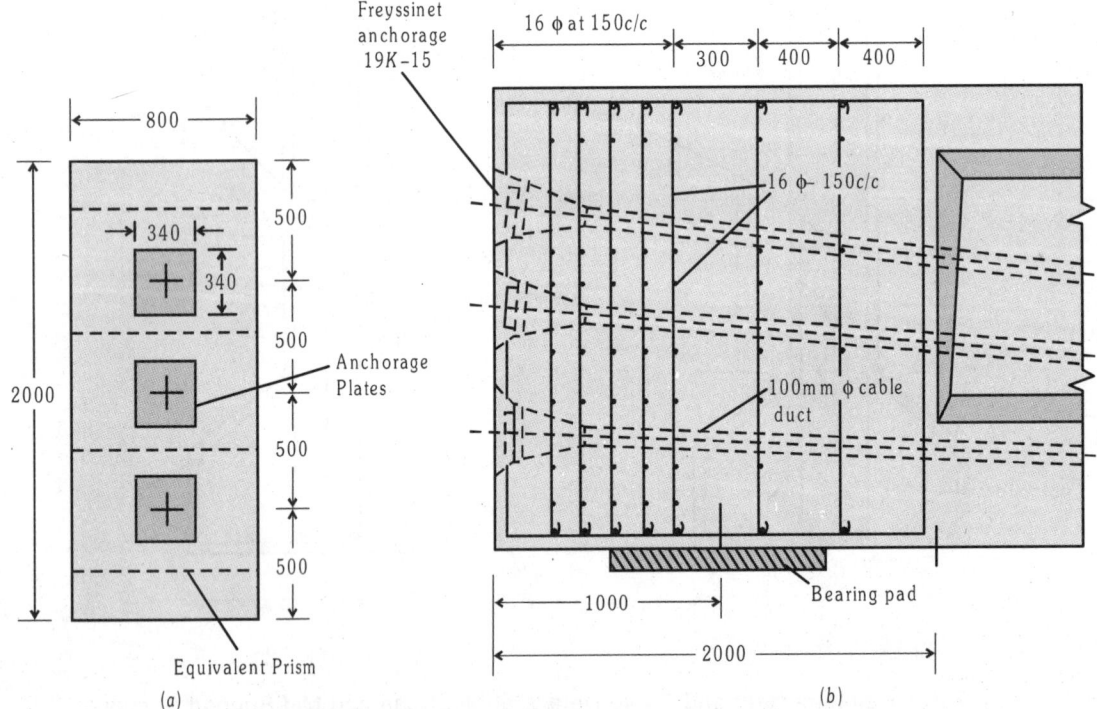

Fig. 7.16: Anchorage Zone Reinforcements in End Block

11. **Cross Girders:** At intervals of 5 m along the span, cross girders of width 200 mm and depth 1600 mm are provided with nominal reinforcements of 0.18 % of the cross-section

comprising 12 mm diameter bars spaced at 300 mm centres, distributed near both vertical faces of cross girder. 10 mm diameter ties are also provided. Two cables housing 12 numbers of 7 mm diameter high tensile wires are positioned at mid third points along the depth with a nominal prestress to provide lateral stiffness to the bridge deck system.

REFERENCES

1. LIN, T.Y., *Prestressed Concrete Structures*. John Wiley and Sons Inc, New York, 1960, pp. 288-292.
2. VISVESVARAYA, H. C. and RAGHAVENDRA, N., Continuity in Prestressed Concrete Construction, Seminar on Problems of Prestressing, Madras, 1970, pp. I-91-94.
3. FINISTERWALDER, U., Modern Designs for Prestressed Concrete Bridges. *Concrete and Constructional Engineering*, London, Vol. 60, No. 3 March 1965, pp. 99-103.
4. KRISHNA RAJU, N., *Prestressed Concrete* (Fourth Edition). Tata McGraw Hill Publishers, New Delhi, 2007.
5. RAINA, V. K., *Concrete Bridge Practice, Analysis, Design and Economics*. Tata McGraw Hill Publishing Co, New Delhi, 1991, pp. 554-577.
6. KRISHNA RAJU, N., *Design of Bridges* (Third Edition), Oxford and I.B.H, Publishing Co, New Delhi, 1998, p. 26.
7. KRISHNA RAJU, N., *Advanced Reinforced Concrete Design* (IS: 456-2000), Second Edition, C.B.S. Publishers, New Delhi, 2005, p. 345.
8. REYNOLDS and STEEDMAN, J., *Reinforced Concrete Designers Hand Book*, Eighth Edition, Cement and Concrete Association, London, 1976.

EXERCISES

1. A post tensioned prestressed concrete continuous beam and slab deck is to be designed for a National Highway crossing to suit the following data:
   Two continuous spans of 30 m each
   Width of carriage way = 15 m
   Kerbs: 600 mm wide on either side
   Thickness of wearing coat = 100 mm
   Live load: IRC Class AA tracked vehicle
   For R.C.C. Slab adopt M-30 Grade concrete and Fe-415 HYSD bars
   For Prestressed Concrete Girders, adopt M-50 Grade concrete and 12K-15 type
   Freyssinet cables conforming to IS: 6006-1983
   Loss Ratio = 0.85
   Spacings of cross girders = 5 m
   Design the deck slab and the continuous girder as Class 1 type structure conforming to IRC: 6-2000, IRC: 18-2000 and IRC: 21-2000. Sketch the details of reinforcements and cables at critical sections.

2. A post tensioned P.S.C. continuous beam and slab deck has been proposed for a National Highway crossing having the following data:
   Three equal spans of 40 m each
   Width of carriage way = 7.5 m
   Foot paths of 1 m on each side

Thickness of wearing coat = 80 mm

Live load: IRC Class AA tracked vehicle

Adopt M-25 Grade concrete and Fe-415 HYSD reinforcements for R.C.C. slab

Adopt M-60 Grade concrete with cube strength at transfer as 40 N/mm² for prestressed beams.

Loss ratio = 0.8

High tensile strands of 15.2 mm diameter conforming to IS: 6006-1983 are available for use in prestressed girders

Design the bridge deck as Class 1 type structure conforming to the specifications of the Indian Standard Codes IS: 21-2000, IRC: 18-2000 and IRC: 6-2000.

Sketch the details of reinforcements in the deck slab and the cables in the prestressed concrete girders.

3. A three span post tensioned prestressed concrete continuous beam and slab deck is to be designed for a National Highway crossing to suit the following data:

Span details: Two side spans of 30 m each with a central span of 40 m

Width of carriage way: Four lane carriage way 15 m wide

Foot paths: 1.5 m on either side

Thickness of wearing coat = 100 mm

Live load: IRC Class AA tracked vehicle or Class A whichever is severe.

Grade of concrete: M-30 for reinforced concrete deck slab and M-60 for prestressed girders

Loss ratio = 0.8

Type of steel: Fe-415 HYSD bars for deck slab and supplementary reinforcements. High tensile strands conforming to IS: 6006-1983.

Design the bridge deck as Class-1 type structure conforming to the specifications of codes IS: 6-2000, 21-2000 and IS: 18-2000.

Sketch the details of reinforcements in the deck slab and continuous girders showing the cable profile along the span.

DESIGN OF PRESTRESSED CONCRETE RIGID FRAME BRIDGES

8.1 GENERAL FEATURES

Rigid frame bridges comprising of portal frames are widely used in the span range of 10 to 30 m for Highway crossings in embankments. Solid slab type rigid frames are ideally suited for spans up to 20 m. The transom comprises of a thick horizontal slab of constant or variable cross-section. The column supports are generally hinged at the base. A typical rigid frame bridge of the solid barrel or slab type is shown in Fig. 8.1 (a).

Rigid frames comprising beams or ribs spaced at 3 to 4 m intervals connected by a slab are more economical for large spans in the range of 20 to 30 m[1]. Figure 8.1 (b) shows the salient features of a typical rigid frame bridge with tee beam ribs and slab.

Reinforced concrete rigid frame bridges[2] require larger concrete sections with large quantities of steel reinforcement when adopted for spans exceeding 10 m. However prestressed concrete rigid frames are suitable for long spans since the concrete cross sections can be reduced in comparison with reinforced concrete rigid frames. Also the quantity of untensioned steel reinforcement is reduced since only the minimum quantity of steel in the form of supplementary reinforcement is used in prestressed concrete rigid frames. However high tensile steel in the form of cables are required for transom and legs of the rigid frame. Prestressing has facilitated the use of longer spans for rigid frame bridges with the added advantage of slender sections of the portal frames.

8.2 ADVANTAGES OF RIGID FRAME BRIDGES

Prestressed concrete rigid frame bridges have the following advantages in comparison with the traditional type of bridges.

1. Rigid frame bridges are essentially made up of portal frames having essentially the structural action of an arch.
2. Prestressing of the portal frame helps to contain the line of thrust or pressure within the middle third zone of the section, thus eliminating the tension and ensuring compressive stresses in the section.
3. Rigid frame bridge comprising of a portal frame being a monolithic structure eliminates the use of separate abutments. The vertical side of the rigid frame serves as retaining walls to retain earth in road crossings of embankments.

4. Slab type rigid frames can be easily cast *in situ* since plain moving form work can be used for rapid construction work.
5. Rigid frame bridges can be advantageously adopted in fly over crossings where roads intersect at different levels
6. Rigid frame bridges do not require expensive bearings at the supports since concrete hinged bearings can be provided with structural advantage.
7. According to Raina[3], large spans of up to 100 m or more is possible with prestressed concrete rigid frames.
8. The depth to span ratio at crown can be as low as 1/50 when variable cross-section is used in the frame.

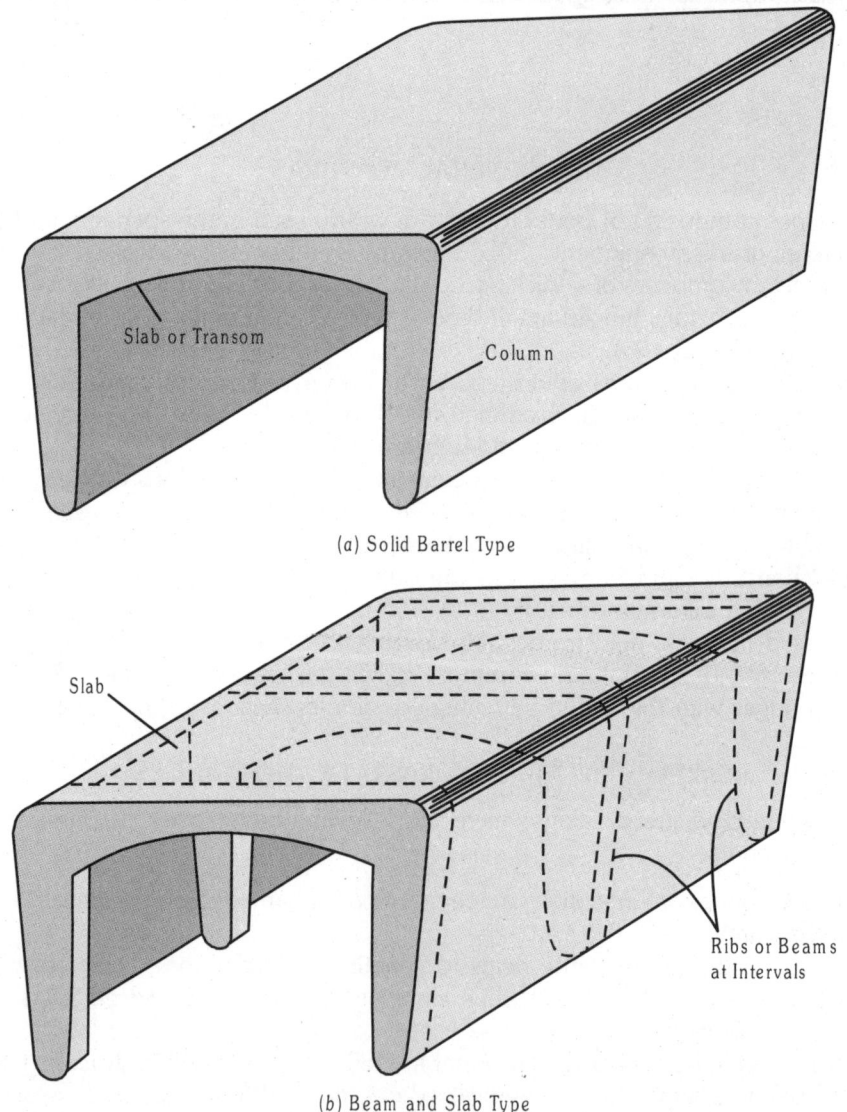

(a) Solid Barrel Type

(b) Beam and Slab Type

Fig. 8.1: Types of Rigid Frame Bridges

Design of Prestressed Concrete Rigid Frame Bridges

9. The stability of the supports is much greater in comparison with independent piers
10. Rigid frame bridges have high stability against lateral forces like wind, earth quake and soil pressure.
11. Rigid frame bridges have greatly reduced horizontal thrust
12. Rigid frame bridges are aesthetically superior and consume less quantity of materials compared to reinforced concrete beam and slab bridges.

8.3 DESIGN PRINCIPLES OF PRESTRESSED CONCRETE PORTAL FRAMES

Two dimensional structures like portal frames develop secondary moments due to prestressing and also the effect of axial shortening is also considered in the design of rigid portal frames. The preliminary dimensions of the structural members are estimated based on empirical relations evolved from past practical experience. The dimensions selected for the critical sections of the portal frame are based on the span of the transom.

Effective length of transom (span of the bridge) = L
Overall depth at centre of transom = (L/30) to (L/40)
Depth at junction of column and transom = (L/15) to (L/20)

In the case of hinged supports, the thickness of column should be at least 500 to 750 mm at the junction of column and beam elements, the depth at the junction is larger in comparison with that at the centre of the beam. In the case of reinforced concrete rigid frames, the members are of variable depth depending upon the moment variation in the frame.

The structural analysis of a rigid frame of variable cross-section involves the determination of frame constants and influence lines for horizontal thrust at supports and bending moment at crown and at salient critical sections in the transom. The author has presented a detailed design of a reinforced concrete rigid frame bridge of variable cross-section to support IRC Class AA loading in a separate monograph[4].

The design of prestressed concrete portal frames involve the derivation of relations between the variation of moments at critical section, permissible compressive stress in concrete, cross-sectional dimensions and the thrusts developed due to maximum and minimum moment conditions. The theoretical procedure recommended by Marshall and Nelson[5] is detailed below:

Let
N_g = Thrust developed due to dead load
N_p = Thrust due to prestress
N_1 = Thrust developed for maximum moment M_1
N_2 = Thrust developed for minimum moment M_2
M_g = Dead load moment
N_A = Total thrust under maximum moment condition ($M_1 + M_g$)
N_B = Total thrust under minimum moment condition ($M_2 + M_g$)

Then
$$N_A = (N_p + N_g + N_1) \qquad \ldots(8.1)$$
$$N_B = (N_p + N_g + N_2) \qquad \ldots(8.2)$$

Assuming that the resultant thrust N_A and N_B pass through the kern points of the section for the maximum and minimum moment conditions resulting in zero tension and maximum compressive stress at the extreme fibres of the section as shown in Fig. 8.2.

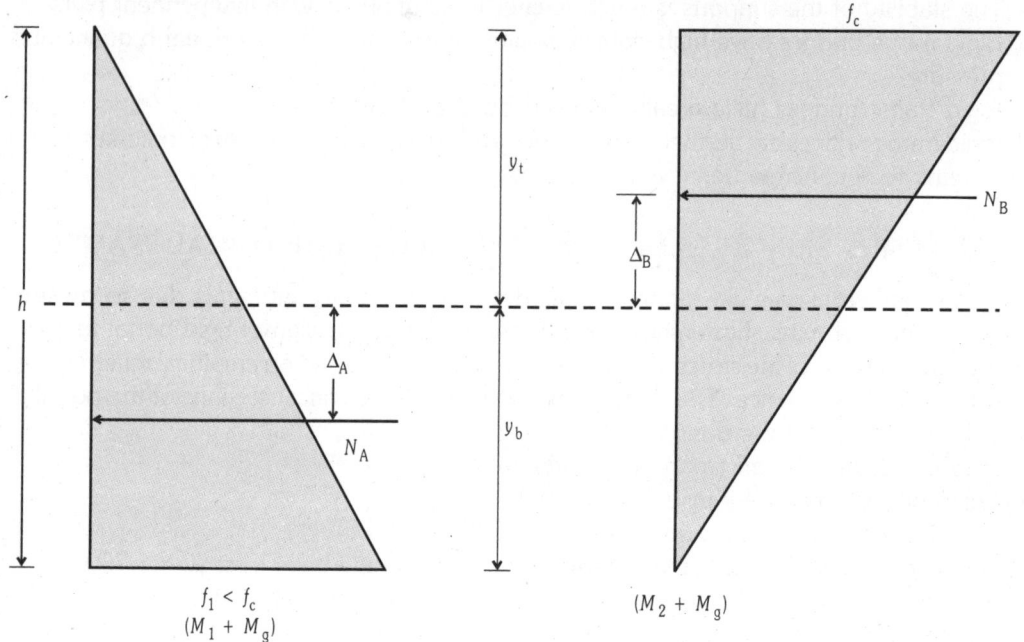

Fig. 8.2: Stress Distribution Under Maximum and Minimum Moment Conditions

If M_L = Maximum moment variation, we have
$$M_L = (N_A \Delta_A + N_B \Delta_B) \quad \ldots(8.3)$$

where Δ_A and $-\Delta_B$ are the core limits (I/Ay_1) and $-(I/Ay_2)$ respectively
and A = cross-sectional area of section
I = second moment of area
y_1 and y_2 = extreme fibre distances

Substituting the values of N_A and N_B from Eq. 8.1 and 8.2 in Eq. 8.3 and assuming a rectangular section of overall depth 'h', we have

$$M_L = (N_p + N_g)\frac{h}{3} + (N_1 + N_2)\frac{h}{6} \quad \ldots(8.4)$$

But $N_B = 0.5 f_c A = (N_p + N_g + N_2) \quad \ldots(8.5)$

If $A = bh$

Then $(N_p + N_g) = [0.5 f_c A - N_2] \quad \ldots(8.6)$

Hence $M_L = \frac{1}{6} f_c bh^2 + (N_1 - N_2)\frac{h}{6} \quad \ldots(8.7)$

Similarly if $N_2 > N_1$

Then $M_L = \frac{1}{6} f_c bh^2 + (N_2 - N_1)\frac{h}{6} \quad \ldots(8.8)$

In this relation, the second term being always negative, the section required will generally be larger than that for beams required to support the same moment variation. Since 'b' and 'h' are the only unknown quantities in the above equation, the depth of the section can be determined by choosing a suitable width 'b'. The dead load moment M_g and thrust N_g may now be calculated by the normal methods and by using equation 8.6, the required prestressing force can be estimated.

Design of Prestressed Concrete Rigid Frame Bridges

The limiting zone for thrust can now be determined easily by plotting the ratios of $[(M_1 + M_g)/N_p]$ in each member on lines at a distance of (N/N_p) times the core distances from the line of centroids using the appropriate value of N for the two moment conditions. The plotting lines in rectangular sections correspond to the values of $[(h/6)/(N_A/N_p)]$ and $[(-h/6)/(N_A/N_p)]$ respectively.

The typical bending moments developed in a portal frame hinged at supports subjected to dead load and distributed live load together with the limiting zone are shown in Fig. 8.3. After determining the limiting zone, it is necessary to locate a concordant cable profile lying within the limiting zone. The cable profile if required may be linearly transformed to suit the practical requirements.

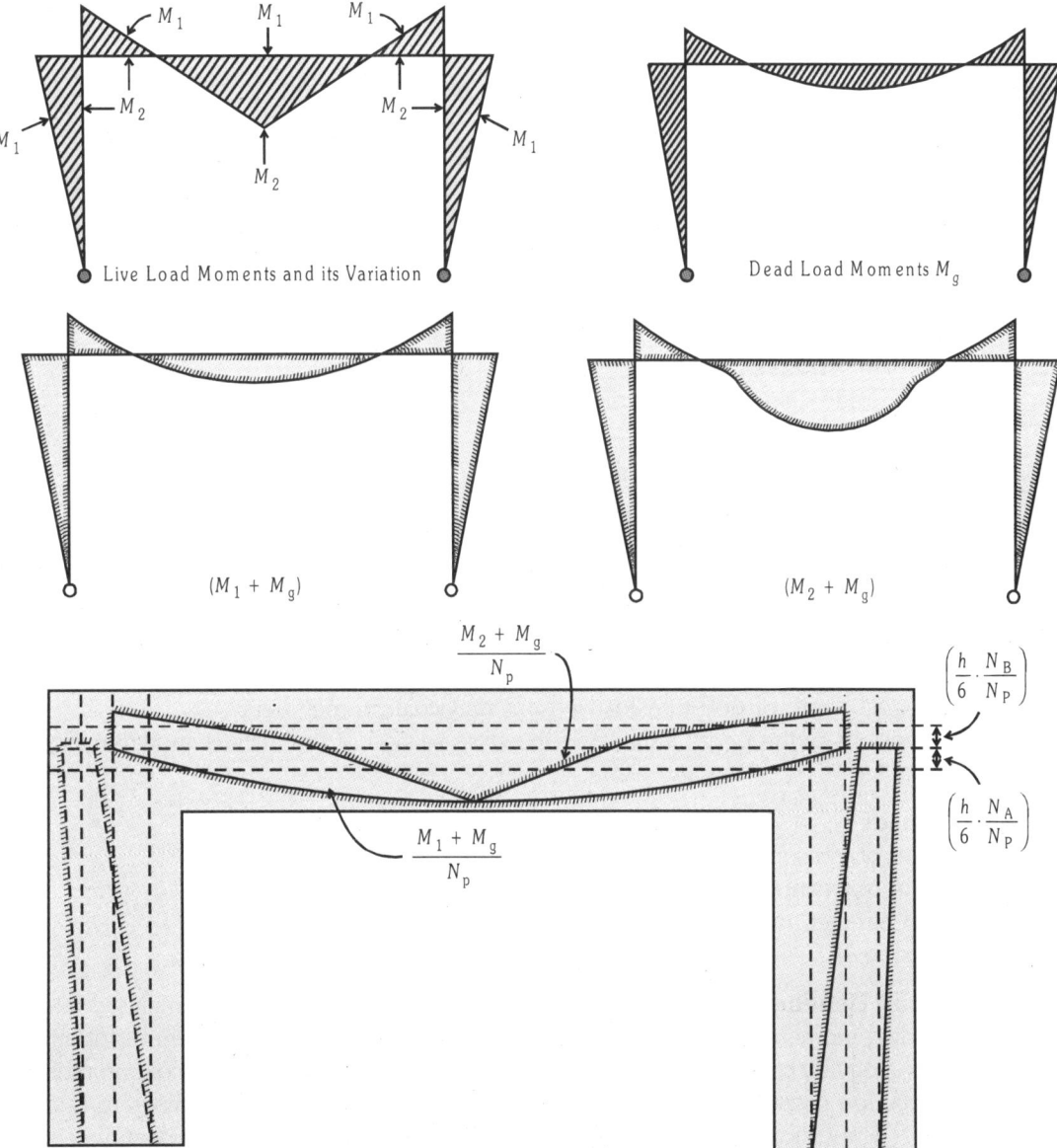

Fig. 8.3: Bending Moments and Limiting Zones

8.4 DESIGN EXAMPLE

Design a prestressed concrete rigid frame bridge with hinged supports to suit the following data:

1. **Data**
   Type of super structure: Two pinned portal frame
   Span (Length of transom) = 15 m
   Height of column legs = 6 m
   Width of road way = 7.5 m
   Foot path: one metre on either side
   Thickness of wearing coat = 80 mm
   Live load: IRC Class AA Tracked vehicle
   Grade of concrete: M-45 for portal frame and M-20 for footing
   Loss ratio = 0.85
   Safe bearing capacity of soil at site = 250 kN/m^2
   Adopt 7K-15 type Freyssinet high tensile steel strands conforming to IS: 6006-1983 for prestressing the members of portal frame.
   Fe-415 HYSD bars for supplementary reinforcements and footing foundation.
   Design the rigid frame as Class 1 type structure conforming to the specifications of Indian Road Congress codes IRC: 18-2000[6] and IRC: 21-2000[7].

2. **Permissible Stresses**
   For M-45 Grade Concrete (According to IRC: 18-2000)
   $$f_{ck} = 45 \text{ N/mm}^2$$
   $$f_{ct} = 0.33 f_{ck} = (0.33 \times 45) = 15 \text{ N/mm}^2$$
   $$f_{tt} = f_{tw} = 0 \text{ (Class 1 type structure)}$$

3. **Dimensions of Portal Frame and Cross Section of Deck:** Rigid rectangular portal frame comprising of transom and column members is proposed to have a uniform cross-section.
   Span of transom = L = 15 m
   Thickness of transom = $(L/30)$ = $[(15 \times 1000)/30]$ = 500 mm
   Adopt overall depth of 600 mm for transom and column members.
   The dimensions of the portal frame and the cross-section of the deck is shown in Figs. 8.4 and 8.5.

4. **Dead Loads**
   Self weight of transom ... = (0.6 × 24) =   14.40 kN/m^2
   Self weight of wearing coat = (0.08 × 22) =   1.76
   Weight of parapet, railing, etc. (Lumpsum) =   1.84
   Total dead load ... =   18.00 kN/m^2

5. **Dead Load Bending Moments**: The frame ABCD shown in Fig. 8.6 is analysed for dead load bending moments by any of the well established methods like moment distribution or column analogy by computing the stiffness coefficients of the transom and column members and distribution coefficients at the joints. The resulting bending moment diagram is shown in Fig. 8.7. The critical moments for design are the maximum positive moment at centre of transom and the negative moment developed at the junction of column and transom.

Design of Prestressed Concrete Rigid Frame Bridges

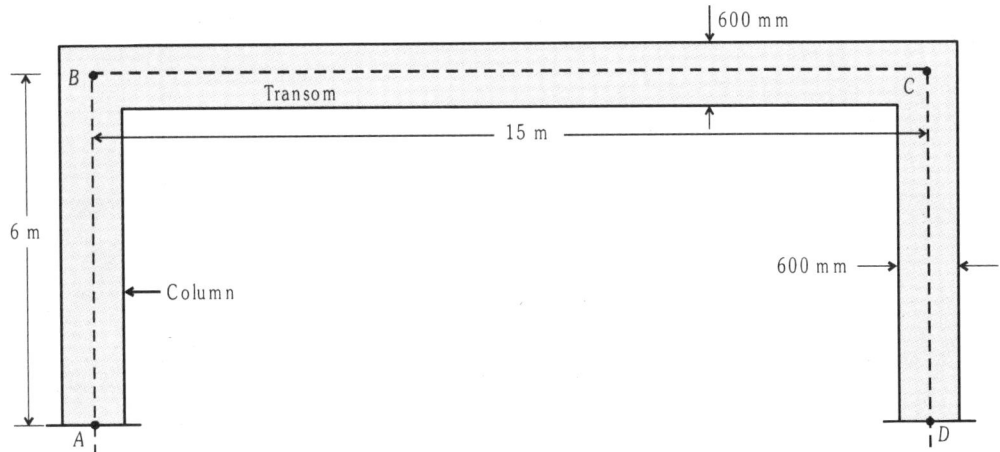

Fig. 8.4: Dimensions of Portal Frame

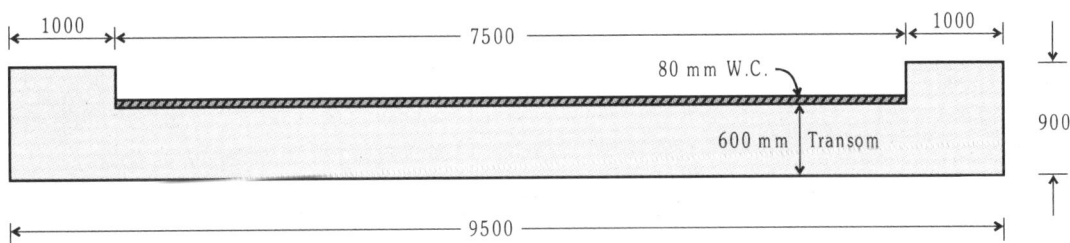

Fig. 8.5: Cross-section of Bridge Deck

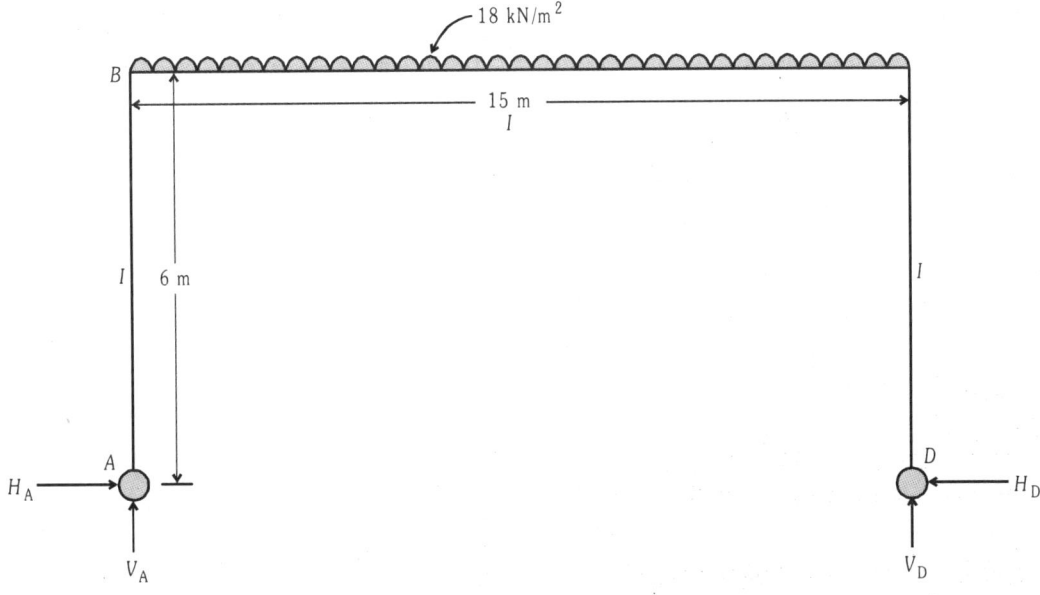

Fig. 8.6: Dead Loads on Frame

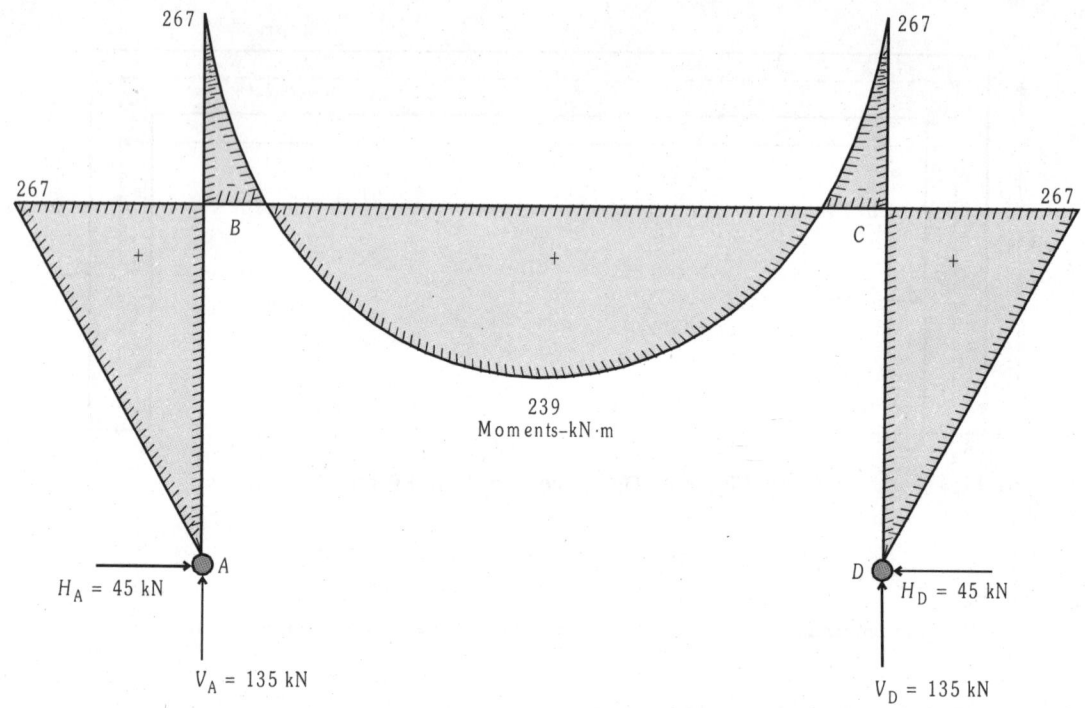

Fig. 8.7: Dead Load Bending Moments

6. **Live Load Distribution on Deck:** The deck is subjected to IRC Class AA tracked vehicle loading comprising an army tank of 700 kN spread over two tracks.
Impact Factor = 10 per cent for a span of 15 m
The tracked vehicle is placed symmetrically at centre of span on the deck.
Effective length of load = [3.6 + 2(0.6 + 0.08)] = 4.96 m
Effective width of slab perpendicular to the span is expressed as

$$b_e = Kx\left(1 - \frac{x}{L}\right) + b_w$$

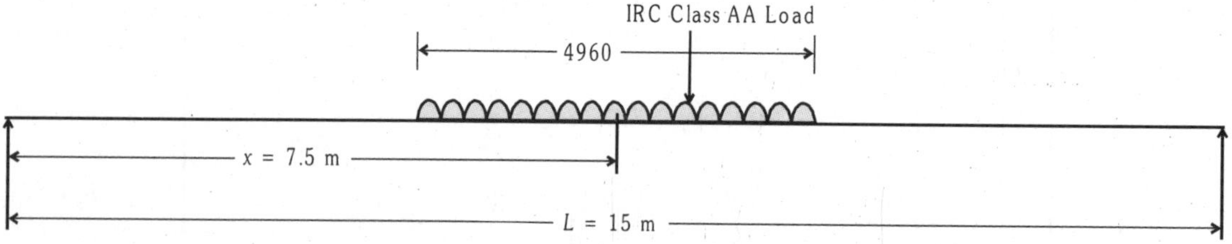

Fig. 8.8: Position of Load for Maximum Bending Moment

Referring to Fig. 8.8
$x = 7.5$ m, $L = 15$ m, $B = 9.5$ m
Ratio $(B/L) = (9.5/15) = 0.63$

$b_w = [0.85 + (2 \times 0.08)] = 1.01$ m

From Table 5.1 for $(B/L) = 0.63$ and continuous slab read out the value of $K = 1.88$

$$\therefore \quad b_e = (1.88 \times 7.5)\left[1 - \frac{7.5}{15}\right] + 1.01 = 8.06 \text{ m}$$

The IRC Class AA tracked vehicle is placed close to the kerb with the required minimum clearance as shown in Fig. 8.9.

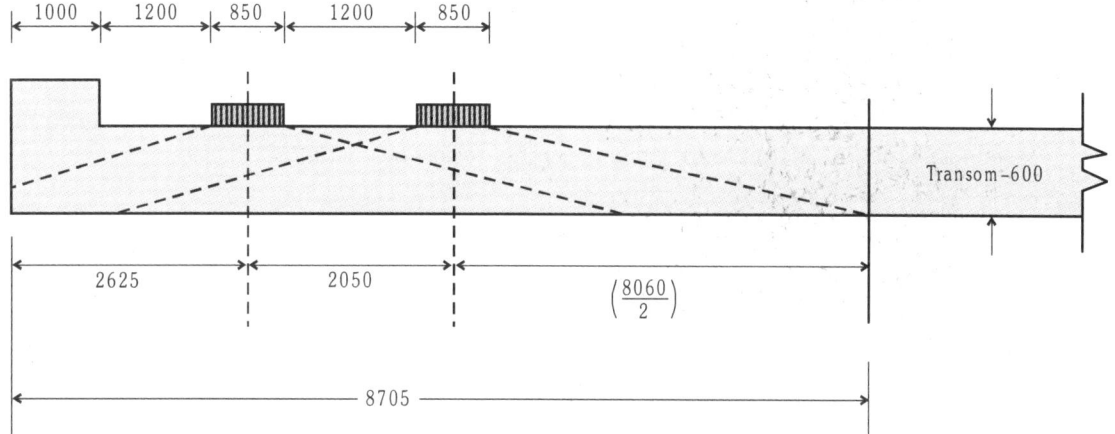

Fig. 8.9: Effective Width of Dispersion for IRC Class AA Tracked Vehicle

Net effective width of dispersion = 8.705 m
Total load of two tracks with impact = $(700 \times 1.10) = 770$ kN

$$\text{Average intensity of load} = \left[\frac{770}{(4.96 \times 8.705)}\right] = 17.8 \text{ kN/m}^2$$

7. **Live Load Bending Moments in Portal Frame:** The portal frame is analysed for maximum positive and negative bending moments when the live load is in the centre of transom. The corresponding live load bending moment diagram for the frame is shown in Fig. 8.10.

8. **Design Bending Moments in Frame**
   Maximum positive moment at mid point of transom = $(150 + 239) = 389$ kN·m
   Maximum negative moment at knee = $(127 + 267) = 394$ kN·m
   Maximum horizontal thrust at hinge = $(21 + 45) = 66$ kN
   Maximum vertical reaction in legs = $(45 + 135) = 180$ kN

9. **Dimensions of Critical Section of Transom**
   For transom, maximum positive design moment at centre is $M_L = 389$ kN·m
   and $(N_1 - N_2) = -66$ kN
   Assuming 1 m width of deck and using Eq. 8.8, we have
   $$M_L = (1/6) f_c b h^2 + (1/6)(N_1 - N_2)h$$
   In this example, $f_c = 15$ N/mm^2

$M_L = (389 \times 10^6)$ N·mm

h = overall depth of the section

$b = 1000$ mm

Substituting these values in the moment equation, we have

$(389 \times 10^6) = [(1/6) \times 15 \times 1000 \times h^2] + [(1/6)(-66 \times 10^3)h]$

Solving, $h = 401$ mm

Similarly for the column members, $M_L = 394$ kN·m and $(N_2 - N_1) = -180$ kN

$(394 \times 10^6) = [(1/6) \times 15 \times 1000 \times h^2] + [(1/6)(-180 \times 10^3)h]$

Solving, $h = 409$ mm

The section adopted for transom and column has an overall depth of 600 mm. Hence the section assumed is adequate to safely resist the design moments.

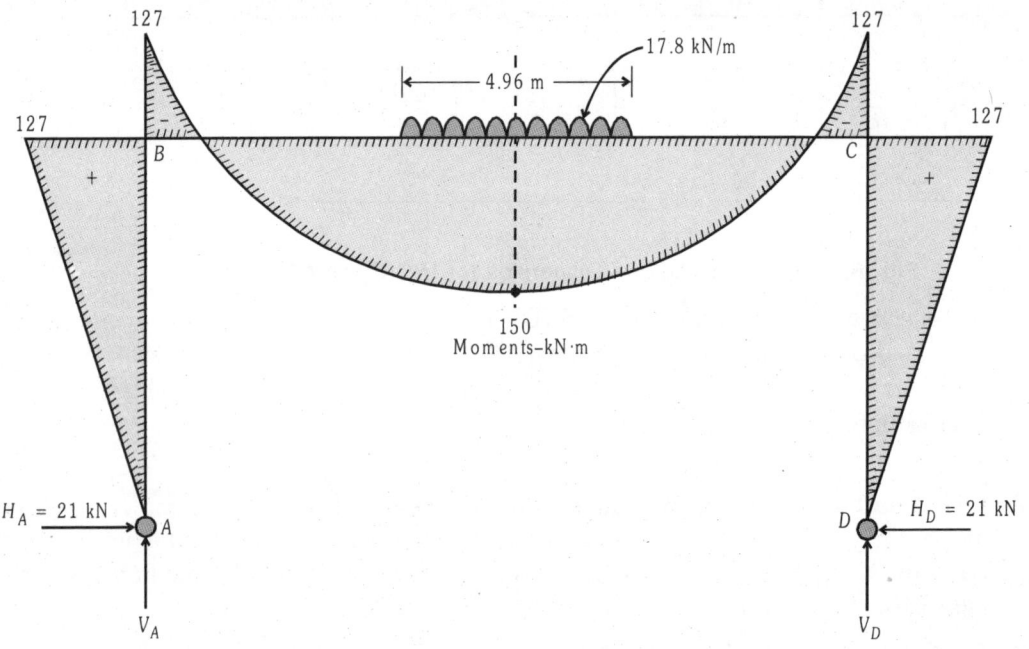

Fig. 8.10: Live Load Bending Moments in Frame

10. **Design of Prestressing Force**: The prestressing force required in the transom and legs is computed using Eq. 8.1 and 8.2.

For transom,
$$N_p = [(1/2) f_c b h - N_g - N_1]$$

where $N_g = 45$ kN and $N_1 = 21$ kN

Hence $N_p = \left(\dfrac{1}{2} \times \dfrac{15}{1000} \times 1000 \times 600\right) - 45 - 21$

$= 4434$ kN

For legs,
$$N_p = [(1/2) f_c b h - N_g - N_1]$$
where $N_g = 135$ kN and $N_1 = 45$ kN

$$N_p = \left(\frac{1}{2} \times \frac{15}{1000} \times 1000 \times 600\right) - 45 - 135$$
$$= 4320 \text{ kN}.$$

11. **Determination of the Cable Zone**: The permissible tendon zone for each member is determined for various load conditions. The core limits for a beam of rectangular section of overall depth 'h' is ($\pm h/6$), but due to the presence of applied loading thrusts, they have to be modified by multiplying them by the factor (N/N_p).
The factor (N/N_p) for the various members is computed as detailed below:

(a) **Transom (Under dead load only)**
$$\frac{N}{N_p} = \left[\frac{4434 + 45}{4434}\right] = 1.01$$

(b) **Transom (Under live load + dead load)**
$$\frac{N}{N_p} = \left[\frac{4434 + 45 + 21}{4434}\right] = 1.014$$

(c) **Top of Legs (Under dead load only)**
$$\frac{N}{N_p} = \left[\frac{4320 + 135}{4320}\right] = 1.031$$

(d) **Top of Legs (Under dead load + live load)**
$$\frac{N}{N_p} = \left[\frac{4320 + 135 + 45}{4320}\right] = 1.041$$

The low values of the ratio of (N/N_p) indicate that the effect of axial thrust is negligible in transom and legs.

12. **Limiting Zone for Prestressing Force:** The limiting zone can be obtained as the ratio of the moments to the thrust and are evaluated for transom and legs as detailed below:

   1. **Leg Foot Zone Limits:** Since the column foot is hinged at supports, there are no moments and hence the core limit is computed as
   $$\frac{h}{6} = \frac{600}{6} = 100 \text{ mm}$$
   The distance measured from the centroidal axis are as follows:
   (a) Upper limit (towards outside of frame) = $(-100 \times 1.031) = -103$ mm
   (b) Lower limit (inside of frame) = $(100 \times 1.041) = 104$ mm

   2. **Leg Top Zone Limits**
   (a) Upper (outer limit)

$$\left[\frac{M_2 + M_g}{N_p} - \frac{h}{6} \cdot \frac{N}{N_p}\right] = \left[\frac{-267 \times 10^3}{4320} - (100 \times 1.03)\right] = -165 \text{ mm}$$

(b) Lower (inner limit)

$$\left[\frac{M_1 + M_g}{N_p} + \frac{h}{6} \cdot \frac{N}{N_p}\right] = \left[\frac{-394 \times 10^3}{4320} + (100 \times 1.04)\right] = 13 \text{ mm}$$

3. Transom Mid Span Zone Limits

(a) Upper limit

$$\left[\frac{M_2 + M_g}{N_p} - \frac{h}{6} \cdot \frac{N}{N_p}\right] = \left[\frac{389 \times 10^3}{4434} - (100 \times 1.014)\right] = -13.6 \text{ mm}$$

(c) Lower limit

$$\left[\frac{M_1 + M_g}{N_p} + \frac{h}{6} \cdot \frac{N}{N_p}\right] = \left[\frac{239 \times 10^3}{4434} + (100 \times 1.014)\right] = 155.3 \text{ mm}$$

4. Transom End Zone Limits

(a) Upper limit

$$\left[\frac{M_2 + M_g}{N_p} - \frac{h}{6} \cdot \frac{N}{N_p}\right] = \left[-\frac{267 \times 10^3}{4434} - (100 \times 1.014)\right] = -161.6 \text{ mm}$$

(b) Lower limit

$$\left[\frac{M_1 + M_g}{N_p} + \frac{h}{6} \cdot \frac{N}{N_p}\right] = \left[-\frac{394 \times 10^3}{4434} + (100 \times 1.014)\right] = 12.6 \text{ mm}$$

13. **Determination of Bending Concordant Profile:** The bending concordant profile lying within the cable zone is determined by considering the moments developed for the two cases of loading. The eccentricity of the cable can be determined by the relation,

$$e = \left[\frac{M_g}{P} + \frac{M_1 + M_2}{2P}\right]$$

At transom mid span,

$$e = \left[\frac{239 \times 10^3}{4434} + \frac{150 \times 10^3}{2 \times 4434}\right] = 70.80 \text{ mm}$$

At transom end,

$$e = \left[-\frac{267 \times 10^3}{4434} - \frac{127 \times 10^3}{2 \times 4434}\right] = -74.52 \text{ mm}$$

Leg top zone,

$$e = \left[-\frac{267 \times 10^3}{4320} - \frac{127 \times 10^3}{2 \times 4320}\right] = -76.50 \text{ mm}$$

Design of Prestressed Concrete Rigid Frame Bridges

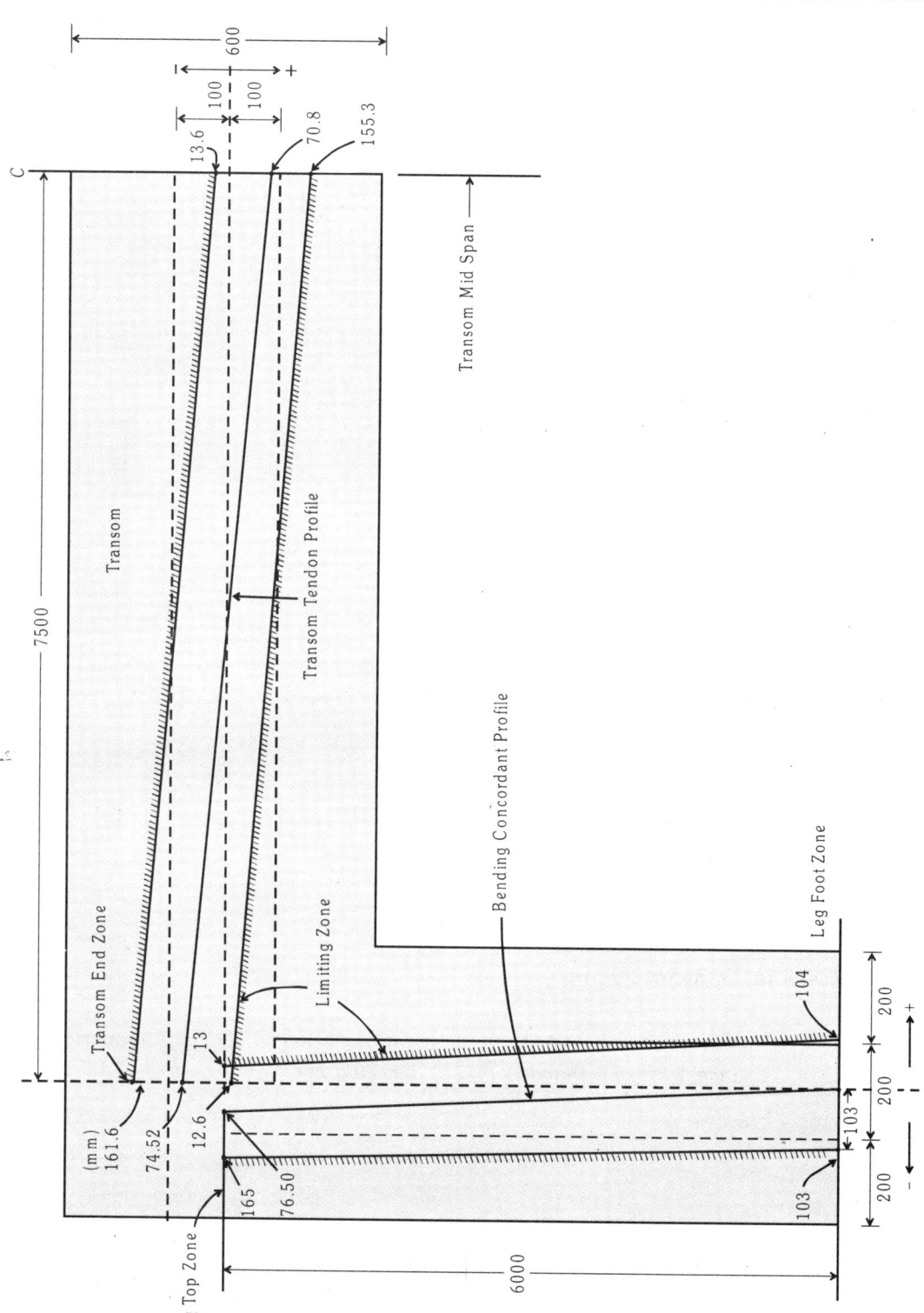

Fig. 8.11: Limiting Zone and Concordant Cable Profile in Two Pinned Portal Frame

The cable profile conforming to these values of eccentricity is straight in the column portion and parabolic in the transom. The limiting zone and bending concordant profile of the cable is shown in Fig. 8.11.

14. Design of Prestressing Force in Cables

(a) Transom

Prestressing force = 4434 kN/m
Using Freyssinet 7K-15 type seven wire strands in 65 mm diameter cable ducts conforming to IRC: 6006-1983[8],
Force in each cable = $(7 \times 0.8 \times 260.7) = 1459$ kN

$$\text{Number of cables/metre} = \left(\frac{4434}{1459}\right) = 3.04$$

$$\text{Spacing of cables} = \left(\frac{1000}{3.04}\right) = 329 \text{ mm}$$

Adopt cables at 300 mm centres

(b) Column

Prestressing force = 4320 kN/m
Adopt the same type of cables and spacing since the prestressing force is more or less similar in magnitude.

15. Check for Stresses at Service Loads

The stresses developed at the extreme fibres of the critical sections of the transom and columns are checked for various loading conditions.

(a) Properties of cross-section

$b = 1000$ mm
$h = 600$ mm
$A = (1000 \times 600) = (6 \times 10^5)$ mm²

$$Z_b = Z_t = Z = \left(\frac{bh^2}{6}\right) = \left(\frac{1000 \times 600^2}{6}\right) = (60 \times 10^6) \text{ mm}^2$$

Loss ratio = $\eta = 0.85$

(b) Stresses in Transom (N/mm²)

P (N)	A (mm²)	(P/A) (N/mm²)	Centre of Span e = 70.8 mm	Supports e = –74.5 mm
(4434×10^3) $(\eta P/A) = 6.28$ N/mm²	(6×10^5)	7.35	$(Pe/Z) = 5.23$ $(\eta Pe/Z) = 4.44$ $M_g = (239 \times 10^6)$ $M_L = (150 \times 10^6)$ $(M_g/Z) = 3.98$ $(M_L/Z) = 2.50$	$(Pe/Z) = 5.50$ $(\eta Pe/Z) = 4.67$ $M_g = (-267 \times 10^6)$ $M_L = (-127 \times 10^6)$ $(M_g/Z) = 4.45$ $(ML/Z) = 2.11$

Design of Prestressed Concrete Rigid Frame Bridges

Resultant Stresses (Transom) Case-1 (Prestress + Dead Load)(+ Compression and – Tension)

Stress at	Centre of Span	Supports
Soffit	$f_b = [(P/A) + (Pe/Z) - (M_g/Z)]$ $= [7.39 + 5.23 - 3.98]$ $= 8.64 \text{ N/mm}^2$	$f_b = [(P/A) - (Pe/Z) + (M_g/Z)]$ $= [7.39 - 5.50 + 4.45]$ $= 6.34 \text{ N/mm}^2$
Top	$f_t = [(P/A) - (Pe/Z) + (M_g/Z)]$ $= [7.39 - 5.23 + 3.98]$ $= 6.14 \text{ N/mm}^2$	$f_t = [(P/A) + (Pe/Z) - (M_g/Z)]$ $= [7.39 + 5.50 - 4.45]$ $= 8.44 \text{ N/mm}^2$

Case-2 (Effective Prestress + Dead Load + Live Load)(+Compression and – Tension)

Stress at	Centre of Span	Supports
Soffit	$f_b = \left(\dfrac{\eta P}{A} + \dfrac{\eta Pe}{Z} - \dfrac{M_g}{Z} - \dfrac{M_L}{Z}\right)$ $= 6.28 + 4.44 - 3.98 - 2.5$ $= 4.24 \text{ N/mm}^2$	$f_b = \left(\dfrac{\eta P}{A} - \dfrac{\eta Pe}{Z} + \dfrac{M_g}{Z} + \dfrac{M_L}{Z}\right)$ $= (6.28 - 4.67 + 4.45 + 2.11)$ $= 8.17 \text{ N/mm}^2$
Top	$f_t = \left(\dfrac{\eta P}{A} - \dfrac{\eta Pe}{Z} + \dfrac{M_g}{Z} + \dfrac{M_L}{Z}\right)$ $= (6.28 - 4.44 + 3.98 + 2.5)$ $= 8.32 \text{ N/mm}^2$	$f_t = \left(\dfrac{\eta P}{A} + \dfrac{\eta Pe}{Z} - \dfrac{M_g}{Z} - \dfrac{M_L}{Z}\right)$ $= (6.28 + 4.67 - 4.45 - 2.11)$ $= 4.39 \text{ N/mm}^2$

Resultant stresses (Column top)

$P = (4320 \times 10^3) \text{ N}$ $\quad Z = (60 \times 10^6) \text{ mm}^3$
$A = (6 \times 10^5) \text{ mm}^2$ $\quad M_g = (267 \times 10^6) \text{ N·mm}$
$e = -76.50 \text{ mm}$ $\quad M_L = (127 \times 10^6) \text{ N·mm}$

$\left(\dfrac{\eta P}{A}\right) = 6.12 \text{ N/mm}^2 \quad \left(\dfrac{Pe}{Z}\right) = 5.50 \text{ N/mm}^2$

$\left(\dfrac{\eta Pe}{Z}\right) = 4.68 \text{ N/mm}^2 \quad \left(\dfrac{M_g}{Z}\right) = 4.45 \text{ N/mm}^2$

$\left(\dfrac{M_L}{Z}\right) = 2.11 \text{ N/mm}^2$

Resultant Stresses (N/mm²)

Location	Prestress + Dead Load	Prestress + Dead Load + Live Load
Outside	$f_o = \left(\dfrac{P}{A} + \dfrac{Pe}{Z} - \dfrac{M_g}{Z}\right)$ $= (7.20 + 5.50 - 4.45)$ $= 8.25 \text{ N/mm}^2$	$f_o = \left(\dfrac{\eta P}{A} + \dfrac{\eta Pe}{Z} - \dfrac{M_g}{Z} - \dfrac{M_L}{Z}\right)$ $= (6.12 + 4.68 - 4.45 - 2.11)$ $= 4.24 \text{ N/mm}^2$
Inside	$f_i = \left(\dfrac{P}{A} + \dfrac{Pe}{Z} - \dfrac{M_g}{Z}\right)$ $= (7.20 - 5.50 + 4.45)$ $= 6.15 \text{ N/mm}^2$	$f_i = \left(\dfrac{\eta P}{A} - \dfrac{\eta Pe}{Z} + \dfrac{M_g}{Z} + \dfrac{M_L}{Z}\right)$ $= (6.12 - 4.68 + 4.45 + 2.11$ $= 8 \text{ N/mm}^2$

All the stresses in concrete under different loading conditions are well within the safe permissible limits.

16. Check for Ultimate Flexural Strength

(a) Centre of Transom
Consider one metre width of section
Overall depth $(h) = 600$ mm
Prestressing steel comprises of 7K-15 type seven wire strands in 65 mm cables
Provided at 300 mm at an eccentricity (e) of 70.8 mm
Effective depth $= d = (300 + 70.8) = 370.8$ mm
A_p (each cable) $= (7 \times 140) = 980$ mm^2
Number of cables per metre $= (1000/300) = 3.33$
A_p/metre $= (3.33 \times 980) = 3263$ mm^2
$f_p = 1862$ N/mm^2
Dead load moment $= G = 239$ kN·m
Live load moment $= Q = 150$ kN·m
According to IRC: 18-2000
Required $M_u = [1.5\ G + 2.5\ Q]$
$\quad\quad = [(1.5 \times 239) + (2.5 \times 150] = 733.5$ kN·m

Case-1 Failure by yielding of steel (Tension failure)
$$M_{us} = [0.9\ d\ A_p f_p]$$
$\quad\quad = [0.9 \times 370.8 \times 3263 \times 1862]$
$\quad\quad = (2027 \times 10^6)$ N·mm
$\quad\quad = 2027$ kN·m $> M_u$

Case-2 Failure by crushing of concrete (Compression failure)
$$M_{uc} = [0.176\ b\ d^2\ f_{ck}]$$
$\quad\quad = [0.176 \times 1000 \times 370.8^2 \times 45]$
$\quad\quad = (1089 \times 10^6)$ N·mm
$\quad\quad = 1089$ kN·m $> M_u$

(b) Transom Support Section
$A_p = 3263$ mm^2/m
$f_p = 1862$ N/mm^2
$e = 74.5$ mm
$d = (300 + 74.5) = 374.5$ mm
$G = 267$ kN·m
$Q = 127$ kN·m
M_u (Required) $= [1.5\ G + 2.5\ Q] = [(1.5 \times 267) + (2.5 \times 127] = 718$ kN·m

Case-1 Failure by yielding of steel
$$M_{us} = [0.9\ d\ A_p\ f_p]$$
$\quad\quad = [0.9 \times 374.5 \times 3263 \times 1862]$
$\quad\quad = (2047 \times 10^6)$ N·mm
$\quad\quad = 2047$ kN·m $> M_u$

Design of Prestressed Concrete Rigid Frame Bridges

Case-2 Failure by crushing of concrete
$$M_{uc} = [0.176\, bd^2 f_{ck}]$$
$$= [0.176 \times 1000 \times 374.5^2 \times 45]$$
$$= (1110 \times 10^6)\ \text{N·mm}$$
$$= 1110\ \text{kN·m} > M_u$$

The critical sections in the rigid frame satisfy the code provisions for flexural strength criteria.

17. **Check for Ultimate Shear Strength:** At transom ends, the dead and live load shear forces are computed as
$$V_g = 135\ \text{kN/m}$$
$$V_q = 44\ \text{kN/m}$$
Required $V_u = [1.5\, V_g + 2.5\, V_q]$
$$= [(1.5 \times 135) + (2.5 \times 44)]$$
$$= 312.5\ \text{kN}$$

The support section is uncracked in flexure, hence
$$V_{co} = 0.67 bh \sqrt{f_t^2 + 0.8 f_{cp} f_t}$$
where $\quad b = 1000$ mm
$\quad\quad\quad h = $ Overall depth $= 600$ mm
$$f_t = 0.24\sqrt{45} = 1.6\ \text{N/mm}^2$$
$$f_{cp} = \left(\frac{P}{A}\right) = \left(\frac{4320 \times 1000}{600 \times 1000}\right) = 7.2\ \text{N/mm}^2$$
$$V_{co} = \left[0.67 \times 10^3 \times 600\sqrt{1.6^2 + 0.8 \times 7.2 \times 1.6}\right]$$
$$= (1379 \times 10^3)\ \text{N}$$
$$= 1379\ \text{kN} > V_u\ \text{(Hence safe)}$$

18. **Design of Hinged Footing**

    (a) **Maximum Live load Reaction:** The live load position for maximum support reaction is shown in Fig. 8.12. Effective width of dispersion is computed as
    $$b_e = Kx(1 - x/L) + b_w$$
    where $\quad x = 1.8$ m
    $\quad\quad\quad L = 15$ m
    $\quad\quad\quad B = 9.5$ m
    $\quad\quad\quad b_w = 1.01$ m
    For $(B/L) = (9.5/15) = 0.63$, Read out from Table 5.1 the value of K as 1.88
    $$b_e = [(1.88 \times 1.8)\{1 - (1.8/15)\} + 1.01]$$
    $$= 3.98\ \text{m}$$

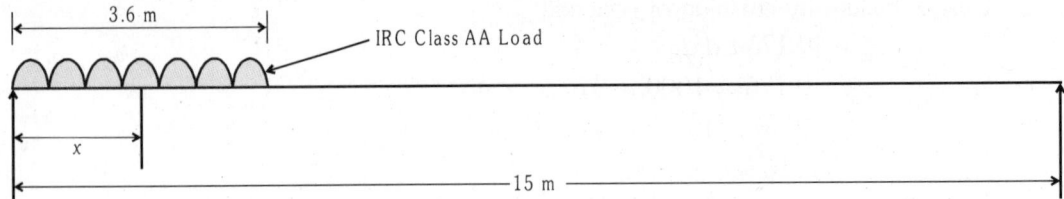

Fig. 8.12: Position of Load for Maixmum Shear

For 2 wheels (Refer Fig. 8.9),
Effective width of dispersion = [(2625 + 2050 + (3980/2)] = 6665 mm = 6.665 m

Intensity of live load = $\left(\dfrac{700 \times 1.1}{6.665 \times 3.6}\right)$ = 32 kN/m²

Live load reaction at support = $\left[\dfrac{32 \times 3.6 \times 13.2}{15}\right]$ = 101 kN

(b) Dead Load Reaction
Reaction due to dead load of frame/metre = 0.5(6 + 6 + 15) (0.6 × 1 × 24) = 194 kN
Total load on footing = (101 + 194) =   295 kN
Self weight of footing (Lump sum) =   30 kN
Total load ... =   325 kN
Assuming safe bearing capacity of soil at site as 250 kN/m²

Area of footing required = $\left[\dfrac{325}{250}\right]$ = 1.3 m²

Adopt a footing of 1.5 m wide throughout the length of 9.5 m,
Referring to Fig. 8.13

Intensity of soil pressure = $\left(\dfrac{325}{1.5 \times 1.0}\right)$ = 217 kN/m²

Maximum bending moment at centre of footing is
  M = (217 × 0.75 × 0.5 × 0.75) = 61 kN·m

Using M-20 Grade concrete

$$d = \sqrt{\dfrac{M}{Qb}} = \sqrt{\dfrac{61 \times 10^6}{0.762 \times 10^3}} = 283 \text{ mm}$$

Since depth required from shear considerations will be higher, adopt d = 350 mm and overall depth = D = 400 mm

$$A_{st} = \left(\dfrac{M}{\sigma_{st} jd}\right) = \left(\dfrac{61 \times 10^6}{200 \times 0.91 \times 350}\right) = 958 \text{ mm}^2$$

Use 16 mm diameter bars at 200 mm centres (A_{st} = 1005 mm²)

(c) Check for Shear Stresses in Footing (IRC: 21-2000)
Maximum shear force at a distance equal to '*d*' is computed as
  V = (217 × 0.1 × 1) = 22 kN

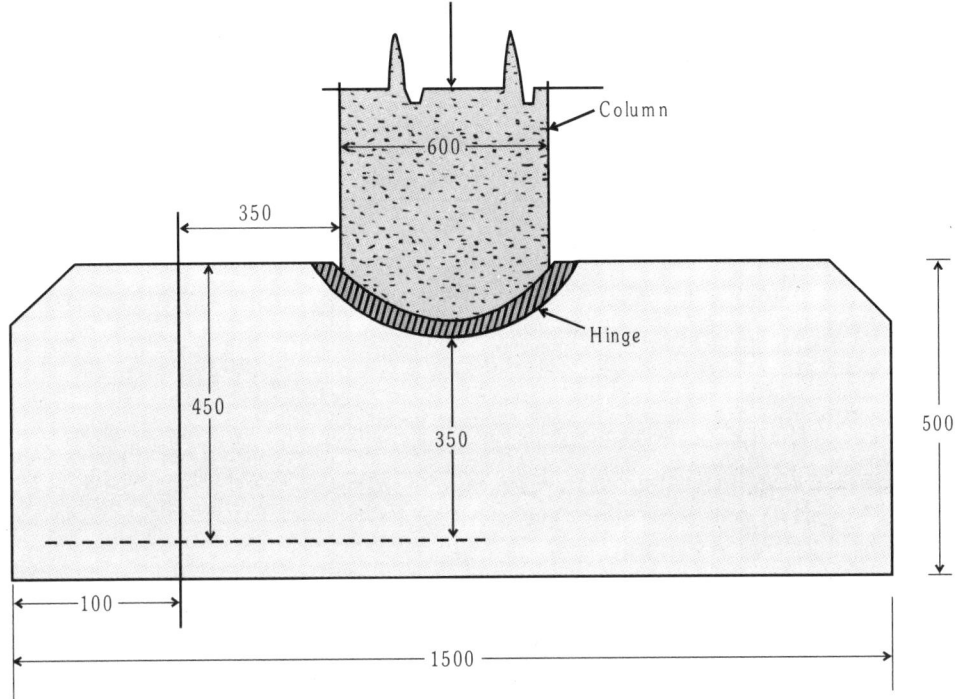

Fig. 8.13: Hinged Footing for Column of Portal Frame

Nominal shear stress $= \tau_v = \left(\dfrac{V}{bd}\right) = \left(\dfrac{22 \times 10^3}{1000 \times 450}\right) = 0.048$ N/mm^2

Ratio $\left(\dfrac{100 A_{st}}{bd}\right) = \left(\dfrac{100 \times 1005}{1000 \times 450}\right) = 0.22$

From Table-12B of IRC: 21-2000 read out the permissible shear stress for M-20 Grade concrete as $\tau_v = 0.20$ N/mm$^2 > \tau_v$

Hence shear stresses are well within safe permissible limits.

19. Supplementary Reinforcements

Using Fe-415 HYSD bars and M-45 Grade concrete

$A_{st} = 0.15$ per cent

$= (0.0015 \times 1000 \times 600)$

$= 900$ mm^2/m

Provide 12 mm diameter bars at 200 mm centres distributed both ways near top and bottom faces of transom and column. Also provide 10 mm diameter links at 200 mm centres in transom and columns.

The details of reinforcements provided in the rigid frame are shown in Fig. 8.14.

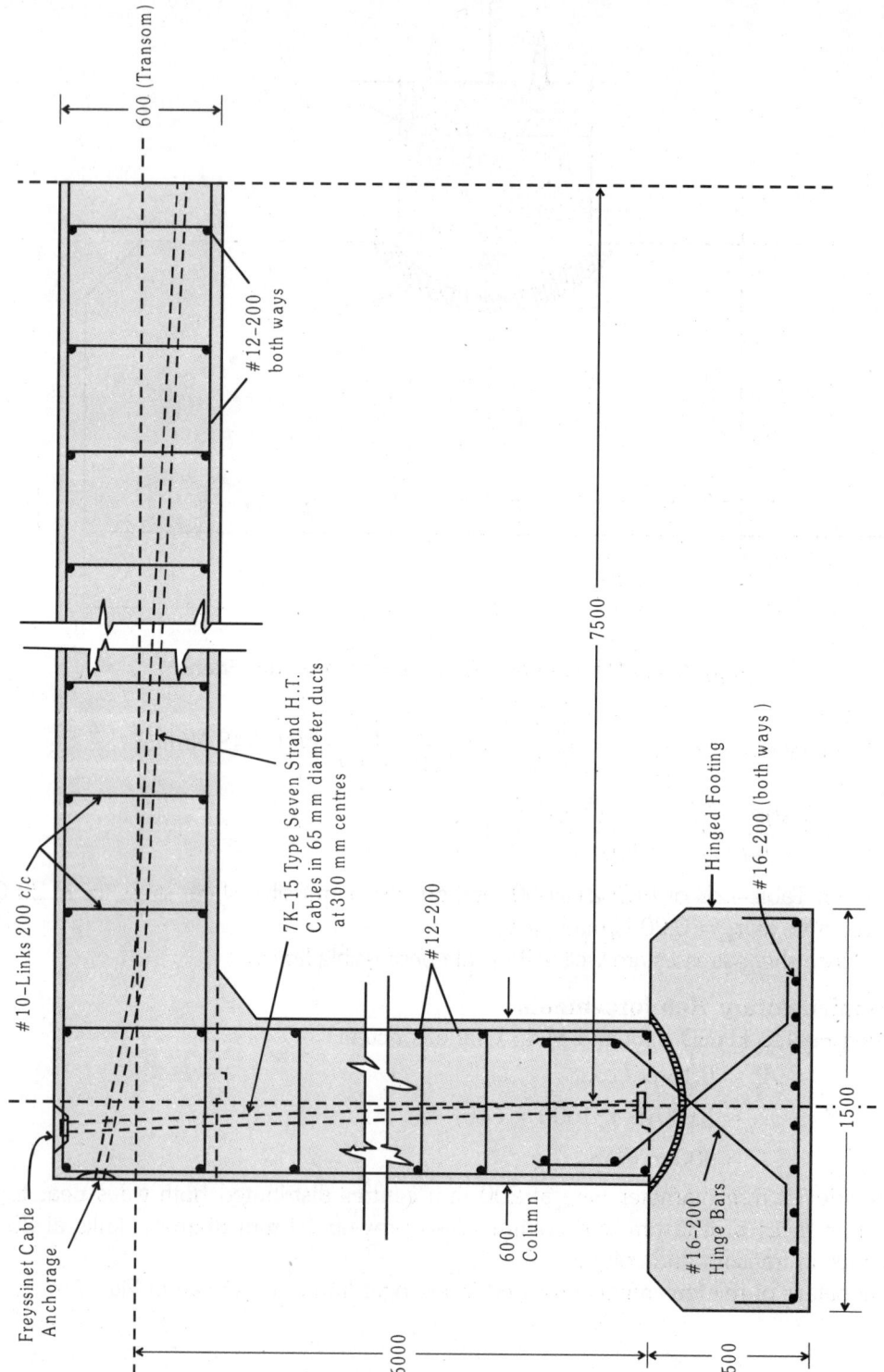

Fig. 8.14: Reinforcement Details in Prestressed Concrete Portal Frame

Design of Prestressed Concrete Rigid Frame Bridges

REFERENCES

1. TAYLOR, F. W., THOMPSON, S. E., SMULSKI, E., *Reinforced Concrete Bridges*. John Wiley and Sons, New York, 1955, p. 456.
2. CHETTOE, C. S. and ADAMS, H. C., *Reinforced Concrete Bridge Design*. Chapman and Hall, London, 1952.
3. RAINA, V. K., *Concrete Bridge Practice, Analysis, Design and Economics*. Tata McGraw Hill Publishing Co, New Delhi, 1991, pp. 189-193.
4. KRISHNA RAJU. N., *Design of Bridges* (Third Edition). Oxford and I.B.H. Publishers, New Delhi, 1998, pp. 202-231.
5. MARSHALL, W. T. and NELSON, H. M., *Structures*. Pitman, London, 1970 pp. 1-442.
6. I.R.C: 18-2000, Design Criteria for Prestressed Concrete Road Bridges (Post Tensioned Concrete), Third Revision, Indian Roads Congress, New Delhi 2000, pp. 1-61.
7. I.R.C: 21-2000, Standard Specifications and Code of Practice for Road Bridges, Section-III, Cement Concrete (Plain and Reinforced), Third Revision, Indian Roads Congress, New Delhi, 2000, pp. 1-80.
8. IS: 6006-1983, Indian Standard Specifications for Uncoated Stress Relieved Strand for Prestressed Concrete (First Revision), Indian Standards Institute, First Reprint, October, 1989.

EXERCISES

1. A prestressed concrete rigid frame has been proposed for a major National Highway crossing. The salient details of the proposed bridge are listed below:

   Type of bridge: Two pinned portal frame
   Span of crossing: 20 m
   Height of column members: 6 m
   Width of Road way: 15 m
   Foot path: 1 m on either side
   Thickness of wearing coat = 100 mm
   Live load: IRC Class AA tracked vehicle
   Grade of concrete: M-50 for portal frame and M-20 for footing
   Loss ratio: 0.85
   7K-15 type Freyssinet high tensile steel strands conforming to IRC: 6006-1983 are available for use.
   Fe-415 HYSD bars are to be used for foundation footing
   Safe bearing capacity of soil at site = 200 kN/m^2
   Design the rigid frame as Class-1 type structure conforming to IRC: 18-2000 and IRC: 21-2000. Sketch the details of cables and reinforcements in the rigid frame.

2. A prestressed concrete rigid frame bridge is required for the crossing of a National highway to suit the following particulars:

   Span of transom = 25 m
   Road width = 15 m between kerbs
   Foot path = 1.5 m on each side.
   Height of column members = 7.5 m

End conditions: Hinged or Pinned

Loading: IRC Class AA or A whichever produces the worst effect

Concrete: M-40 Grade for columns and transom members, M-20 Grade for foundation footing.

Adopt suitable prestressing cables with high tensile steel strands conforming to IS: 6006-1983 and Fe-415 HYSD bars for the footing.

Safe bearing capacity of soil at site = 250 kN/m^2

Design the rigid frame as Class 1 type structure conforming to IRC: 18-2000 and IRC: 21-2000. Sketch the details of cables and reinforcements in the frame.

9

PRESTRESSED CONCRETE CELLULAR BOX GIRDER BRIDGES

9.1 GENERAL FEATURES

Cellular box girder bridge decks with multiple cells are being increasingly adopted for urban fly overs and long span bridges in preference to the traditional tee beam and slab bridge decks due to their inherent advantages. The development of cantilever construction techniques by Finisterwalder[1] using pre cast or cast *in situ* elements has paved the way for use of multi-celled modular units for bridge decks. This method eliminates the use of expensive form work and scaffolding especially for bridges in deep ravines and rivers with large depth of water.

Segmental bridge deck construction by free cantilever method for long spans compensates for the high cost of tall piers and deep foundations. According to Raina[2], cantilever construction is ideally suited for box girder decks and the method of construction is elegantly convenient and cost effective in comparison with other traditional methods.

The cellular segments 2.5 to 3 m long can either be cast *in situ* on traveling gantries or can be precast in a yard and erected by launching truss or floating cranes. In the case of bridges with fewer spans, it is more economical to adopt *in situ* construction. In the case of long span bridges, progress of work is speeded up by using precast elements.

9.2 ADVANTAGES OF SEGMENTAL BOX GIRDER CONSTRUCTION FOR LONG SPAN BRIDGE DECKS

The use of precast single or multi-celled modular units in segmental construction of long span bridges has resulted in several advantages which are listed below:

1. The quality of concrete can be maintained at specified requirements.
2. Precast segments manufactured at a plant near the site of the bridge with a high degree of quality control exhibit superior characteristics.
3. Good dimensional control of segments will result in positive assembling of the units at bridge site.
4. The system is amenable to new techniques and speed of erection.
5. Proper curing of precast segments will result in the elimination of shrinkage problems.
6. Use of high quality concrete reduces the creep effects.

7. Match casting of the consecutive segments and epoxy jointing with precisely time controlled prestressing results in a monolithic integrated structure.
8. Segmental box girders have superior torsional resistance and exhibit non distortional characteristics.
9. Cellular box girders are ideally suited for curved bridge decks due to their superior torsional rigidity.
10. Segmental construction can be adopted by using precast or cast in situ units depending upon the site and availability of mechanical equipment.
11. Bridge construction is faster when precast segments are assembled and prestressed at bridge site.
12. Bigger cables with high tensile strands carrying large forces can be conveniently adopted in segmental construction.

9.3 TYPICAL CROSS-SECTIONS OF CELLULAR BOX DECKS

The use of precast segments for long span bridge decks has been well established by the advent of cantilever construction techniques developed by Finisterwalder[1]. Depending upon the width of the carriageway and span considerations, different types of cellular box decks are selected to suit the structural, constructional and aesthetic requirements of the bridge deck.

Typical cross-sections of cellular box girders for highway traffic on long span prestressed concrete bridges are compiled in Fig. 9.1. Single cell precast box girders of the type shown in Fig. 9.1 (a) have been used for the construction of the bridge between Oleron Island and the continent (France). The precast segmental bridge is made up of 26 main spans of 79 m each built by using single cell precast box girders[3]. Notable examples of single cell[4] segmental construction include the Barak bridge at Silchar in Assam (India) having main span of 130 m and the Lubha bridge in Assam built using variable depth single celled box girders over a span of 172 m.

Precast single cell box girders were used in the construction of Boussen's bridge in France having spans varying from 49 to 96 m, to cross the Garrone River. Mahatma Gandhi Sethu[5] at Patna is a classic example of single celled segmental construction with girders of varying depth. Being the longest bridge on record extending over 5.575 km, it is made up of 46 spans of 122 m with a two lane deck.

Two and three celled box decks shown in Fig. 9.1 (b and c) are generally preferred for multilane carriage way in major Highway bridges. Twin celled cellular box girders have been used for 80 m span sections of the bridge between Rio and Niteroi across Gunabara bay in Brazil. Two celled box units have been used for the Bhaghirathi bridge[6] in West Bengal for spans of 78 m. The bridge is of balanced cantilever type extending over three main spans supported on deep foundations on soft soil.

Precast multicell box units have been used for the Castejon bridge in Spain. Four celled precast units shown in Fig. 9.1 (d) have been used for the Sirsi Circle fly over at Bangalore[7,8]. The super structure comprises of precast post tensioned continuous box girders with curved soffit extending over eight spans each of 36 m. M-40 Grade concrete was used for the precast segments of length 2.9 m. The box girder segmental units were precast adopting the long line match casting technique of producing one full span of 12 segmental units at a time.

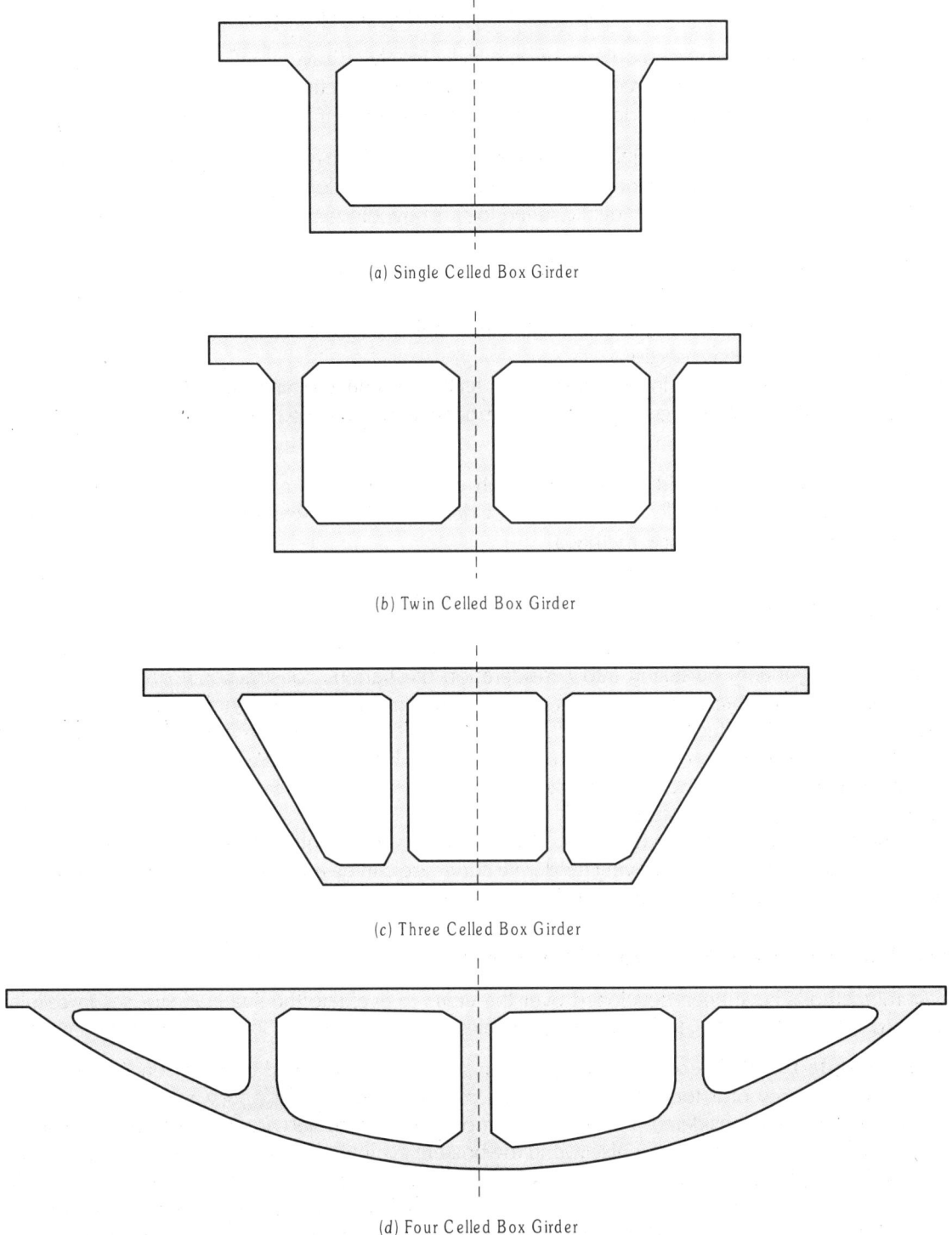

(a) Single Celled Box Girder

(b) Twin Celled Box Girder

(c) Three Celled Box Girder

(d) Four Celled Box Girder

Fig. 9.1: Typical Cross-sections of Cellular Box Girders for Bridge Decks

9.4 ANALYSIS OF BOX GIRDER BRIDGE DECKS

Rigorous analysis of cellular box girder bridge decks involving several influencing parameters is a complex task requiring extensive technical skill and construction experience. Box girders are preferred for long span cellular decks mainly due to the following structural advantages:

1. Hollow box sections are highly efficient as a structural form allowing optimum use of materials with considerable reduction in self weight especially for long span bridges.
2. The cellular box section is ideally suited to resist bending and particularly torsional moments in curved bridge decks.
3. The shape of the cross-section can be varied along the span to suit the flexural and torsional moments.
4. The hollow box section provides high lateral stability coupled with superior torsional rigidity compared to beam sections.
5. Recent developments in the construction techniques have overcome the problem of casting box sections and at present it is more competitive over a wide range of spans and invariably preferred for long spans.

Analysis of box girder bridge decks although complex is very much simplified by the use of computers, which can handle large number of design variables while using Finite Strip[9], Finite Plate[10, 11] and Finite Element[12] methods.

Dead Load Stresses in Box Girder Bridge Decks

In the case of straight bridges, the dead load stresses developed is generally evaluated using the simple bending theory by taking into consideration the various constructional stages. The shear lag[13] generally encountered in reinforced concrete sections in zones of high shear forces at support sections, reduces the effective flange width and the effective stiffness of the member in addition to modifying the distribution of stresses across the section. However in prestressed concrete sections the shear lag effect may not be very significant since a large percentage of the dead load is counterbalanced by the prestress.

In the case of curved bridge decks, the dead load stresses are influenced by the torsion induced at critical sections. Box girders having moderate curvature can be analyzed by the method proposed by Witecki[14] for horizontally curved bridge decks.

Live Load Stresses in Box Girder Bridge Decks

Several methods have been developed over the years to evaluate the live load stresses in cellular box girder decks. According to Raina[2], the principles of the five methods listed below are noteworthy.

1. **Simple Beam Theory:** The longitudinal stresses are estimated by pure bending or simple beam theory and torsion of thin walled members. Distortion and warping of the cross-section is not considered. Local bending moments in deck slab are evaluated from influence surfaces. Transverse moments around the box are evaluated by moment distribution method. The live loads are distributed at each node and the structure is treated as a grillage and analyzed by using computer programmes. This method is suitable only when the distortional stiffness of the box section is high and the span length is such that warping stresses are negligibly small.

2. **Torsional Analysis of Box Girders:** The simple beam theory and grillage analysis can be supplemented by using the torsional analysis of thin walled box girders proposed by Richmond[15]. In the case of box sections where warping and distortion significantly influences the stresses, this method improves the accuracy of results obtained from simple beam theory.

3. **Box Girder Analysis by Folded Plate Theory:** Thin walled box girder decks are treated as folded plates with rigid joints along the longitudinal edges according to Chuk-h, E et al.[11]. Using matrix methods and classical plate theory, overall stiffness matrix for edge loading is formulated. By using Fourier series and harmonic loading applied to the structure, the final results are obtained by superposition of the results. This method can be used only for constant cross-sections and is suitable for straight or curved bridge decks.

4. **Finite Strip Method:** The finite strip method proposed by Cheung[9] is more versatile in which the loaded surfaces are divided into finite strips as shown in Fig. 9.2. Analysis is carried out by applying loads directly to the longitudinal edges between the plates. The degrees of freedom chosen are the same but in this case the variation of these displacements across the strip is approximated by polynomials. Harmonic analysis by superposition of loading yields the final results. Curved bridge decks, continuous structures with variable depth girders can be analyzed. However this method is rarely used.

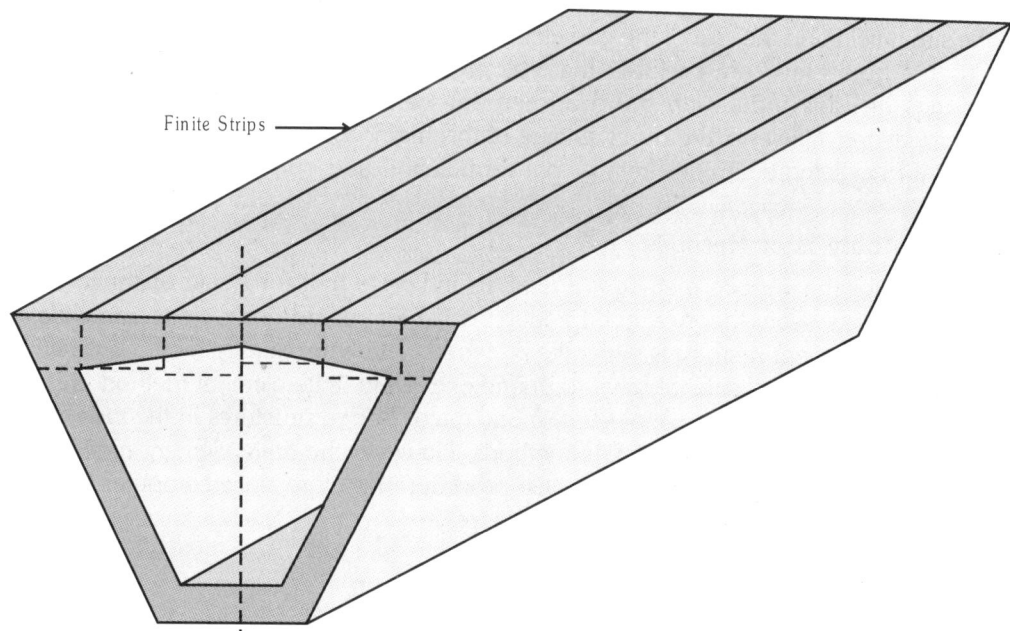

Fig. 9.2: Finite Strip Method

5. **Finite Element method:** The finite element method first developed by Zienkiwicz[16] for complex stress analysis problems is considered as a powerful tool which can handle all types of structures. This method has been applied for the analysis of concrete box girder decks in the pre and post cracking stages by Trikha and Edwards[17]. The accuracy of the finite element method depends upon the number and size of elements used in the analysis of the structure as shown in Fig. 9.3.

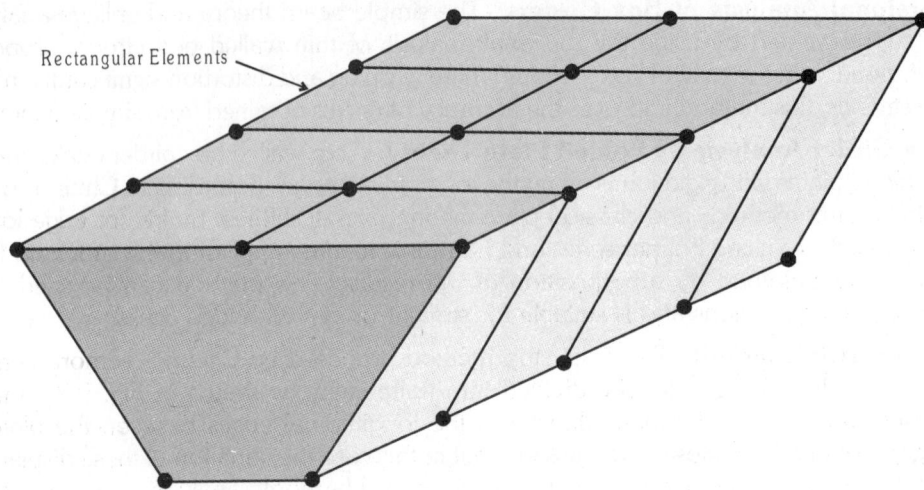

Fig. 9.3: Finite Element Method of Analysis

Compatibility of deformation is ensured at the nodes and the accuracy of the representation between nodes depends on the type of displacement function assumed for the strain field and the size of the grid elements. Research investigations by Sisodiya et al.[12], Cusens[18] and Lim[19] are useful in the type of finite elements to be chosen for box girder bridges besides indicating good agreement between experimental and theoretical investigations. However the disadvantage of this method lies in the large computer storage required involving time consuming input formulation and output analysis.

Comparative Analysis of Methods

A critical survey of the various methods of analysis indicates that for single or multi celled box girder bridges, simple beam theory is good enough while for curved decks which are wide and or shallow, the finite element method is generally preferred. The approximate methods outlined in 1 and 2 are recommended for straight forward bridge decks and finite element method presented in 5 for more complex bridge decks. Analysis of many major bridge structures in the past have been successfully attempted by the approximate methods. However the finite element method is likely to be adopted as the most general and versatile procedure in future since computer programmes are available for different types of box girder sections.

9.5 DESIGN PRINCIPLES

The selection of final cross-sectional dimensions and design of reinforcements in prestressed concrete box girders involves several iterative steps based on past experience, availability of constructional facilities, skilled labour and many other factors. Recently the Indian Roads Congress has brought out a special publication[20] covering the important guide lines for design and construction of segmental bridges. The salient steps in the design process are summarized below:

1. **Working Stress Design:** The tentative sectional details of the cellular box girders are selected based on past experience using span/depth ratios, lanes of traffic and the nature of loads on the deck. Reinforcements are designed for service load bending moments and

shear forces using the codal specifications of working stress design. The section selected should satisfy the limit state of deflection prescribed in the National Codes of Practice. Prestressing force is designed to control the longitudinal stresses using the conventional design procedure. Prestressing cable lay outs are arranged such that the centroid of the total prestressing force required at the section has the designed eccentricity and lies within the permissible tendon zone.

2. **Limit State of Collapse in Flexure:** Critical sections of bridge deck are checked for ultimate moment and shear strength capacity as specified in IRC: 18-2000 code requirements. If the high tensile steel area provided is insufficient to develop the required flexural strength, additional untensioned steel is provided in the tension zone to resist the balance moment. Normally the centre of span and support sections of continuous structures are checked for flexural strength requirements.

3. **Limit State of Collapse in Shear:** The behaviour of large cellular prestressed concrete box girders in shear is not clearly established. However the ultimate shear strength of the critical sections may be estimated with reasonable confidence using codal procedures specified in various National codes. According to FIB-CEB[21] and Leonhardt[22], shear design can be omitted under working stress design if a number of specified criteria for ultimate load design are satisfied. However, minimum shear reinforcements in the form of vertical links with a cross sectional area of 0.18 percent of the web should be provided according to IRC: 18-2000 when HYSD bars are used for stirrups.

4. **Continuous Prestressed Concrete Bridge Decks:** In the case of continuous prestressed bridge decks, the range of moment at critical sections such as support and span sections is determined by computing the maximum positive and negative moments due to live and dead loads. The overall cross-sectional requirement is checked using permissible compressive stress in concrete.

    The limiting zone for thrust is computed as the ratio of (M_{max}/P) and (M_{min}/P) at each section measured from the upper and lower kern respectively. The profile of a cable lying within the limiting zone and suitable for a concordant profile is determined. If necessary, the cable profile may be linearly transformed to reduce the slopes at supports with due regard to cover requirements. The reader may refer to a separate manograph[18] for details of theory and design of continuous prestressed concrete beams.

5. **Transverse Bending:** In cellular box girders, reinforcements are designed for transverse bending moment developed due to I.R.C. loads. The deck slab between the vertical ribs bends in the transverse direction generating bending moments due to live and dead loads. The transverse diaphragms are provided to limit the warping stresses in the box girder and according to Camp Bell et al.[23] and Sisodiya[24], transverse diaphragms are required only at the supports and it suits also the constructional considerations.

    The minimum dimensions of concrete box girders are usually dictated by detailing requirements and the rigidity of the cross-section will generally be sufficient to resist the effects of distortional resonance and instability. In cellular box girders, the transverse bending strength of the deck is improved by provision of proper fillets at the junction of flanges and web with extra corner reinforcements as shown in Fig. 9.4.

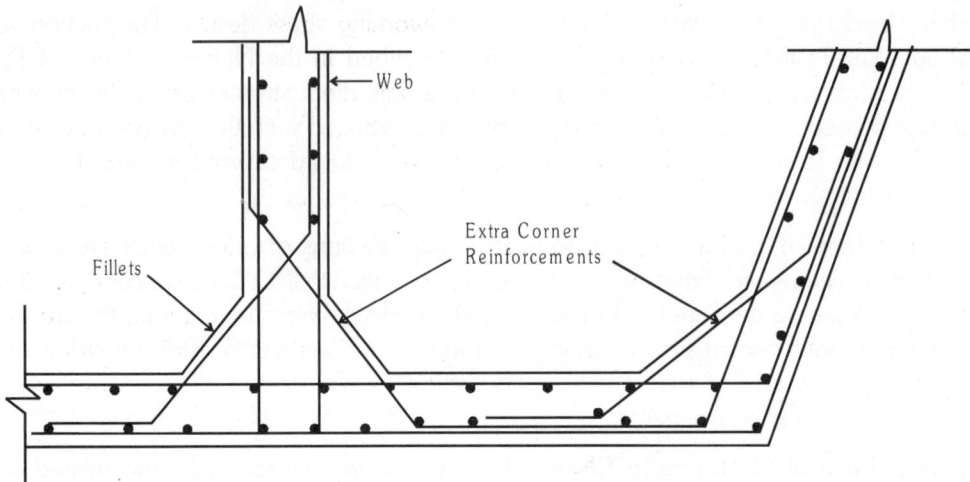

Fig. 9.4: Corner Reinforcements in Box Girders

6. **Control of Shear and Principal Tensile Stresses:** A critical analysis of the shear and principal tensile stresses developed in prestressed concrete beams has been presented in a separate monograph[25] by the author. The principal tensile stresses in the webs can be controlled by the following methods:

   1. Increasing the thickness of the web
   2. Vertical prestressing (Fig. 9.5) which induces compression but not advisable due to short cables and constructional problems.

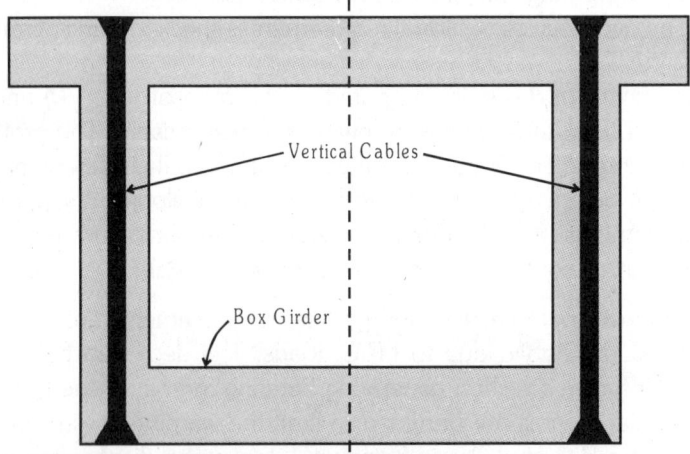

Fig. 9.5: Vertical Prestressing of Webs

   3. Providing longitudinal cap cables in cantilever construction (Fig. 9.6) which reduces the effective shearing force acting at support sections.
   4. Use of haunched girder units near supports as shown in Fig. 9.7 in which the vertical component of the longitudinal prestressing force counteracts the shear.

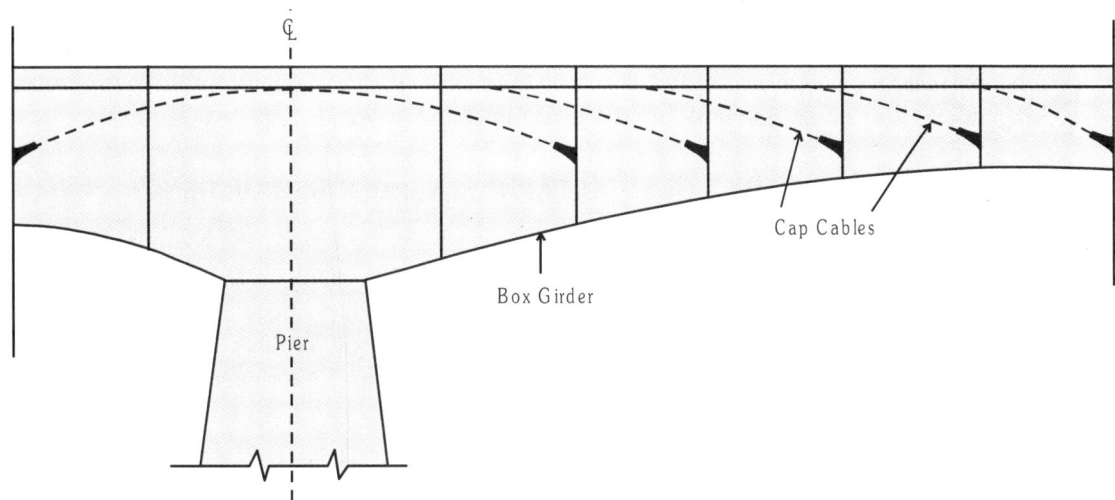

Fig. 9.6: Cap Cables in Cantilever Construction

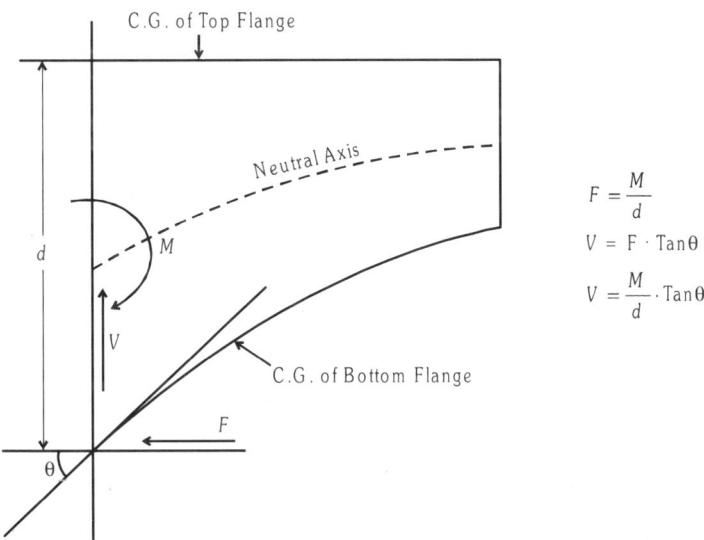

$$F = \frac{M}{d}$$
$$V = F \cdot \text{Tan}\theta$$
$$V = \frac{M}{d} \cdot \text{Tan}\theta$$

Fig. 9.7: Shear in Haunched Girders

The common practice is to use curved cables in continuous box girders. At supports where shear is maximum due to dead loads, the vertical component of the cable force reduces the magnitude of shear force. The horizontal component of the prestressing force will further induce compressive stress which in turn reduces the principal tension thus avoiding diagonal tension cracks near support sections.

7. **Torsion:** Cellular box sections are ideally suited for resisting torsion arising in curved bridge decks. Particular care is essential when thin flanged sections are used. In such cases principal

tension developed due to combined torsion and shear should be limited. IRC: 18-2000 specifies that calculations for torsion are required only for ultimate loads. Computation of torsional resistance of prestressed sections and the design of torsion reinforcements comprising longitudinal and transverse links are discussed in detail in section 4.6.

9.6 DESIGN EXAMPLE OF CELLULAR BOX GIRDER BRIDGE DECK

A cellular multi celled prestressed concrete box girder deck is to be designed for a National Highway crossing. The proposed bridge deck is made up of two continuous spans each of 50 m. The road width is 7.5 m with foot paths 1.25 m on each side. The box girder is proposed to have 4 cells 2 m wide by 2 m deep and should support IRC Class AA tracked vehicle loading. Design the cellular bridge deck adopting M-60 Grade concrete, Fe-415 HYSD bars and high tensile steel strands of 15.2 mm diameter conforming to the relevant Indian Standards.

1 Data
Span = 50 m
Cross section: Multi celled box girder
Cell dimensions = 2 m wide by 2 m deep
Road width = 7.5 m
Foot paths: 1.25 m wide on either side of road way
Wearing coat = 80 mm
Thickness of web = 300 mm to house 27 K-15 Freyssinet type anchorages
 (27 strands of 15.2 mm diameter in 110 mm diameter cables)
 (Refer Table A.3.3 in Appendix-3 for details of Freyssinet anchorages)
Thickness of top and bottom slabs = 300 mm
Concrete grade: M-60
Loss ratio = 0.80
Type of tendons: High tensile strands of 15.2 mm diameter conforming to
 IRC: 6006-2000
Type of supplementary reinforcements: Fe-415 HYSD bars
Design the bridge deck as Class-1 type structure conforming to the codes
IRC: 6-1966, IRC: 18-2000 and IRC: 21-2000.

2. Permissible Stresses:
For M-60 Grade concrete and Fe-415 HYSD bars, the permissible stresses according to IRC: 21-2000 are:

σ_{cb} = 11.5 N/mm² $\qquad$ j = 0.879
σ_{st} = 200 N/mm² $\qquad$ Q = 1.844
m = 10

For M-60 Grade concrete, the permissible stresses according to IRC: 18-2000 are:

f_{ck} = 60 N/mm²
f_{ci} = 45 N/mm²
f_{ct} = 0.45 f_{ci} = (0.45 × 45) = 20 N/mm²
f_{cw} = 0.33 f_{ck} = (0.33 × 60) = 20 N/mm²

$f_{tt} = f_{tw} = 0$ (Class-1 type member)

$E_c = 5000\sqrt{f_{ck}} = 5000\sqrt{60} = 38730 \text{ N/mm}^2 = 38.7 \text{ kN/mm}^2$

3. Cross-section of Box Girder

Overall depth of the box girder $= \left(\dfrac{\text{Span}}{25}\right) = \left(\dfrac{50}{25}\right) = 2$ m

Width of road way = 7.5 m
Width of foot paths = (2 × 1.25) = 2.5 m
Total width of box girder at road level = (7.5 + 2.50) = 10 m
Spacing between webs = 2 m
4 celled box girder is adopted
Thickness of web as per Clause 9.3.2.1 of IRC: 18-2000 is computed as
$\quad\quad t_w$ = [200 + diameter of cable duct for housing 27 K-15 strands]
$\quad\quad\quad$ = [200 + 100] = 300 mm
At end supports where anchorages are located, web thickness increased to 400 mm
Thickness of top and bottom slabs = 300 mm
The multi celled box girder section selected is shown in Fig. 9.8
Section properties of the symmetrical I-girder shown in Fig. 9.9 are as follows:
Cross sectional area = $A = 1.62 \text{ m}^2$
Second moment of area = $I = 0.94 \text{ m}^4$
Distance of the extreme fibre from centroid = $y = y_t = y_b = 1$ m
Section modulus = $Z = Z_t = Z_b = (I/y) = 0.94 \text{ m}^3$

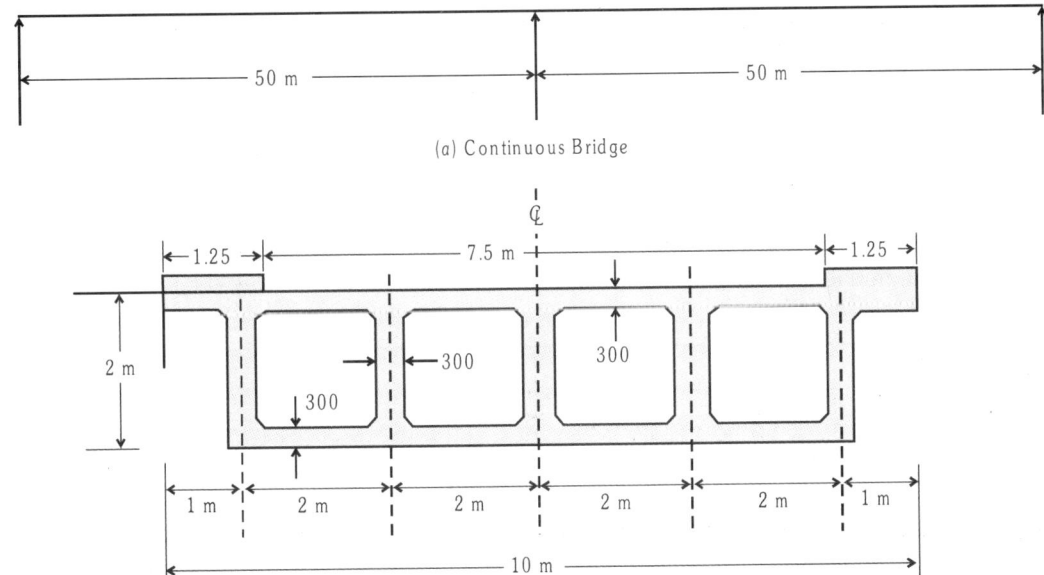

(a) Continuous Bridge

(b) Cross-section of Deck

Fig. 9.8: Two Span Box Girder Bridge Deck

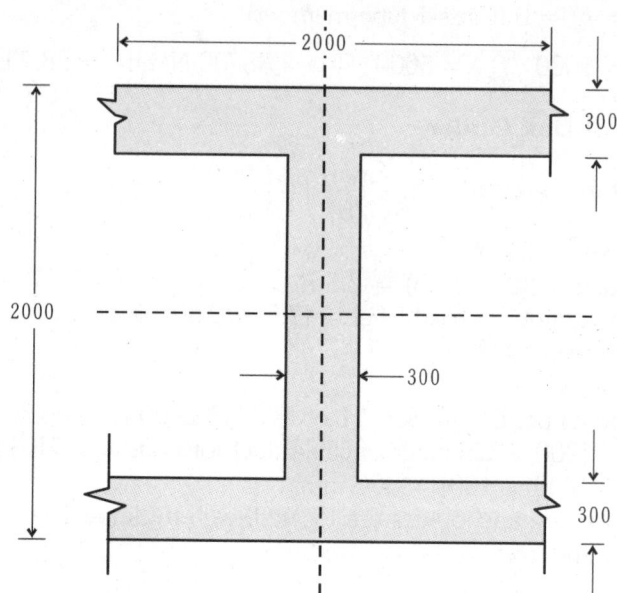

Fig. 9.9: Cross-section of Web Girder

4. Design of Slab Panel

(a) Dead Load Bending Moments

Dead weight of slab = (1 × 1 × 0.3 × 24) =   7.20 kN/m²
Dead weight of W.C. = (0.08 × 22) ... =   1.76
Total dead load = g ...= 8.96 =   9.00 kN/m²

Referring to the bending moment coefficients compiled in a separate monograph by the author[26] and the Fig. 9.10,

Maximum negative bending moment due to dead load at supports = (0.107 × gL)
= (0.107 × 9 × 2) = 1.93 kN·m/m
Maximum positive bending moment at centre of span = (0.077 × gL)
= (0.77 × 9 ×2) = 1.38 kN·m/m
Maximum shear force = (0.60 × gL) = (0.60 × 9 × 2) = 10.8 kN

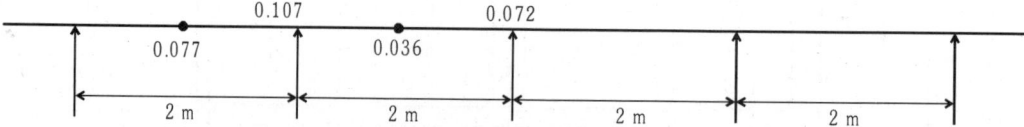

Fig. 9.10: Dead Load Bending Momenet Coefficients in Four Span Continuous Slab

(b) Live Load Bending Moments:
The slab panel is continuous over webs in the transverse direction and free in the longitudinal direction. The slab spanning in the transverse direction is designed for IRC Class AA tracked loading using the procedure specified in IRC: 21-2000. When IRC Class AA tracked vehicle traverses on the deck, maximum bending moment in the transverse direction of the slab will develop when one tracked wheel occupies the centre of slab as shown in Fig. 9.11.

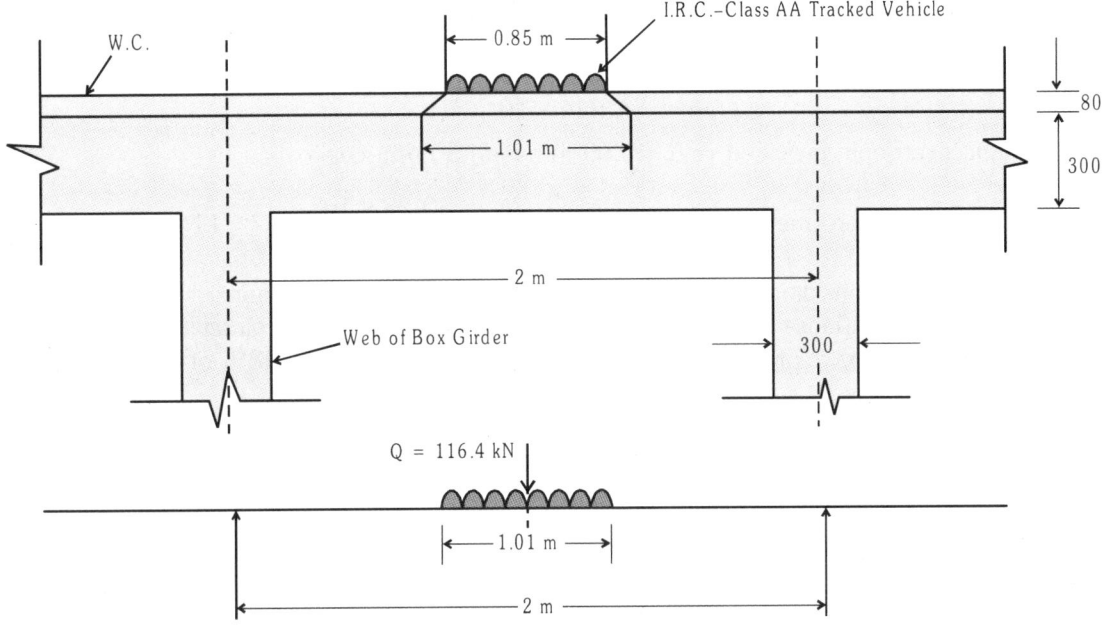

Fig. 9.11: Position of IRC Class AA Load for Maximum B.M. in slab

The effective width of dispersion of the wheel through the wearing coat is computed as
$$u = [0.85 + (2 \times 0.08)] = 1.01 \text{ m}$$
$$v = [3.60 + (2 \times 0.080] = 3.76 \text{ m}$$

Average intensity of wheel load with Impact Factor $= \left[\dfrac{1.25 \times 350}{3.76 \times 1.01}\right] = 115.20$ kN/m²

Concentrated load acting at the centre of span in the transverse direction is computed as
$$Q = (115.20 \times 1.01) = 116.4 \text{ kN}$$

Referring to the bending moment and shear force coefficients[26] compiled in Fig. 9.12,

Maximum positive B.M. at middle of end span $= [0.210\ QL]$
$$= [0.210 \times 116.4 \times 2] = 48.9 \text{ kN·m}$$

Maximum negative B.M. at penultimate support $= [0.181\ QL]$
$$= [0.181 \times 116.4 \times 2] = 42.13 \text{ kN·m}$$

Maximum shear force $= [0.60 Q] = [0.60 \times 116.4] = 69.8$ kN

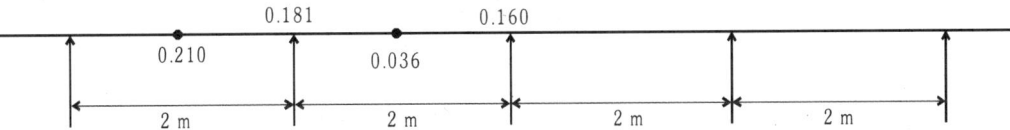

Fig. 9.12: Live Load Bending Moment and Shear Force Coefficients in Four Span Continuous Slab

(c) Design Bending Moments and Shear Forces

Total Positive Bending Moment $= [DLBM + LLBM] = [1.38 + 48.9] = 50.28$ kN·m
Total Negative Bending Moment $= [DLBM + LLBM] = [1.93 + 42.13] = 44.06$ kN·m
Maximum Shear Force $= [DLSF + LLSF] = [10.8 + 69.8] = 80.6$ kN

(d) Design of Top Slab Section and Reinforcements

Effective depth $= d = \sqrt{\dfrac{M}{Qb}} = \sqrt{\dfrac{50.28 \times 10^6}{1.844 \times 10^3}} = 165.12$ mm

Effective depth provided $= d = 250$ mm with 50 mm cover

Area of reinforcement $= A_{st} = \left(\dfrac{M}{\sigma_{st} jd}\right) = \left(\dfrac{50.28 \times 10^6}{200 \times 0.879 \times 250}\right) = 1144$ mm^2

Provide 16 mm diameter bars at 160 mm centres ($A_{st} = 1257$ mm^2)
Referring to Clause 15.4 of IRC: 18-2000, for top slab of box girder decks, minimum supplementary reinforcement in the longitudinal direction is 0.18% of the gross cross-sectional area.
Distribution reinforcement $= (0.0018 \times 1000 \times 300) = 540$ mm^2
Provide 12 mm diameter bars at 160 mm centres ($A_{st} = 707$ mm^2)

(e) Check for Shear Stress in Slab

Design shear force $= V = 80.6$ kN

Design shear stress $= \tau = \left(\dfrac{V}{bd}\right) = \left(\dfrac{80.6 \times 1000}{1000 \times 250}\right) = 0.32$ N/mm^2

$\left(\dfrac{100 A_{st}}{bd}\right) = \left(\dfrac{100 \times 1257}{1000 \times 250}\right) = 0.50$

According to IRC: 21-2000, the permissible shear stress (τ_c) in concrete is obtained from Table 4.2 (Table 12B of IRC: 21-2000) as

$\tau_c = 0.32$ N/mm^2

For solid slabs, permissible shear stress $= K \tau_c$ where K is a constant depending upon the thickness of the slab compiled in Table 4.4 (Table 12C of IRC: 21-2000).
In this case the thickness of the slab $= 300$ mm and hence $K = 1.00$
Permissible shear stress in slab $= (K \tau_c) = (1.00 \times 0.32) = 0.32$ N/mm^2
Since τ does not exceed $K\tau_c$, shear reinforcements are not required

5. Design of Web Girder

(a) **Dead Load Bending Moments and Shear Forces**: The continuous box girder is treated as an assemblage of I-sections with web serving the function of a main girder and flanges of symmetrical size as shown in Fig. 9.9.

Self weight of flanges $= (2 \times 0.3 \times 24) =$ 14.40 kN/m^2
Self weight of W.C. ... $= (1 \times 1 \times 22) =$ 1.76
Total load ... =...   16.16
Self weight of web $= (1.4 \times 0.3 \times 24) =$ 10.08
Total load on each I-girder $= g = [(2 \times 16.16) + 10.08] = 43$ kN/m
The dead load bending moment coefficients[26] for a two span continuous beam is shown in Fig. 9.13. The dead load bending moments at mid support and mid span sections are computed as:

$$M_{gB} = 0.125\,gL^2 = (0.125 \times 43 \times 50^2) = 13438 \text{ kN·m}$$
$$M_{gD} = 0.071\,gL^2 = (0.071 \times 43 \times 50^2) = 7633 \text{ kN·m}$$

Dead load shear is maximum near the mid support section and is computed as
$$V_g = 0.62gL = (0.62 \times 43 \times 50) = 1333 \text{ kN}$$

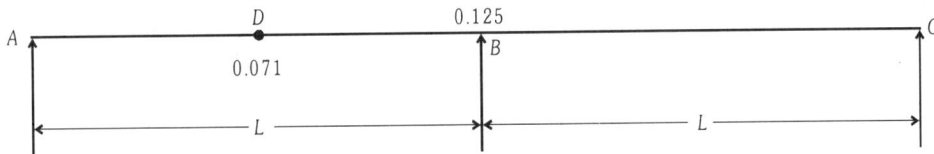

Fig. 9.13: Dead Load Bending Moment Coefficients

(b) Live load Bending Moments in Continuous Web Girder: Maximum live load reaction occurs in the web girder when the transverse disposition of the IRC Class AA tracked vehicle load is arranged to have the maximum eccentricity with respect to the centre of the bridge deck as shown in Fig. 9.14. Maximum reaction due to live loads in girder B is computed as

$$R_B = \left[\frac{W \times 1.1}{2}\right] = 0.55W = (0.55 \times 700) = 385 \text{ kN}$$

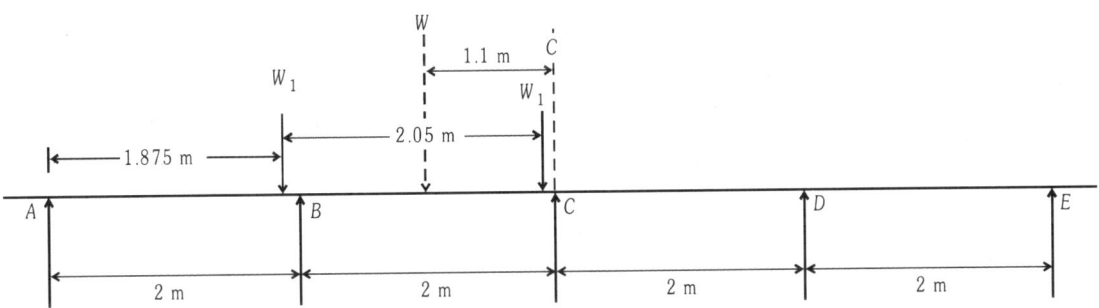

Fig. 9.14: Position of IRC Class AA Live Loads for Maximum Reaction in Girder

Hence the concentrated load = Q = 385 kN

This load acting over a length of 3.6 m in the longitudinal direction is positioned at the centre of span of the two span continuous beam as shown in Fig. 9.15 to compute the maximum positive and negative moments.

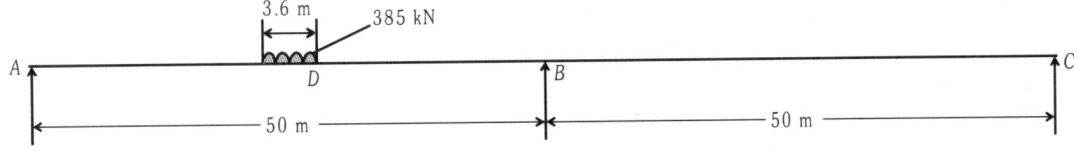

Fig. 9.15: Position of Live Load for Maximum Moments in Two Span Continuous Beam

The live load bending moment coefficients[26] for maximum positive and negative moments in a two span continuous beam are shown in Fig. 9.16.

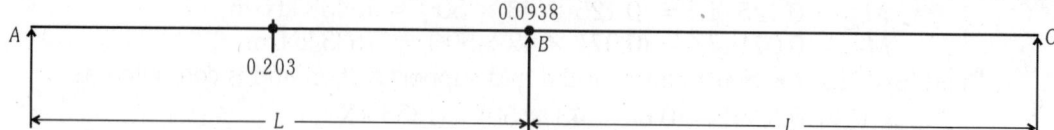

Fig. 9.16: Live Load Bending Moment Coefficients for a Two Span Continuous Girder

Maximum positive live load bending moment with impact factor at centre of span is computed as

$$M_{max}(\text{Positive}) = (I.F.)(0.203QL)$$
$$= (1.10)(0.203 \times 385 \times 50)$$
$$= 4298 \text{ kN·m}$$

Maximum negative live load bending moment with impact factor at mid support is computed as

$$M_{max}(\text{Negative}) = (I.F.)(0.0938QL)$$
$$= (1.1)(0.0938 \times 385 \times 50)$$
$$= 1986 \text{ kN·m}$$

(c) Live Load Shear Force in Girder: The maximum live load shear force develops in the interior webs when the IRC Class AA loads are placed near the mid support as shown in Fig. 9.17.

Reaction of load W on interior girder $= \left(\dfrac{350 \times 48.2}{50}\right) = 338 \text{ kN}$

Maximum live load shear force with impact $= (338 \times 1.1) = 372 \text{ kN}$

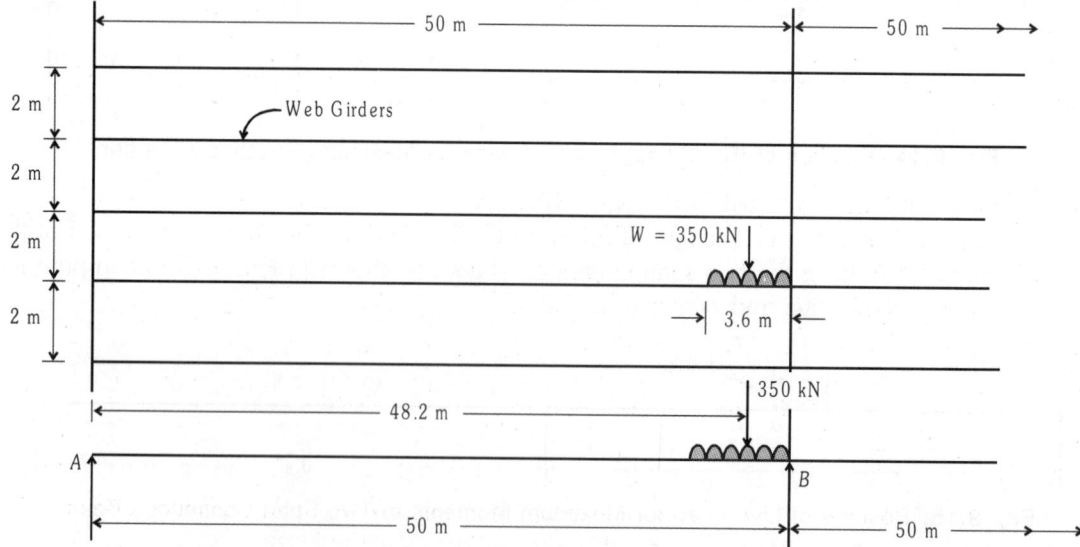

Fig. 9.17: Position of IRC Class AA Loads for Maximum Shear Force in Web Girder

(d) Design Bending Moments and Shear Forces: The design bending moments and shear forces at service and ultimate loads as stipulated in IRC: 18-2000 are compiled in Table 9.1.

Table 9.1: Service and Ultimate Load Moments and Shear Forces in Web Girder

	D.L.B.M.	L.L.B.M.	Total Working B.M.	Required Ultimate B.M.	Units
	(a) Bending Moments (Outer Web Girder)				
Mid Span Section (D)	M_g 7633	M_q 4298	$(M_g + M_q)$ 11931	$M_u = (1.5\,M_g + 2.5\,M_q)$ 22195	kN·m
Mid Support Section (B)	13438	1986	15429	25122	kN·m
	D.L.S.F.	L.L.S.F.	Total Working S.F.	Required Ultimate S.F.	Units
	(b) Shear Forces (Inner Web Girder)				
Near Mid Support Section (B)	V_g 1333	V_q 372	$(V_g + V_q)$ 1705	$V_u = (1.5\,V_g + 2.5\,V_q)$ 2930	kN

(e) Check for Minimum Section Modulus: At the mid support section B, the dead and live load moments are listed as

$M_{gB} = 13438$ kN·m

$M_{QB} = 1986$ kN·m

$M_{dB} = [M_{gB} + M_{QB}] = [13438 + 1986] = 15424$ kN·m

$f_{br} = [\eta f_{ct} - f_{tw}] = [(0.8 \times 20) - 0] = 16$ N/mm^2

$$f_{inf} = \left[\frac{f_{tw}}{\eta} + \frac{M_g}{\eta Z_b}\right] = \left[0 + \frac{13438 \times 10^6}{0.8 \times 0.94 \times 10^9}\right] = 17.86 \text{ N/mm}^2$$

$$Z_b \geq \left[\frac{M_q + (1-\eta)M_g}{f_{br}}\right]$$

$$\geq \left[\frac{(1986 \times 10^6) + (1-0.8)13438 \times 10^6}{16}\right]$$

$= (0.292 \times 10^9)$ mm^3 < (0.94×10^9) mm^3 (section provided)

Hence section provided is adequate.

(f) Prestressing Force: For the two continuous spans AB and BC, a concordant cable profile is selected such that the secondary moments are zero. The cable profile selected with eccentricity at mid support twice that at mid spans is shown in Fig. 9.18.
Providing an effective cover = 300 mm,
Maximum possible eccentricity at support = e = (1000 – 300) = 700 mm
Prestressing force is computed from the relation,

$$P = \left[\frac{Af_b Z_b}{Z_b + Ae}\right] = \left[\frac{(1.62 \times 10^6)(17.86)(0.94 \times 10^9)}{(0.94 \times 10^9) + (1.62 \times 10^6 \times 700)}\right]$$
$$= 13109932 \text{ N}$$
$$= 13110 \text{ kN}$$

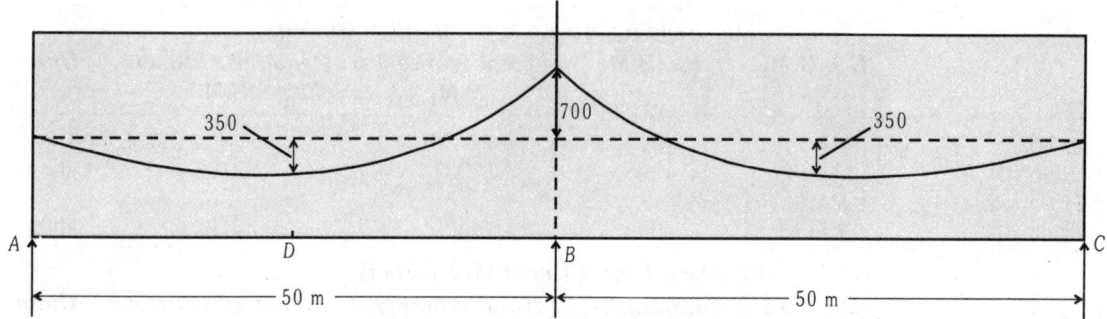

Fig. 9.18: Concordant Cable Profile

Using Freyssinet system with anchorage type 27K-15 (27 strands of 15.2 mm diameter) in 110 mm diameter cable ducts (Refer Table A.3.3 in Appendix),
Force in each cable = $(27 \times 0.8 \times 265) = 5724$ kN
Provide three cables carrying an initial prestressing force of
$$P = (3 \times 5000) = 15000 \text{ kN}$$
Area of each strand of 15.2 mm diameter tendon = 140 mm²
Area of 27 strands in each cable = $(27 \times 140) = 3780$ mm²
Total area in 3 cables = $A_p = (3 \times 3780) = 11340$ mm²
The cables are arranged in a parabolic concordant profile so that the centroid of the group of cables has an eccentricity of 700 mm towards the top fibre at mid support section B and an eccentricity of 350 mm towards the soffit at mid span section D. The centroid of the cables is concentric at the end supports A and C. The selected cable profile is shown in Fig. 9.19.

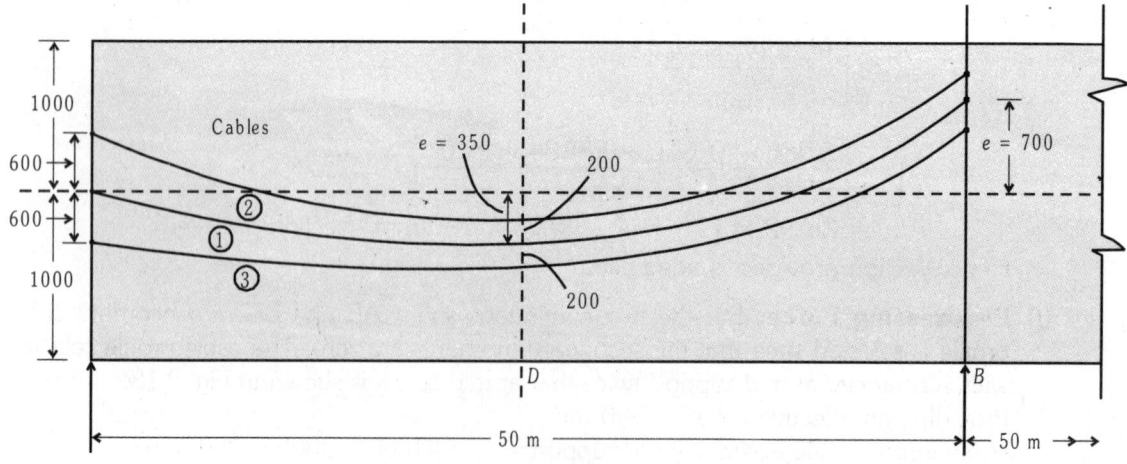

Fig. 9.19: Profiles of Individual Cables in Span

(g) Check for Stresses at Service Loads

1. Centre of span section

   $P = 15000$ kN $\quad\quad h = 0.80$

   $e = 350$ mm $\quad\quad M_g = 7633$ kN·m

   $A = (1.62 \times 10^6)$ mm^2 $\quad\quad M_q = 4298$ kN·m

   $Z = (0.94 \times 10^9)$ mm^3

   $$\left(\frac{P}{A}\right) = \left[\frac{15000 \times 10^3}{1.62 \times 10^6}\right] = 9.25 \text{ N/mm}^2$$

   $$\left(\frac{Pe}{Z}\right) = \left[\frac{15000 \times 10^3 \times 350}{0.94 \times 10^9}\right] = 5.58 \text{ N/mm}^2$$

   $$\left(\frac{M_g}{Z}\right) = \left[\frac{7633 \times 10^6}{0.94 \times 10^9}\right] = 8.12 \text{ N/mm}^2$$

   $$\left(\frac{M_q}{Z}\right) = \left[\frac{4298 \times 10^6}{0.94 \times 10^9}\right] = 4.57 \text{ N/mm}^2$$

   At transfer stage, the stresses at extreme fibres are computed as

   $$\sigma_t = \left[\frac{P}{A} - \frac{Pe}{Z} + \frac{M_g}{Z}\right] = [9.25 - 5.58 + 8.12] = 11.79 \text{ N/mm}^2$$

   $$\sigma_b = \left[\frac{P}{A} + \frac{Pe}{Z} - \frac{M_g}{Z}\right] = [9.25 + 5.58 - 8.12] = 6.71 \text{ N/mm}^2$$

   At service load stage, the stresses at extreme fibres are computed as

   $$\sigma_t = \left[\frac{\eta P}{A} - \frac{\eta Pe}{Z} + \frac{M_g}{Z} + \frac{M_q}{Z}\right]$$

   $$= [0.8(9.25 - 5.58) + 8.12 + 4.57]$$

   $$= 15.62 \text{ N/mm}^2 < 20 \text{ N/mm}^2$$

   $$\sigma_b = \left[\frac{\eta P}{A} + \frac{\eta Pe}{Z} - \frac{M_g}{Z} - \frac{M_q}{Z}\right]$$

   $$= [0.8(9.25 + 5.58) - 8.12 - 4.47]$$

   $$= -0.8 \text{ N/mm}^2 \text{ (Negligible tension)}$$

2. Mid support section

   $P = 15000$ kN $\quad\quad \eta = 0.80$

   $e = 700$ mm $\quad\quad M_g = 13438$ kN·m

   $A = (1.62 \times 10^6)$ mm^2 $\quad\quad M_q = 1986$ kN·m

   $Z = (0.94 \times 10^9)$ mm^3

$$\left(\frac{P}{A}\right) = \left[\frac{15000 \times 10^3}{1.62 \times 10^6}\right] = 9.25 \text{ N/mm}^2$$

$$\left(\frac{Pe}{Z}\right) = \left[\frac{15000 \times 10^3 \times 700}{0.94 \times 10^9}\right] = 11.16 \text{ N/mm}^2$$

$$\left(\frac{M_g}{Z}\right) = \left[\frac{13438 \times 10^6}{0.94 \times 10^9}\right] = 14.29 \text{ N/mm}^2$$

$$\left(\frac{M_q}{Z}\right) = \left[\frac{1986 \times 10^6}{0.94 \times 10^9}\right] = 2.11 \text{ N/mm}^2$$

At transfer stage, the stresses at extreme fibres are computed as
$$\sigma_t = [9.25 + 11.16 - 14.29] = 6.12 \text{ N/mm}^2$$
$$\sigma_b = [9.25 - 11.16 + 14.29] = 12.38 \text{ N/mm}^2$$

At service load stage, the stresses at extreme fibres are computed as
$$\sigma_t = [0.8 (9.25 + 11.16) - 14.29 \; 2.11] = -0.072 \text{ N/mm}^2$$
$$\sigma_b = [0.8 (9.25 - 11.16) + 14.29 + 2.11] = 14.87 \text{ N/mm}^2$$

All the stresses are well within the maximum permissible limits of 20 N/mm² and no tensile stresses develop at transfer and service load stages.

(h) Check for Ultimate Flexural Strength

1. Centre of span section

A_p = 11340 mm² D_f = 300 mm
b = 2000 mm f_{ck} = 60 N/mm²
b_w = 300 mm f_p = 1862 N/mm²
d = 1350 mm M_u (Required) = 22195 kN·m

According to IRC: 18-2000

(i) Failure by yielding of steel
$$M_u = [0.9 \, d A_p f_p]$$
$$= [0.9 \times 1350 \times 11340 \times 1862]$$
$$= [25654 \times 10^6] \text{ N·mm}$$
$$= 25654 \text{ kN·m} > 22195 \text{ kN·m (Hence Safe)}$$

(ii) Failure by crushing of concrete
$$M_u = [0.176 \, b_w \, d^2 \, f_{ck} + 0.67 \times 0.8 \, (b - b_w)(d - 0.5 \, D_f) \, D_f \, f_{ck}]$$
$$= [(0.176 \times 300 \times 1350^2 \times 60] + [(0.67 \times 0.8)(2000 - 300)$$
$$(1350 - 0.5 \times 300)(300 \times 60)]$$
$$= (25454 \times 10^6) \text{ N·mm}$$
$$= 25454 \text{ kN·m} > 22195 \text{ kN·m (Hence safe)}$$

2. Mid support section

$A_p = 11340$ mm² $\quad D_f = 300$ mm
$b = 2000$ mm $\quad f_{ck} = 60$ N/mm²
$b_w = 300$ mm $\quad f_p = 1862$ N/mm²
$d = 1700$ mm $\quad M_u$ (Required) $= 25122$ kN·m

(i) Failure by yielding of steel
$$M_u = [0.9\, d\, A_p\, f_p]$$
$\quad\quad = [0.9 \times 1700 \times 11340 \times 1862]$
$\quad\quad = [32306 \times 10^6]$ N·mm
$\quad\quad = 32306$ kN·m > 25122 kN·m (Hence safe)

(ii) Failure by crushing of concrete
$$M_u = [0.176\, b_w\, d^2 f_{ck} + 0.67 \times 0.8\, (b - b_w)(d - 0.5 D_f) D_f\, f_{ck}]$$
$\quad\quad = [(0.176 \times 300 \times 1700^2 \times 60] + [(0.67 \times 0.8)(2000 - 300)$
$\quad\quad\quad\quad (1700 - 0.5 \times 300)(300 \times 60)]$
$\quad\quad = (34577 \times 10^6)$ N·mm
$\quad\quad = 34577$ kN·m > 25122 kN·m (Hence safe)

The ultimate flexural strength of centre of span and mid support sections are greater than the required design ultimate moment. Hence the design satisfies limit state of collapse.

(i) Check for ultimate shear strength
Design shear force $= V_u = 2930$ kN
According to IRC: 18-2000, the ultimate shear resistance of the support section uncracked in flexure is given by the relation,

$$V_{cw} = 0.67 b_w\, h \sqrt{f_t^2 + 0.8 f_{cp}\, f_t} + \eta P \sin\theta$$

here $b_w = 300$ mm
$\quad h = 2000$ mm

$f_t = 0.24 \sqrt{f_{ck}} = 0.24 \sqrt{60} = 1.85$ N/mm²

$$f_{cp} = \left(\frac{\eta P}{A}\right) = \left(\frac{0.8 \times 15000 \times 10^3}{1.62 \times 10^6}\right) = 7.4 \text{ N/mm}^2$$

$$\theta = \left(\frac{4e}{L}\right) = \left(\frac{4 \times 700}{50 \times 1000}\right) = 0.056$$

$$V_{cw} = \left[0.67 \times 300 \times 2000 \sqrt{1.85^2 + (0.8 \times 7.4 \times 1.85)}\right]$$
$$+ [0.8 \times 15000 \times 10^3 \times 0.056]$$
$\quad\quad = (2195 \times 10^3)$ N
$\quad\quad = 2195$ kN

Balance shear force = (2930 − 2195) = 735 kN
Using 12 mm diameter two legged stirrups, the spacing is calculated as

$$S_v = \left\{\frac{0.87 \times 415 \times 2 \times 113 \times 1900}{735 \times 10^3}\right\} = 211 \text{ mm}$$

Adopt 12 mm diameter two legged stirrups at a spacing of 200 mm centres at supports gradually increased to 300 mm towards the centre of span.

(j) **Supplementary Reinforcement:** Longitudinal reinforcements of not less than 0.18 per cent of the gross cross-sectional area are to be provided according to the specifications of IRC: 18-2000 to safeguard against shrinkage cracking in webs.

$$A_{st} = (0.0018 \times 1000 \times 300) = 540 \text{ mm}^2$$

Provide 12 mm diameter bars at 160 mm centres (A_{st} = 707 mm²).
The reinforcements provided at mid support and centre of span sections are shown in Fig. 9.20.

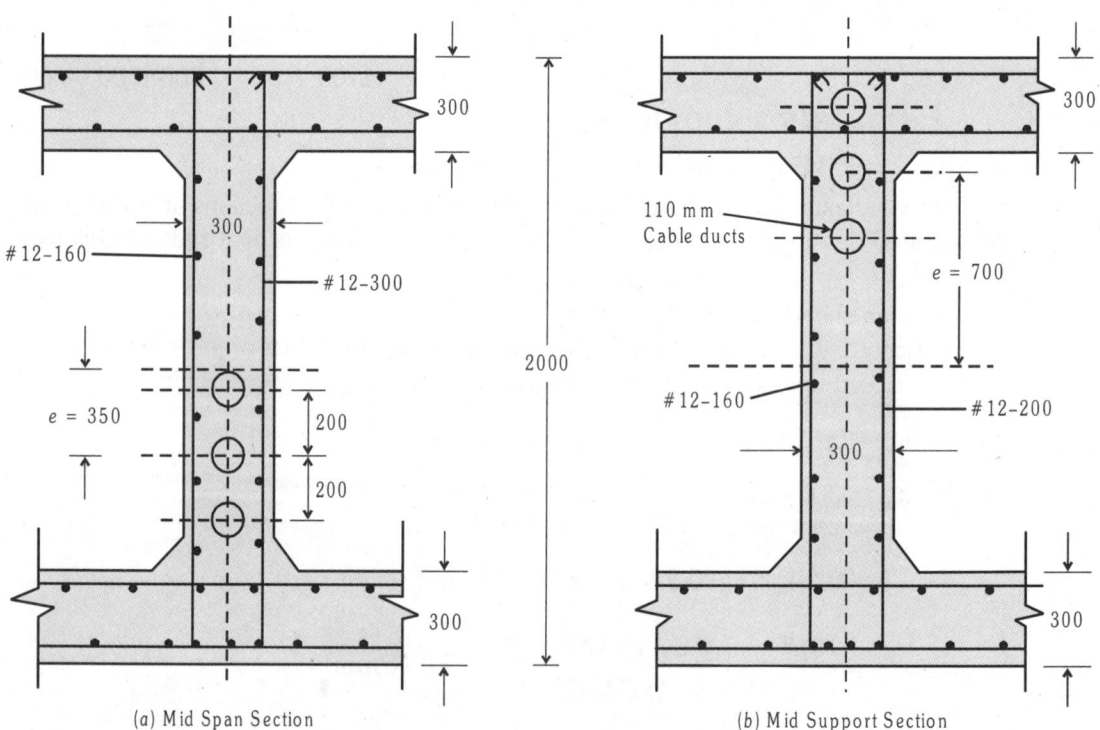

Fig. 9.20: Details of Reinforcements and Cables at Mid Span and Support Sections

(k) **Design of End Blocks:** Near the end supports where anchorages are located, the web thickness is increased to 600 mm to accommodate the anchorage bearing plates of size 400 mm by 400 mm in conjunction with Freyssinet 27K-15 type prestressing strands as shown in Fig. 9.21. The bursting tension is computed using the data in Table 4.6 (Table 8 of IRC: 18-2000). The bursting tension in the horizontal plane is computed using the parameters,

$2y_{po} = 400$ mm

$2y_o = 600$ mm

Ratio $\left(\dfrac{y_{po}}{y_o}\right) = \left(\dfrac{400}{600}\right) = 0.66$

$\left(\dfrac{F_{bst}}{P_k}\right) = 0.12$

Bursting tension = F_{bst} = (0.12 × 5000) = 600 kN
Using Fe-415 HYSD bars, anchorage zone reinforcement required is

$$A_{st} = \left(\dfrac{600 \times 10^3}{0.87 \times 415}\right) = 1662 \text{ mm}^2$$

Provide 16 mm diameter bars at 150 mm centres in the horizontal and vertical planes distributed in the region from $0.2y_o$ to $2y_o$ (60 mm to 600 mm) as shown in Fig. 9.21.

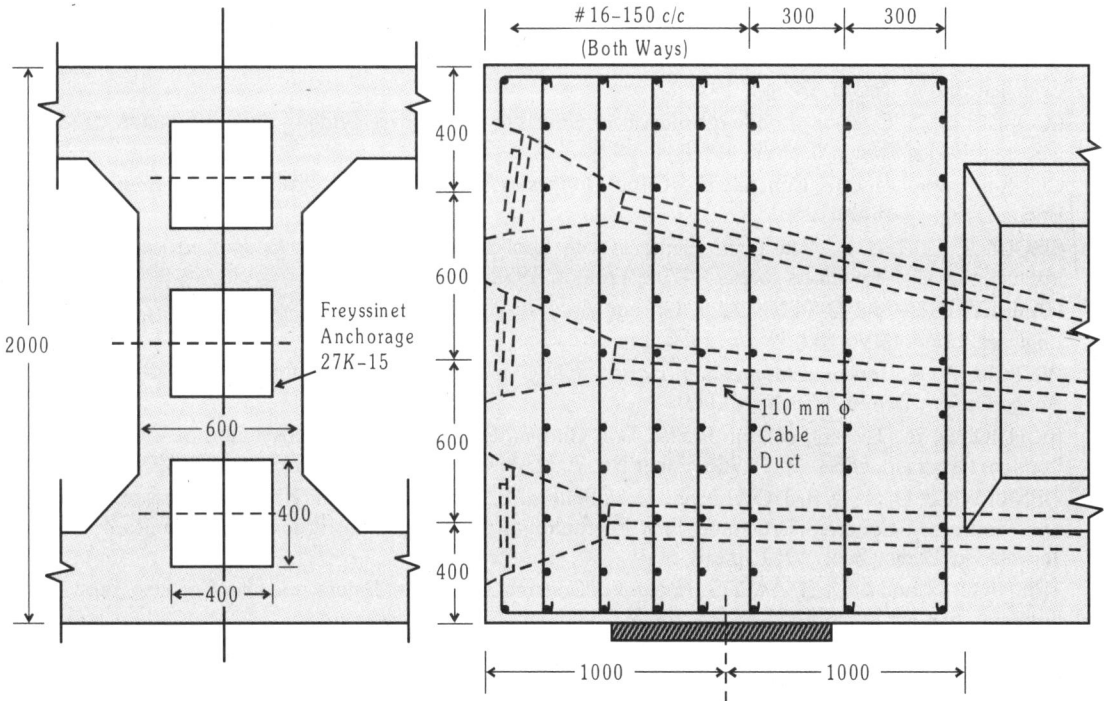

Fig. 9.21: Anchorage Zone Reinforcements in End Block

REFERENCES

1. FINISTERWALDER, U., Modern Designs for Prestressed Concrete Bridges. *Concrete and Constructional Engineering*, London, Vol. 60, No. 3, March 1965, pp. 99-103.
2. RAINA, V. K., *Concrete Bridge Practice, Analysis, Design and Economics*. Tata McGraw Hill Publishing Co, Ltd, New Delhi, 1991, pp. 186-200.
3. JACQUES FAUCHART., Panorama of the Execution Methods of the Prestressed Concrete Bridges now used in France, Proceedings of the Seminar on Problems of Prestressing, Indian National Group of the International Association for Bridge and Structural Engineering, Madras 1970, pp. I-131 to I-157.
4. RAIKAR, R. N., Outstanding Concrete Structures of India, Maharashtra India Chapter of American Concrete Institute. Proceedings at ACI Fall Convention, New York, 1984, p. 146.
5. RAIKAR, R. N., Outstanding Concrete Structures of India, Maharashtra India Chapter of American Concrete Institute. Proceedings at ACI Fall Convention, New York, p. 430.
6. SUBBA RAO, T. N., Long Span Prestressed Concrete Bridges in India, Proc. of The Seminar on Problems of Prestressing, Madras, 1970, pp. I-113 to I-130.
7. NAIDU, M. P. and C. SANKARALINGAM, Precast Segmental Construction of Sirsi Fly Over, Bangalore, Souvenir, IRC Diamond Jubilee Session, Highways Department, Tamil Nadu, January 2000, pp. 97-108.
8. JOHNSON VICTOR, D., *Essentials of Bridge Engineering* (V Edition). Oxford and IBH Publishing Co., New Delhi, 2001, pp. 369-373.
9. CHEUNG, Y. K., *Analysis of Box Girder Bridges by the Finite Strip Method*. ACI Publication SP-26 on Concrete Bridge Design, Detroit, 1971.
10. SCORDELIS, A. C. *et al.*, *Load Distribution in Concrete Box Girder Bridges*, ACI Publication SP-23 on Concrete Bridge Design, Detroit, 1969.
11. CHUK-h, E and DUDNIK, *Concrete Box Girder Bridges as Folded Plates*. ACI Publication SP-23 on Concrete Bridge Design, Detroit, 1969.
12. SISODIYA, R. G, *et al.*, New Finite Elements with Application to Box Girder Bridges, Proceedings of the Institution of Civil Engineers, paper 7479S, London 1972.
13. HAMBLY, E. C. and E. PENNELLS, Grillage Analysis Applied to Cellular Bridge Decks. *The Structural Engineer*, London, Vol. 53, No. 7, 1975, pp. 267-275.
14. WITECKI, A. A., *Torsional Moments in Horizontally Curved Bridges*. ACI Publication SP-23 on Concrete Bridge Design, Detroit, Michigan, 1969.
15. RICHMOND, B., Twisting of Thin Walled Box Girders, Proceedings of the Institution of Civil Engineers, London, Paper No. 6868, April 1966, Paper No. 7031, Jan. 1968, Paper No. 7174 and 7189, August 1969.
16. ZIENKIEWICZ, O. C., D. R. J. OWEN and D. V. PHILLIPS., Finite Element Method in Analysis of Reinforced and Prestressed Concrete Structures. First International Conference on Structural Mechanics in Reactor Technology, Berlin, Sept. 1971, paper M5/1.
17. TRIKHA, D. N. and A. D. EDWARDS, Analysis of Concrete Box Girder Before and after Cracking, Proc. of the Institution of Civil Engineers, London, Paper 7571, 1972.
18. CUSENS, A. R., *Box and Cellular Girder Bridges*, 'A State of the Art Survey'. ACI Publication SP: 26 on Concrete Bridge Design, Detroit, 1971.
19. LIM, P. K. *et al.*, Finite Element Analysis of Curved Box Girder Bridges, Conference on Developments in Bridge Design and Construction, Cardiff, 1971.
20. Guide Lines for Design and Construction of Segmental Bridges, Indian Roads Congress Publication IRC: SP: 65-2005, New Delhi, 2005.
21. FIB-CEB International Recommendations for the Design and Construction of Concrete Structures, Prague, 1970.
22. LEONHARDT, F., Shear and Torsion in Prestressed Concrete, Proceedings of the Sixth F.I.P Congress, Prauge, 1970.

23. CAMPBELL-ALLEN, D. and R. J. L. WEDGWOOD, Need for diaphragms in concrete box girders. *Journal of the Structural Division*, ASCE, March 1971.
24. SISODIYA, R. G., *et al.*, Diaphragms in single and double cell box girder bridges with varying angles of skew. *Journal of the American Concrete Institute*, July 1972.
25. KRISHNA RAJU, N. *Prestressed Concrete* (Fourth Edition). Tata McGraw Hill Publishers, New Delhi, 2007, pp. 423-471
26. KRISHNA RAJU, N., *Advanced Reinforced Concrete Design* (II-Edition), (IS: 456-2000), C.B.S. Publishers, New Delhi, 2005, p. 345.

EXERCISES

1. A cellular prestressed concrete box girder bridge deck is to be designed for a National Highway crossing. Design the bridge deck to suit the following data:
   Effective span of the bridge = 40 m
   Road width = 7.5 m
   Foot paths: 1 m wide on either side
   Wearing coat = 80 mm
   Type of loading: IRC Class AA tracked vehicle
   Grade of concrete: M-50
   Type of high tensile steel: H.T. strands of 15.2 mm diameter conforming to IS: 6006 with Freyssinet anchorages.
   Type of supplementary reinforcement: Fe-415 HYSD Bars
   Design the bridge deck conforming to the Indian Standard IRC: 6, IRC: 18 and 21 codes. Sketch the details of cable profile at salient sections and the details of reinforcements in the anchorage zone.

2. A prestressed concrete multi cell continuous box girder bridge deck is proposed or a National Highway crossing of a river 120 mm wide. Design the bridge deck continuous over two spans of 60 m each to suit the following data:
   Road width = 15 m
   Foot paths: 1 m on either side
   Wearing coat: 100 mm
   Type of loading: IRC Class AA tracked vehicle
   Grade of concrete: M-60
   High tensile steel strands conforming to IS: 6006 in conjunction with Freyssinet K-Range anchorages are available for use. Fe-415 HYSD bars are available for use in deck slab and as supplementary reinforcement in webs. It is proposed to have box cells of 2 m by 2 m vents with 300 mm thick cell walls.
   Design the bridge deck conforming to the Indian Standards IRC: 6-2000, IRC: 18 and IRC: 21-2000. Sketch the details of cable profile along the span and details of reinforcements at critical sections and the anchorage zone.

10

PRESTRESSED CONCRETE CABLE STAYED BRIDGES

10.1 EVOLUTION OF CABLE STAYED BRIDGES

The rebuilding of bridges destroyed during the World War II in Germany together with the progress achieved in steel construction technology such as cables of superior mechanical properties and concrete of high strength paved the way for rapid development of cable stayed bridges during the second half of the 20th century. Cable stayed structural system proved to be economically, structurally and aesthetically superior to the conventional steel suspension bridges for the range of medium to long spans.

Basically the following salient factors helped for the successful development of cable stayed bridges:

1. The development of high strength steels with ultimate tensile strength in the range of 1200 to 1800 N/mm^2 in the form of wires and strands and high strength and high performance concrete of desirable structural properties.
2. The development of orthotropic steel and concrete bridge decks.
3. The development of methods of structural analysis of statically indeterminate structures with high redundancy using electronic computers.
4. Past experience of bridge construction using basic elements of cable stayed bridges.
5. Development of techniques of extradosed prestressed concrete box girder decks supported by cable stays for rapid construction.
6. The ability to analyze complex structures through model studies for aerodynamic stability.

In 1952, Leonhardt designed a cable stayed foot bridge across the river Rhine in Dusseldorf, Germany. However the first modern cable stayed bridge built in Sweeden was the Stromsund bridge[1] designed by Dischinger and constructed in the year 1955. Subsequently more than 200 major bridges of this type with steel or concrete decks have been built or are under design or construction throughout the world. A comprehensive list of prominent cable stayed bridges constructed in various countries and their salient features are compiled in Table 10.1[2].

Table 10.1: World's Prominent Cable Stayed Bridges

S. No.	Name of the Bridge	Material of Deck	Span (m)	Height of Tower (m)	Year
1.	Rafael Urdaneta (Venezuela)	Concrete	235	68	1962
2.	Dnepr Bridge (USSR)	Concrete	144	30	1963
3.	Knie (Germany)	Steel	320	114	1969
4.	Duisburg (Germany)	Steel	350	50	1970
5.	Wadikuf (Libiya)	Concrete	282	50	1972
6.	Mesapatomia (Argentina)	Concrete	340	97	1972
7.	Westgate (Australia)	Steel	336	46	1974
8.	Kholbrand (Germany)	Steel	325	98	1974
9.	Saint-Nazaire (France)	Steel	404	67	1975
10.	Brazo-Largo (Argentina)	Steel	330	48	1976
11.	Zarate (Argentina)	Steel	330	67	1975
12.	Brotonne (France)	Concrete	320	65	1975
13.	Luling (USA)	Steel	376	75	1980
14.	Dames Point (USA)	Concrete	396	92	1995
	Indian Cable Stayed Bridges				
1.	Haradwar (U.P.)	Concrete	130	20	1990
2.	Jogighopa (Assam)	Concrete	286	50	1995
3.	Akkar (Sikkim)	Concrete	154	56	1995
4.	Vidyasagar Sethu (Kolkata)	Concrete and Steel	452	100	1995
5.	Vivekananda Tollway Bridge (Kolkata)	Concrete and Steel	110	40	2008
	Longest and Highest Cable Stayed Bridges				
1.	Messina Straights (Italy)	Steel	1800	–	1982
2.	Normandie Bridge (France)	Concrete and Steel	624	–	1996
3.	Millau Bridge (France)	Concrete and Steel	400	300	2004
4.	Sutong Bridge (China)	Concrete and Steel	1088	–	2008

10.2 ADAVANTAGES OF CABLE STAYED BRIDGE DECKS

The dawn of 21st century has witnessed revolutionary developments in the analysis, design and construction of long span bridges using prestressed concrete box girders in conjunction with steel strand cable stays. The biggest advantage of cable stayed bridge is the simplicity of its form and accessibility of the deck and stays for periodic inspection. In case of damage the cable stays are easily replaceable by design.

It is well established that cable stayed bridge decks are aerodynamically superior in comparison with suspension bridges. The multiple cable stay system ensures additional system damping of large number of cables and facilitates the use of shallower depth of deck. Aerodynamic studies of cable stayed bridge decks has resulted in the simplification of the deck geometry of modern bridges.

According to Raina[1] the construction of bridge decks with spans up to 400 m adopting concrete decks (even longer spans with steel or orthotropic decks) and using cable stay techniques, derived from prestressing that enable simple installation and the eventual replacement of cable stays, have pushed the span limit beyond 500 m, exclusively reserved for suspension bridges.

At present, the world's tallest and longest cable stayed bridge is located in France. The French bridge located near the town of Millau is considered to be an engineering feat since some of the bridge pillars rise gracefully to a height exceeding 300 m in the valley with spans ranging from 60 to 400 m.

10.3 STRUCTURAL COMPONENTS OF CABLE STAYED BRIDGES

The salient structural components of a typical cable stayed bridge are

(a) Towers or pylons
(b) Deck system
(c) Cable system supporting the deck

Figure 10.1 shows the structural components of a typical cable stayed bridge. The deck comprises of reinforced or prestressed concrete or steel or orthotropic system with longitudinal and cross girders. Box girders are also preferred for longer spans. The high tensile cables supporting the deck system may be in single or double planes with their ends anchored to the deck on either side of the pylon. The cables pass over saddles located in the tower.

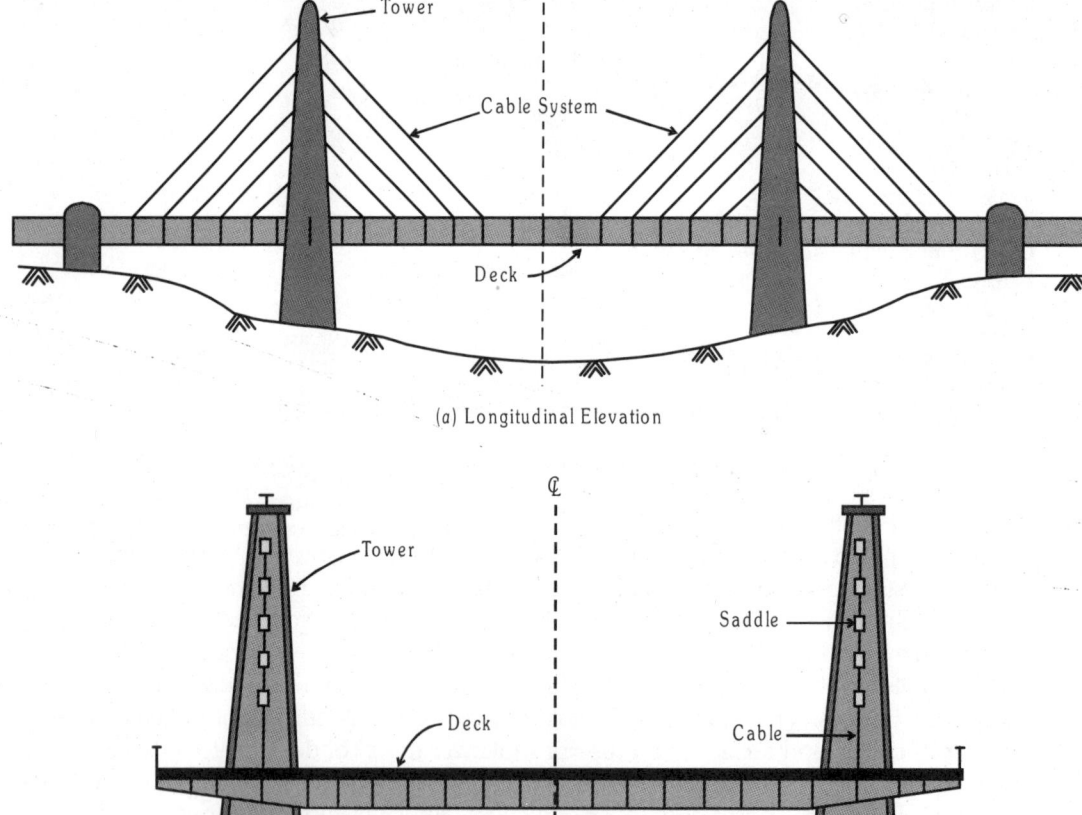

Fig. 10.1: Structural Components of a Typical Cable Stayed Bridge

Prestressed Concrete Cable Stayed Bridges

The live load on the bridge deck is transmitted as tension to the cable stays, which in turn transfer the load to the tower or pylons as compression. In comparison with the traditional suspension bridges, cable stayed bridges require significantly lesser quantity of cable steel resulting in the reduction of overall costs of the bridge.

10.4 TOWERS OR PYLONS

The principal supporting structural element in a cable stayed bridge is the tower or the pylon which transmits the dead and live loads of the bridge deck to the foundations. Pylons are of different configurations and are designed to accommodate different types of cable arrangements. The type of tower is also influenced by bridge site conditions, design features, aesthetic and economical considerations. Generally the arrangement of the cable stays (single or multi layer) determines the design of both the deck and pylons.

The towers can be arranged to support one axial layer of cable stays or two lateral layers of cable stays. The arrangement of single and double layer cable stays are shown in Figs. 10.2 and 10.3 respectively. Generally the towers are massive and are made of reinforced concrete supported on pile foundations.

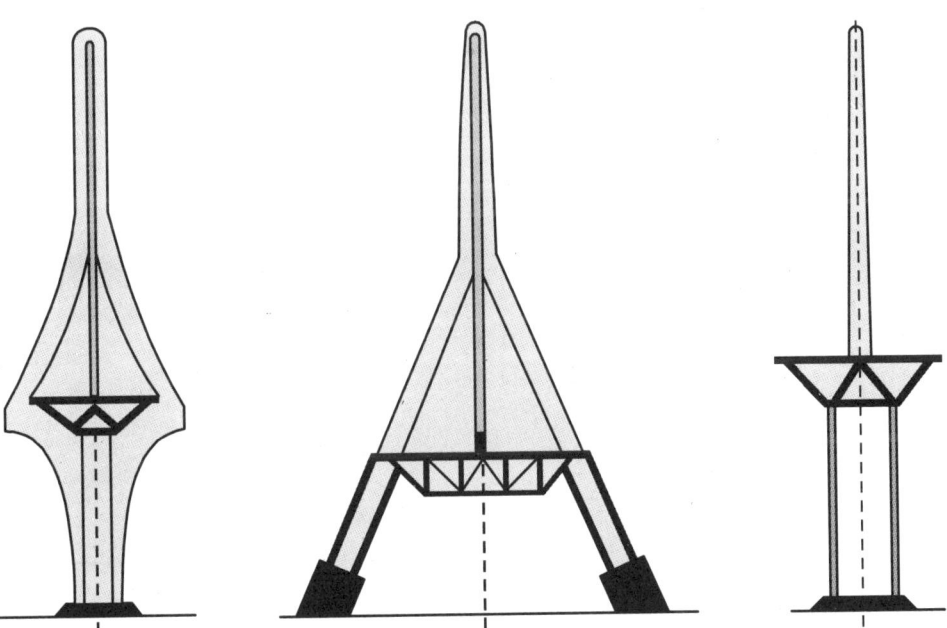

Fig. 10.2: Transverse Arrangement of Pylons (One Axial Layer of Stays)

The forces from the cable stays are transferred to the pylon by any of the following different arrangements:

1. The cables pass over a saddle at the top of the pylon as shown in Fig. 10.4.
2. The cable stays fixed at the top of the tower may cross each other inside the pylon as shown in Fig. 10.5.
3. A relay device assembly fixed to the top of the tower connecting the upper anchorages of associated cable stays as shown in Figs. 10.6 and 10.7.

Prestressed Concrete Bridges

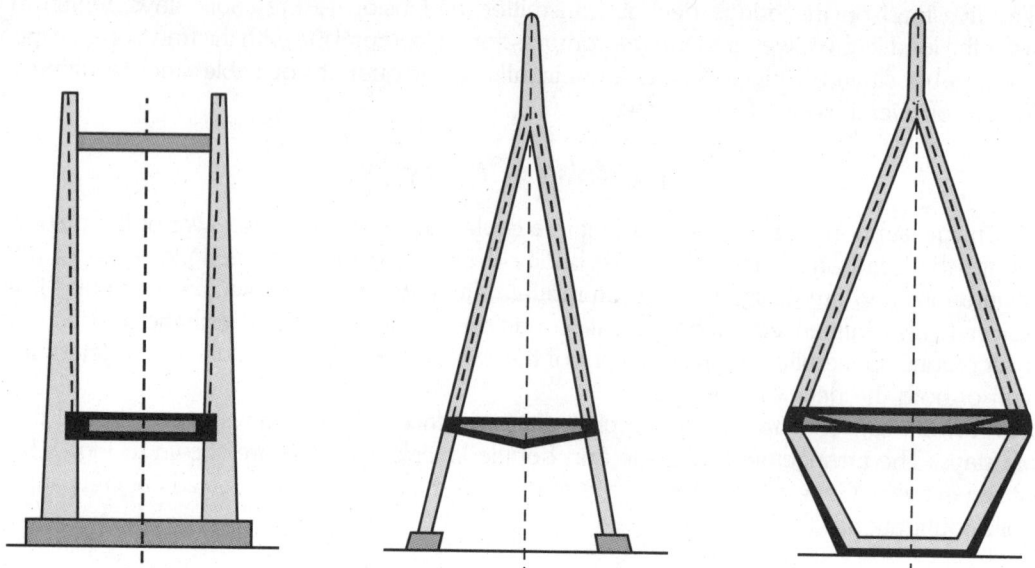

Fig. 10.3: Transverse Arrangement of Pylons (Two Lateral Layers of Stays)

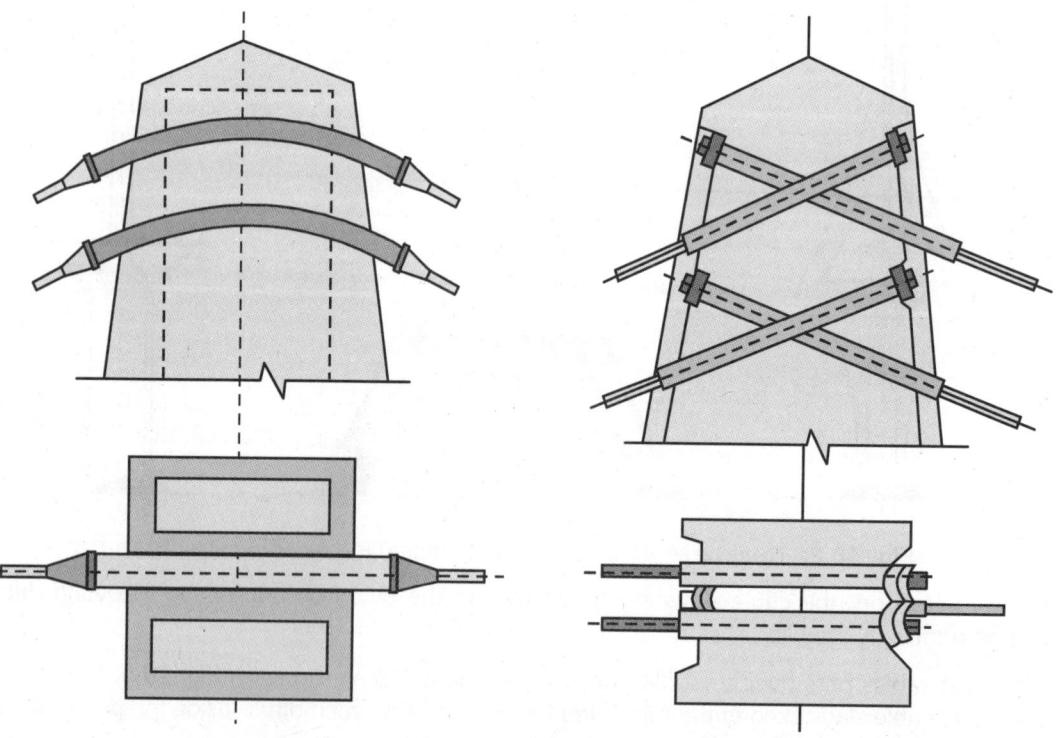

Fig. 10.4: Saddle Housed in Tower **Fig. 10.5:** Crossing of Stay Cables at the Tower

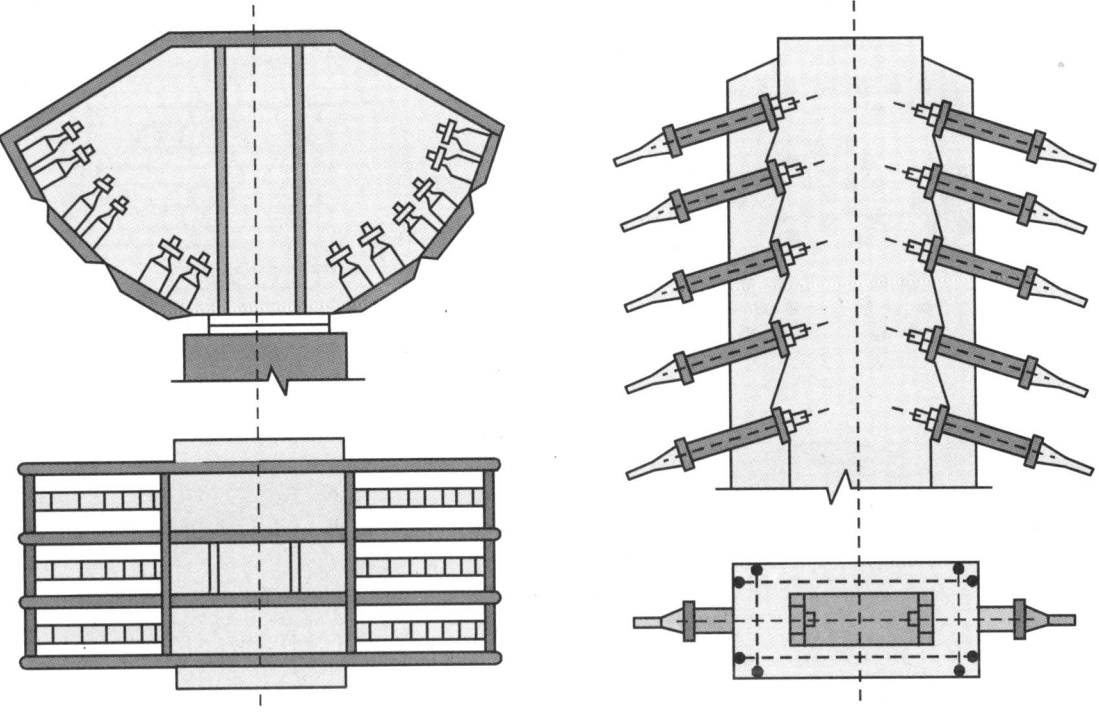

Fig. 10.6: Relay Device (Grouped Anchorages)   **Fig. 10.7:** Relay Device (Distributed Anchorages)

10.5 CABLE STAYS

High tensile steel of different types having an ultimate tensile strength in the range of 1500 to 2000 N/mm^2 is generally used in cable stays. The four main types of cable stays commonly used are

1. Locked coil cable
2. Twisted cable
3. Parallel wires
4. Parallel strands

Figure 10.8 shows the typical cross-sections of different types of cables. Locked coil system generally used in suspension bridges consists of stranded cables with an outer layer of Z-shaped wires interlaced to form a locked unit and this is the first type used in the construction of cable stayed bridges. However the stress variations being very high, this type was found to be unsatisfactory for cable stayed bridges. This problem was overcome by the development of parallel wire cable stays using prestressing techniques. The period from 1960 to 1980 heralded the use of seven wire strands for prestressing and the research investigations on cable stays resulted in the use of strands with higher ultimate tensile strength (1860 N/mm^2) and superior fatigue resistance.

The use of larger diameter high tensile strands resulted in the development of Freyssinet cable stays comprising a bundle of parallel strands of 15 mm diameter enclosed in a polyethylene tube which allows the threading and grouting of the cables. A steel wire spiral spacer inside the polyethylene tube ensures a proper grout cover around the bundle of strands as shown in Fig. 10.9.

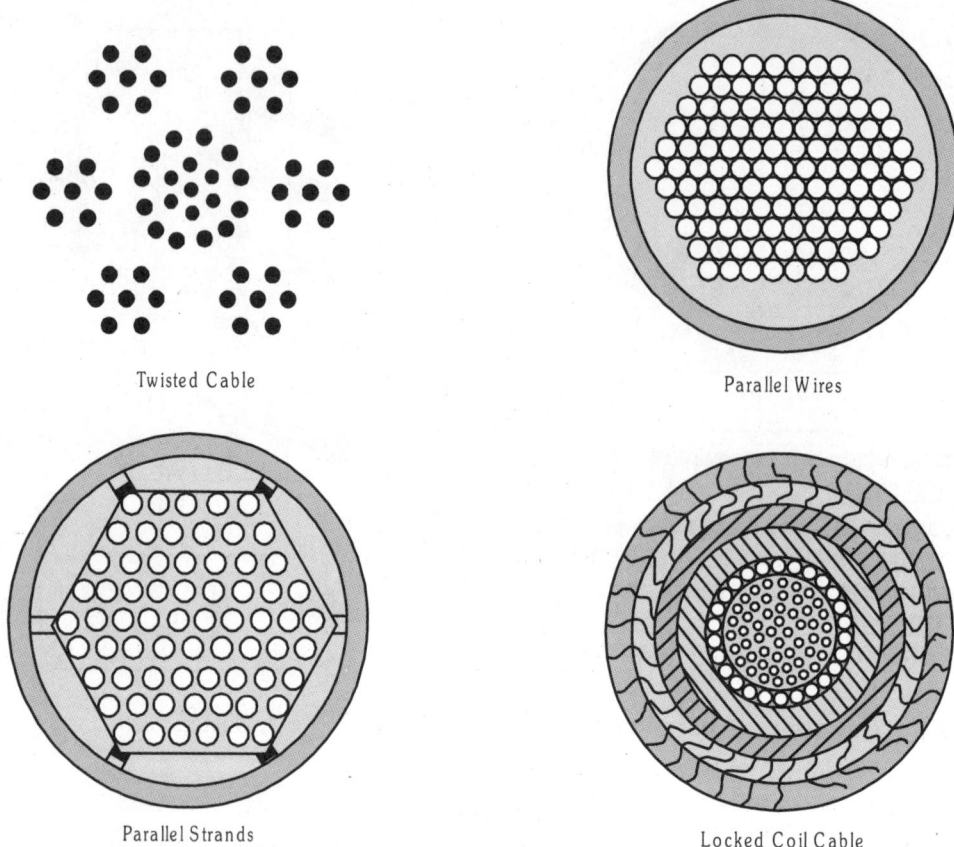

Fig. 10.8: Different Types of Cable Stays

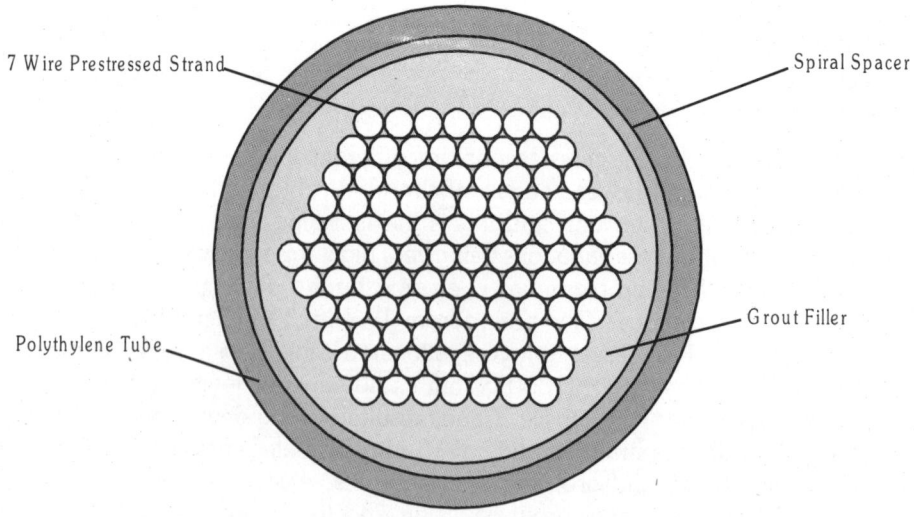

Fig. 10.9: Section of Freyssinet Cable Stay

Prestressed Concrete Cable Stayed Bridges

In Switzerland, B.B.R.-HIAM (High Amplitude) parallel wire cables with a patented anchorage system was developed by the combined efforts of the famous Swiss firm B.B.R. and the German consulting firm Leonhardt and Andra[3, 4]. The B.B.R.-HIAM stay anchorages consist of parallel wire bundles. The wires are provided with a variant B.B.R. button head anchorage with special features to improve their fatigue strength. The Freyssinet and the HIAM-B.B.R. anchorages do not reduce either the static or fatigue strength of the wires. Research investigations have conclusively shown that in both static and dynamic tests, the tendons invariably fail in free length of the wires. The static and dynamic properties of the stay cables are, therefore, determined by the mechanical properties of the individual wires.

The strand used for the stay cables must comply with the specifications prescribed in (a) BS: 5896-1980, (b) ASTM-A-416-80, (c) EURONORM 138-79 or any other equivalent standard. The strands should withstand a stress range of 195 N/mm^2 with the upper stress limit of 0.8 times the breaking stress with a fatigue life of not less than 2 million cycles. The standard specifications of strands used for cable stays are compiled in Table 10.2. Table 10.3 lists the ultimate load characteristics of various types of Freyssinet cable stays of different sizes.

Table 10.2: Standard Specifications of Strands used in Cable Stays

Standard	A.S.T.M. A 416-80	EURONORM 138-79	BS: 5896-1980
Nominal Diameter (mm)	15.24	15.70	15.70
Nominal Tensile Strength (N/mm^2)	1862	1770	1770
Nominal Steel Area (mm^2)	140	150	150
Nominal Weight (kg/m)	1.102	1.180	1.180
Characteristic Breaking Load (kN)	260.7	265.0	265.0
Characteristic 0.1% Proof Load (kN)	–	225.0	225.0
Load at 1 per cent Elongation (kN)	221.5	233.0	233.0
Minimum Elongation at Maximum Load (%)	3.5	3.5	3.5
Length-L (mm)	610	500	500

Table 10.3: Ultimate Load Characteristics of Freyssinet Cable Stays

Type of Strand		Single Strand	19H15	27H15	**37H15**	48H15	**61H15**	75H15	**91H15**
15.24 mm	A	140	2660	3780	5180	6720	8540	10500	12740
Grade 270	Af_{pu}	260.7	4953	7039	9646	12514	15902	19552	23724
A-426	g	1.102	20.94	29.75	40.77	52.90	67.22	82.65	100.28
15.7 mm Super	A	150	2050	4050	5550	7200	9150	11250	13650
Strand to	Af_{pu}	265	5035	7155	9805	12700	16165	19875	24115
Euronorm EU 138	g	1.19	22.42	31.86	43.66	56.64	71.98	88.50	107.28
15.2 mm Class III	A	139	2641	3753	5143	6672	8479	10425	12649
French	Af_{pu}	252.1	4790	6807	9328	12100	15378	18907	22941
Circular N° 73-175	g	1.091	20.73	29.46	40.37	52.37	66.55	81.82	107.28

A—Nominal cross-section (mm^2); Af_{pu} = Ulitimate tensile force in strand (kN)
g—Nominal weight (kg/m)
* Recommended standard sizes are indicated in bold characters.
The other sizes can be envisaged when quantities justify.

The cable stay generally consists of 3 types of zones as shown in Fig. 10.10.

1. The free length zone is the main portion formed by the bundle of parallel wires or strands depending upon the desired capacity.
2. The transmission zone comprises essentially an anchorage tube and form work tube permitting sliding to allow for adjusting, restressing or destressing if necessary both during construction and during the service life of the bridge.
3. Anchorage zone located at the ends of the stay cables where they are anchored against the pylon and the deck. The anchorages should provide perfect distribution of strands eliminating contact between the individual strands anchoring with jaws specially designed for their fatigue endurance.

The stay cables are protected by using either a rigid type or flexible type protection. In the rigid type protection, the strands in a steel or polyethylene sheath are injected with cement grout. In the flexible type protection, the strands housed in a high density polyethylene sheath are grouted with epoxy pitch, grease or wax. Durability of the stay cables under aggressive environmental conditions is a major factor. A more effective protective system comprises of two successive barriers. The first one is the polyethylene tube sheathing and the second barrier being the cement grout injected into the annular space around the wire bundle.

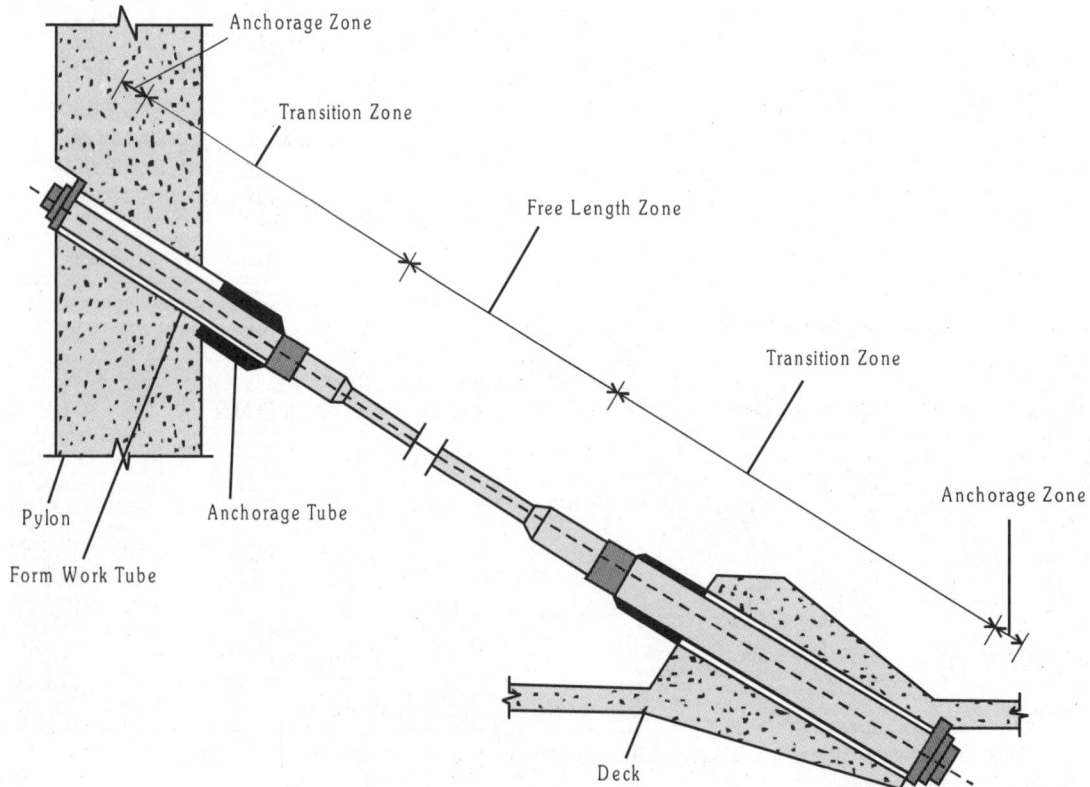

Fig. 10.10: General Layout of Freyssinet Cable Stay

10.6 LONGITUDINAL CABLE PROFILES

The arrangement of stay cable profiles in the longitudinal direction is influenced by several factors such as clear span, height of pylons, width of carriage way, spacing of towers and the level of approach roads. Different types of cable configurations have been used for cable stayed bridges built during the last several decades. For shorter span lengths a single forestay and a backstay with a single pylon is sufficient to satisfy the loading requirements. A typical example of this type used in Severin Bridge at Cologne, Germany is shown in Fig. 10.11.

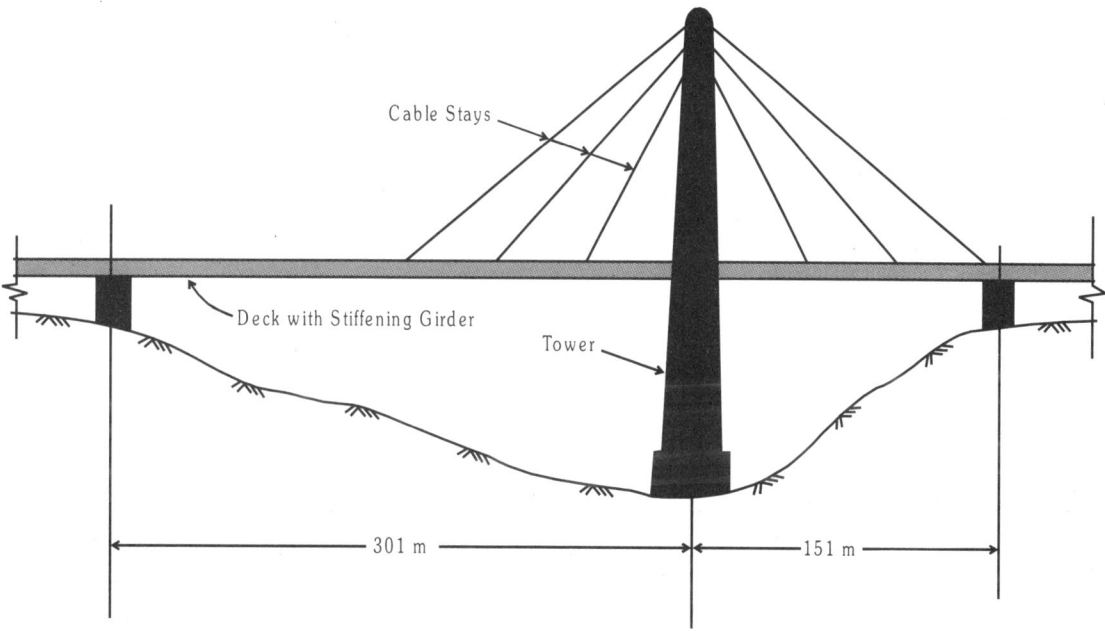

Fig. 10.11: Severin Bridge at Lologne, Germany

A brief survey indicates that basically there are four major types of cable configurations commonly used for cable stayed bridge decks. They are grouped as, (1) Fan type, (2) Harp type, (3) Mixed type, (4) Star type. The four types are illustrated in Fig. 10.12. The selection of the type depends upon the span, width of carriage way, type of loading, height of pylons, economy and aesthetic aspects.

The fan type is aesthetically superior and as a rule, the most economical choice for a tower of slenderness ratio $(h/L) \leq 0.3$. For an equal tower height, the average inclination of the cable stay is lower. However the cable stays are longer and converge towards a single point at the top of the tower posing problems of anchoring arrangement and replacement of damaged cable stay is difficult. To overcome this problem, a mixed type solution involving the fan and harp type is adopted by spreading the anchorages along the depth of the tower according to their dimensions closer to the top of the mast.

In the double plane system, the harp type is preferred as it minimizes the intersection of cables when viewed from an oblique angle. From the motorist's point of view, the harp system is more attractive with the cable connections distributed throughout the height of the tower and hence

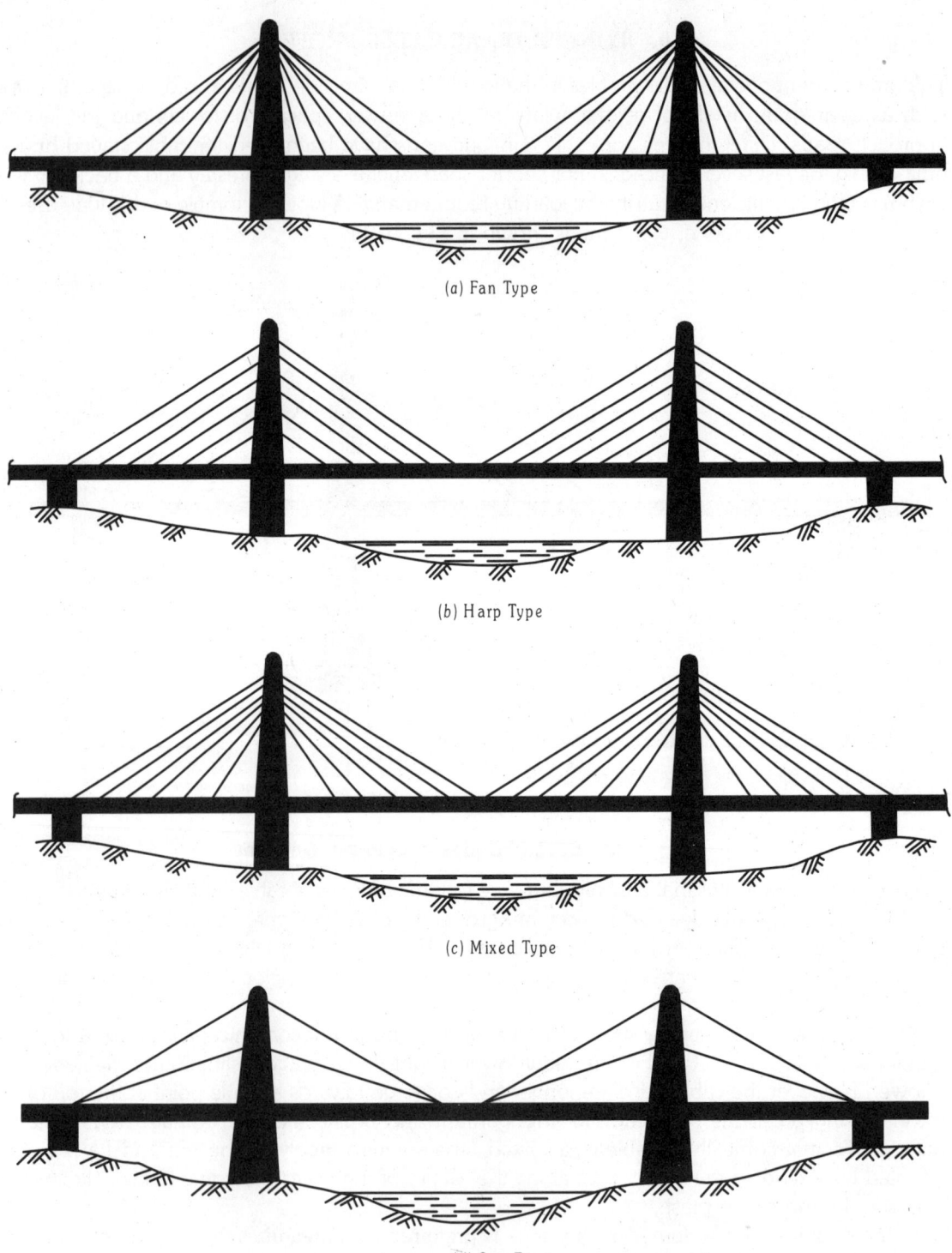

(a) Fan Type

(b) Harp Type

(c) Mixed Type

(d) Star Type

Fig. 10.12: Types of Longitudinal Cable Arrangements

results in an efficient tower design in comparison with the fan type. The structural behaviour of the pylon is influenced by the type of cable system. The harp type induces more bending moments while the fan type increases buckling problems in the tower.

The mixed type where the anchorages of the cables are distributed over the height of tower represents a compromise between the extremes of fan and the harp systems and it is ideally suited when it is difficult to accommodate all the cables at the top of the tower. In the case of bridges with shorter span and lighter loads, the star system with fewer cables may be preferred due to its unique aesthetic appearance.

10.7 SUPERIORITY OF CABLE STAYED BRIDGES OVER CONVENTIONAL BRIDGES

Comparative analysis conducted by Leonhardt[4] has conclusively proved the economical advantages of cable stayed bridges. Cable stayed bridge is an innovative structure and is preferred to conventional steel suspension bridges for long spans mainly due to the reduction in bending moments in the stiffening girder resulting in smaller sections of the girders leading to economy in overall costs. Figrue 10.13 shows the comparison of bending moments in a five span conventional continuous girder system with that of cable stayed girder for the same span. The ratio of maximum bending moment in the cable stayed girder is nearly one tenth of that of the conventional continuous girder system. In addition, the moments can be controlled to make them more uniformly distributed along the girder length resulting in efficient material utilization even with a very low depth to span ratio of 1/90. Cable stayed bridges with their sleek appearance are aesthetically superior in comparison with steel suspension bridges.

10.8 BASIC PRINCIPLES OF STRUCTURAL ANALYSIS

1. **Approximate Methods:** Cable stayed bridge system can be considered as a structure with a high degree of statically indeterminacy requiring complex analysis for reasonably accurate solutions. The deflections of the basic system under service loads may be determined by the application of the classical theory of structures by neglecting the deformations of the system when formulating the equilibrium conditions.

    In the case of statically determinate basic structure, the resulting equations are linear involving loads and internal forces and linear superposition is applicable for the internal forces caused by different types of loads. If Hooke's law is assumed to be valid, linear superposition applies also to the displacements leading to the determination of the stresses in cable stayed bridge system.

    Broadly the design process for a cable stayed bridge with a predetermined geometrical lay out may be examined under the following stages:

    (a) Preliminary dimensions are assumed for the various structural elements such as the deck, cable stays and towers and sectional properties are evaluated.
    (b) The assumed sectional system is analyzed by any of the statical methods of analysis. Stresses and displacements under the given load system are determined and compared with the maximum permissible unit stresses and displacement/span ratios prescribed in the codal specifications.
    (c) If required, a new set of sectional properties are chosen to satisfy the requirements of the specifications.

224   **Prestressed Concrete Bridges**

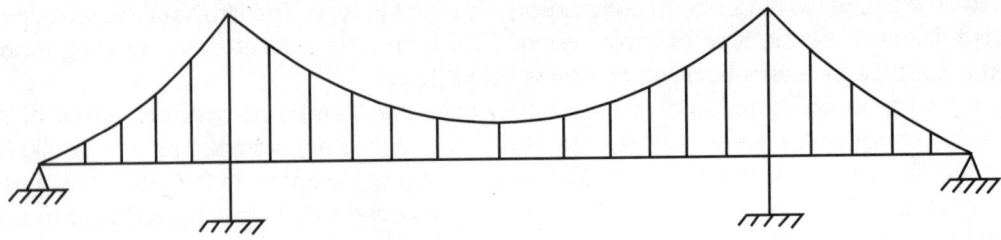

(a) Suspension Bridge

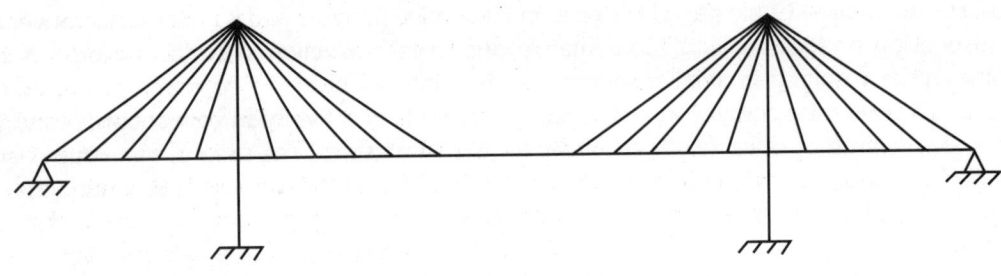

(c) Cable Stayed Bridge

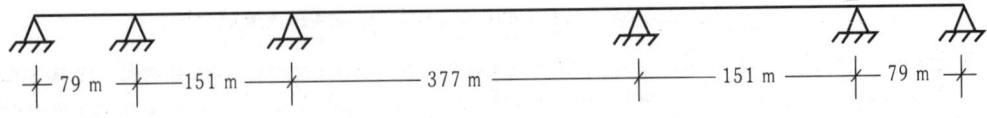

(c) Continuous Girder System

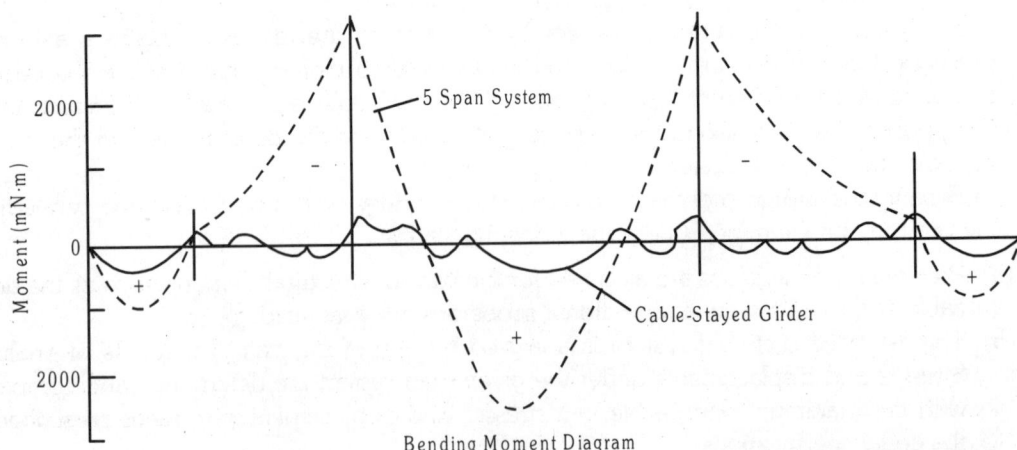

Bending Moment Diagram

Fig. 10.13: Variation of Moments in Cable Stayed and Continuous Bridge Systems

The above stages are repeated until compatibility is achieved between the sectional properties and the actual stresses and displacements and the associated codified values. Several general methods for the linear structural analysis of a cable stayed bridge system are available. These methods do not require extensive use of computers and are termed as 'Classical Methods'.

2. **Rigorous Methods of Structural Analysis:** The modern cable stayed bridge can be considered as a structure with a high degree of statical indeterminacy. The stiffening girder deck behaves as a continuous beam supported elastically at the points of cable attachments. The analysis of cable stayed bridge system requires the use of computer to achieve satisfactory results. The use of computer is primarily in the analysis rather than in design application.

Influence diagrams for cable forces, bending moments and shear forces in the stiffening girder, reactions in towers and piers are generated by using computer programmes. The rapid solution of various parametric efforts and loadings facilitates in achieving a reasonably efficient design using the computer. Probably the most significant problem is the determination of optimum section of the stiffening girder and stay cable size and its configuration. The solution to the complex problem is simplified by assuming a linear elastic system which may be analyzed using the standard stiffness or flexibility method[5, 6].

Based on this approach, several general computer programmes are available such as FRAN, STRESS, STRUDL. The non linear behaviour of the cable stayed bridge system is accounted by introducing the concept of a straight chord member with a modified or ideal modulus of elasticity substituted for the actual cable member. The inclusion of this concept allows for the application of a plane frame computer programme properly adopted to account for the non linearity by an iteration procedure. However for rigorous analysis and design of cable stayed bridges, the reader may refer for the specialist literature and monographs of Troitsky[7], Smith[8], Tang[10] and Shreier[11].

10.9 STRUCTURAL ANALYSIS OF CABLE STAYED BRIDGE

The computations of the following basic items are essential for the approximate structural analysis of the cable stayed bridge system.

1. Cable configuration and optimum inclination of cables
2. Height of pylon and length of panels
3. Forces in cables
4. Approximate self weight of stiffening girder
5. Self weight of cables
6. Degree of redundancy

The approximate analysis of a cable stayed bridge deck with a predetermined geometrical lay out may be classified into the following stages:

(a) Assumption of preliminary set of sectional properties of the structural components of the cable stayed bridge
(b) Analysis of the assumed system by any one of the statical methods of analysis
(c) Determination of stresses and displacements under the given load system and comparison with the maximum permissible codified displacements/span ratios.
(d) Based on results, new set of sectional properties are chosen and the analysis repeated to satisfy the specification requirements.

The iterative procedure is repeated until satisfactory convergence is obtained between the sectional properties assumed and the required agreement with the codified specifications.

The classical methods being simpler are suitable for preliminary structural analysis. However the disadvantage of this method is that a linear elastic behaviour of the bridge system although not true is to be assumed.

1. Determination of Optimum Inclination of Cables

The stiffness of the bridge deck system comprising the stiffening girder, deck and cables is significantly influenced by the height of the tower. The inclination of the cables is an important parameter influencing the stresses in the cable. As the angle of inclination of the cable with respect to the stiffening girder increases, the stresses in the cable decrease, as does the required cross-section of the tower. However, as the height of the tower increases, the length of the cables and their axial deformations also increase together with the amount of metal in the cables.

Based on a simplified bridge system hinged at the locations of the cable connections to the stiffening girder, it is possible to derive empirical relations between the optimum amount of material and the inclination of the cables. The following notations are used in deriving the relations:

A_n = Cross-sectional area of cable (mm²)
a = Length of each panel
E = Modulus of elasticity of the cable material (kN/m²)
F_n = Force in the cable (kN)
f = Permissible stress in the cable (kN/m²)
L_n = Length of the cable (m)
α_n = The angle of inclination of cables
n = The corresponding number of panels
$na = L_n \cos \alpha_n$ = Horizontal projection of cable length
P_n = Vertical component of the cable force (kN)
W = Weight of the cable (kN)
y = Specific weight of the cable material (kN/m³)

Referring to Fig. 10.14, the weight of the cable is computed as

$$W = A_n L_n y \qquad \ldots(10.1)$$

and

$$A_n = \left(\frac{F_n}{f}\right), \text{ also } L_n = \left(\frac{na}{\cos \alpha_n}\right)$$

$$F_n = \left(\frac{P_n}{\sin \alpha_n}\right)$$

∴

$$A_n = \left(\frac{P_n}{f \sin \alpha_n}\right) \qquad \ldots(10.2)$$

By substituting Eq. (10.1) in Eq (10.2), we have the relation,

$$W = \left[\frac{P_n n a y}{f \sin \alpha_n \cos \alpha_n}\right]$$

Assuming $P_n = 1$ and designating $(nay/f) = C$, we have the relation

$$W = \left[\frac{C}{\sin\alpha_n \cos\alpha_n}\right] \qquad \ldots(10.3)$$

Hence the weight of the cable is a function of $[1/(\sin\alpha_n \cos\alpha_n)]$.

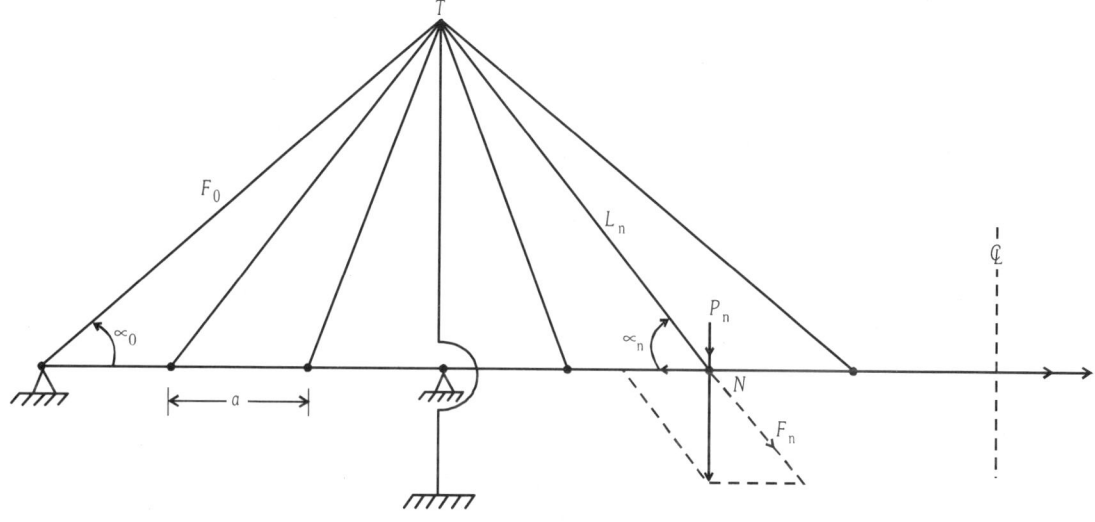

(a) Forces at joint-N

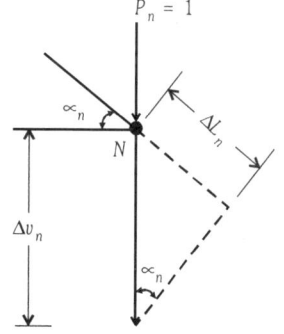

(b) Vertical deflection of joint due to elongation of cable ΔL_n

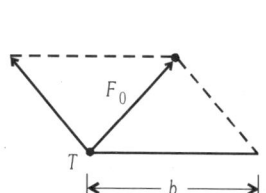

(c) Displacement of top of tower-T

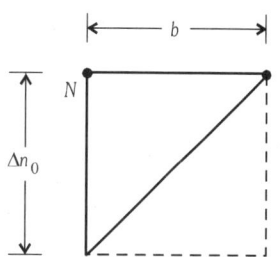

(d) Vertical deflection of joint N due to elongation of cable ΔL_n and displacement of the top of the tower b under load $P_n = 1$

Fig. 10.14: Analysis of Simplified Cable Stayed Bridge System

The force in the cable P_n due to the load $P_n = 1$ at the joint N is expressed as

$$F_n = \left[\frac{P_n}{\sin\alpha_n}\right] \qquad \ldots(10.4)$$

The corresponding elongation of the cable is expressed as

$$\Delta L_n = \left[\frac{P_n L_n}{EA_n \sin\alpha_n}\right] \qquad \ldots(10.5)$$

The force in the upper cable F_o transferred by the tower from the cable is

$$F_o = \left[\frac{F_n \cos\alpha_n}{\cos\alpha_n}\right] = \left[\frac{P_n \cot\alpha_n}{\cos\alpha_n}\right] \qquad \ldots(10.6)$$

The corresponding elongation of the cable F_o is

$$\Delta L_o = \left[\frac{P_n L_o \cot\alpha_n}{EA_o \cos\alpha_o}\right] \qquad \ldots(10.7)$$

and the corresponding displacement of the top of the tower is

$$b = \left(\frac{\Delta L_o}{\cos\alpha_o}\right) \qquad \ldots(10.8)$$

The vertical deflection of the joint N due to the elongation Δv_n of the cable is expressed by the relation

$$\Delta v_n = \left[\frac{\Delta L_n}{\sin\alpha_n}\right] = \left[\frac{P_n L_n}{EA_n \sin^2\alpha_n}\right] \qquad \ldots(10.9)$$

The vertical deflection of the joint N due to the elongation L_o of the cable and displacement b of the top of the tower due to the load P_n at the joint N is

$$\Delta n_o = b \cot\alpha_n = \left(\frac{\Delta L_o \cot\alpha_n}{\cos\alpha_n}\right) = \left(\frac{P_n L_o \cot^2\alpha_n}{EA_o \cos^2\alpha_n}\right) \qquad \ldots(10.10)$$

Therefore the total vertical deflection of the joint N under the load P is given by

$$\Delta_{tot} = (\Delta v_n + \Delta n_o) = \left[\frac{P_n L_n}{EA_n \sin^2\alpha_n}\right] + \left[\frac{P_n L_o \cot^2\alpha_n}{EA_o \cos^2\alpha_n}\right] \qquad \ldots(10.11)$$

After substituting in expression (10.11), the values of

$$P_n L_n = F_n na \tan\alpha_n = F_n n_1 a \left(\frac{\sin\alpha_n}{\cos\alpha_n}\right)$$

The expression for total deflection is given by

$$\Delta_{tot} = \left[\frac{C_1}{\sin\alpha_n \cos\alpha_n}\right] + \left[\frac{C_2 \cos^2\alpha_n}{\sin\alpha_n \cos^3\alpha_n}\right] \qquad \ldots(10.12)$$

where the constants are expressed as

$$C_1 = \left(\frac{F_n na}{EA_n}\right) \text{ and } C_2 = \left(\frac{F_n n_1 a}{EA_o}\right)$$

If the angle and the number of panels are equal, then
$$na = n_1 a \text{ also } \alpha_n = \alpha_o = \alpha \text{ and } C_1 = C_2$$

The final expression for the total deflection is given by

$$\Delta_{tot} = \left[\frac{C_3}{\sin\alpha \, \cos\alpha} \right] \qquad \ldots(10.13)$$

where $(C_1 + C_2) = 2C_1 = 2C_2 = C_3$

Comparison of the expressions (10.3) and (10.13) indicates that the displacement of the joint in the stiffening girder, and therefore the bending moment, follows the same pattern as the change in the weight of the cable. The values of Δ_{tot} as a function of the angle α_n are shown in Fig. 10.15.

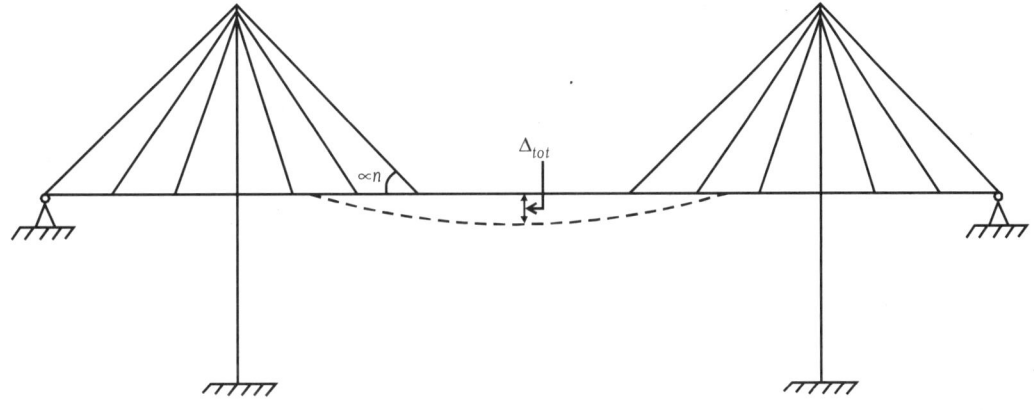

(a) Deflection of Deck

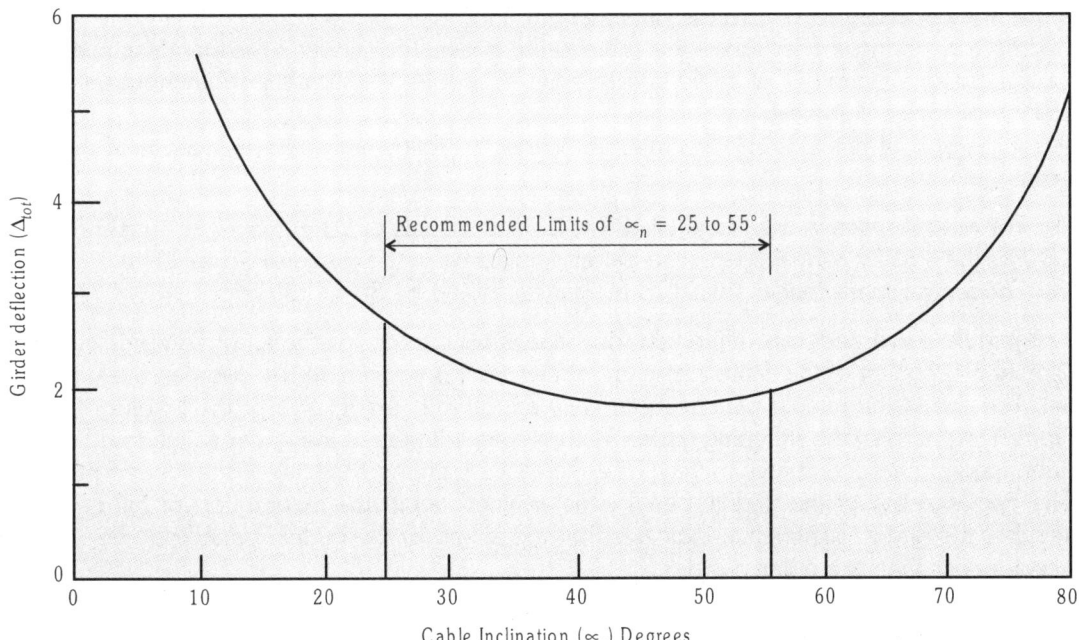

Fig. 10.15: Relation Between Cable Inclination and Deflection of Stiffening Girder

The relation between the cable inclination and deflection of joint indicates that the optimum angle of the cable inclination is 45° and may vary in the reasonable limits of 25 to 55°. The low values of the angle of inclination correspond to the external cables while the greater values correspond to the cable nearest to the tower.

2. Determination of Height of Tower and Length of Panels

The height of the tower as a function of the panel length 'na' may be expressed as follows:
$$h = na \tan 25° = 0.465 na \qquad \ldots(10.14)$$
Using three cables on each side of the tower, $n = 3$
$$h = (0.465 \times 3a) = 1.4a \qquad \ldots(10.15)$$
and with four cables, we have
$$h = (0.465 \times 4a) = 1.86a \qquad \ldots(10.16)$$

The middle panel is usually longer than the remaining panels and may be taken as $1.3a$. In that case, the ratio of the tower height to the length of the mid span considering a total of six panels is computed as

$$\left(\frac{h}{L}\right) = \left[\frac{1.4a}{(6+1.3)a}\right] = \left(\frac{1}{5.2}\right)$$

Hence the height of the tower, $h = \left(\dfrac{L}{5.2}\right) \qquad \ldots(10.17)$

The number and length of the panels are basically determined by the bridge system and its structural characteristics. It is possible to reduce the moment of inertia of the girder and for this purpose it is necessary to reduce the panel length. However the reduction of the girder depth is limited because of the connection of the cable to the girder. Technically it is certainly convenient to have the minimum number of cable connections to reduce the number of anchorages and for regulation of forces in the cables.

A comparison of the existing structures indicates the following optimum values of the panel lengths:

1. For central spans in the range of 137 to 150 m, panels of 20 m length are recommended.
2. For the smaller central spans, the panels should be in the range of 15 to 17 m.
3. For central spans longer than 170 m, panels should be 30 m in length.

The middle panel performs differently from the other panels since it is not compressed by the horizontal component of the cable forces and therefore it is possible to use comparatively a longer panel. The size of the middle panel substantially affects the distribution of the loadings between the remaining parts of the stiffening girder. With greater stiffness of the girder at the middle panel, the non loaded part contributes more to the increased carrying capacity of the loaded part.

The optimum size of the middle panel is determined under the assumption of full use of the material of the girder. Experience indicates that the length of the middle panel may be 20 to 30 per cent longer than the other panels.

3. Determination of Cable Forces

For the preliminary design it is possible for a cable system with five equal panels to use the empirical relation,

$$M_{max} = 0.007 w L^2$$

and for one with seven panels

$$M_{max} = 0.006 w L^2$$

where w = Total load (dead + live load) kN/m
L = length of the panel (m)

The maximum bending moment in the stiffening girder can be estimated using these empirical equations. The cable stay forces depend on factors such as the length of span, number and size of panels and angles of inclination of cables, dead weight of deck and live loads. Referring to the Fig. 10.16 the force in the cable is computed using the following notations:

P = Force in the cable
S = Spacing of cable
w = Total distributed load per metre of deck
α = Angle of inclination of cable with horizontal
R = Vertical reaction at cable stay node = $S \cdot w$.

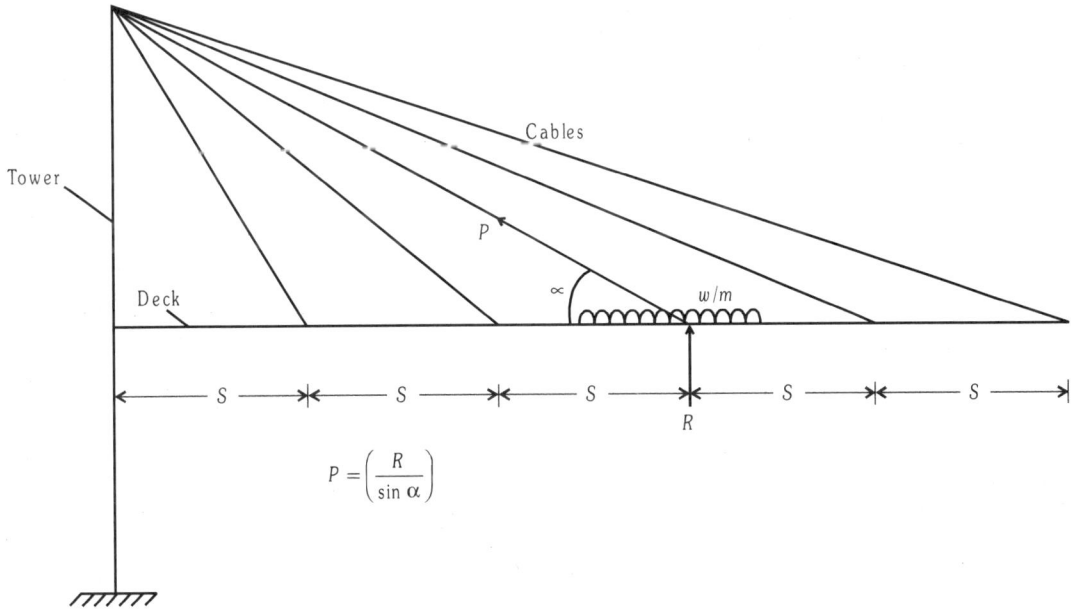

Fig. 10.16: Forces in Stay Cables

Equilibrium of forces at the node yields the force in the cable as

$$P = \left(\frac{R}{\sin \alpha}\right)$$

For example, assuming the total reaction at the node = R = 1000 kN and $\sin \alpha$ = 0.65. The force in the inclined cable is computed as

$$P = \left(\frac{1000}{0.65}\right) = 1538 \text{ kN}$$

The size of the high tensile cable is designed to resist this cable force.

4. Determination of Approximate Self Weight of Stiffening Girders

The problem of the determination of the approximate self weight of stiffening girders for the preliminary design is not yet properly developed. Technical literature provides very little information on this subject. However on the basis of the analysis performed in the previous section it is possible to estimate the approximate value of the self weight of the girder using the formulae for the maximum bending moments in the five and seven panel bridge systems.

The maximum bending moment developed in the stiffening girder including the effect of change of temperature may be expressed by the empirical relation,

$$M = 0.05(\psi g + p + q)L^2$$

where
L = Span length
g = The theoretical weight of the stiffening girder per unit length of span
p = The uniformly distributed weight of deck per unit length of span
q = The uniformly distributed live load carried by a single girder per unit length
ψ = The construction coefficient of the stiffening girder.

If the permissible flexural stress is denoted as f, the required section modulus of the girder is expressed as

$$Z = \left[\frac{0.05(\psi g + p + q)L^2}{f}\right]$$

The section modulus of the stiffening girder may be expressed by its cross-sectional properties. If the girder consists of an I or box section, assuming equal areas of the top and bottom chords, it is possible to express the section modulus as

$$Z = \left[\frac{bh^2}{6} + \frac{2A_c(h/2)^2}{(h/2)}\right] = A_w(h/6) + A_c h$$

Assuming that the cross-sectional area of a single chord and web are equal, we have $A_c = A_w$ and therefore the total area can be expressed as $A = 3A_w$. Hence the section modulus is written as

$$Z = 1.17A_w h = 1.17(A/3)h$$

$\therefore \quad A = (2.5Z/h)$

Substituting the value of the section modulus, we have

$$A = \left[\frac{0.0125(\psi g + p + q)L^2}{fh}\right]$$

By multiplying the theoretical cross-sectional area of the girder by the specific weight of the material γ, the theoretical weight of the stiffening girder per unit length can be expressed as

$$g = \left[\frac{0.0125(\psi g + p + q)L^2 \gamma}{fh}\right]$$

and after transformation, this becomes

$$g = \left[\frac{p + q}{(fh/(0.0125L^2\gamma)) - \psi}\right]$$

Prestressed Concrete Cable Stayed Bridges

Assuming the depth of the girder as one-hundredth of span ($h = L/100$), the theoretical weight is expressed as

$$g = \left[\frac{p+q}{(f/1.25L\gamma) - \psi} \right]$$

This relation has been developed without taking into consideration the axial force acting on the stiffening girder. However in the middle panel there is no axial force and at sections where large axial forces exist, the bending moments as a rule are relatively small. Therefore, the relation developed for the weight of the girder may be used as the first approximation assuming the construction coefficient $\psi = 1.4$. The empirical relation corresponds to the five panel system. In the case of seven panels, the corresponding formula is given by

$$g = \left[\frac{p+q}{(f/1.071L\gamma) - \psi} \right]$$

A comparative analysis of the empirical relations for the self weight of stiffening girders indicate that for the larger spans it is more economical to divide the span into greater number of panels. The empirical formulas may provide good estimates of the weight of the girders only for spans in the range of 270 to 400 m. Larger spans should be divided into nine, eleven and even greater number of panels to obtain a relatively light stiffening girder.

5. Determination of Self Weight of Cables

It is possible to determine the approximate weight of the cable by using the empirical formulas developed for expressing the weight of the stiffening girder. The forces developed in the cables depend upon the number of cables supporting the deck. Reduction in the number of cables increases the forces developed in the cables. Hence it may be assumed that the weight of the cables depend to some extent on the number of cables.

Referring to Fig. 10.17 and using the following notations,

F_o = Force developed in the backstay cable
F_1 = Force developed in the first cable of length L_1
F_2 = Force developed in the second cable of length L_2
Q_1 = Weight of first cable
Q_2 = Weight of second cable
γ = Specific weight of the cable material
L = Span of the stiffening girder

α_1 and α_2 are angles subtended by first and second cables with the horizontal plane.

The weight of the cables for a five panel bridge is determined by using the following empirical relations:

$$F_1 = \left[\frac{0.237(g+p+q)L}{\sin\alpha_1} \right]$$

$$F_2 = \left[\frac{0.174(g+p+q)L}{\sin\alpha_2} \right]$$

The lengths of the first and second cables are obtained as

$$L_1 = \left(\frac{0.2L}{\cos\alpha_1}\right) \text{ and } L_2 = \left(\frac{0.4L}{\cos\alpha_2}\right)$$

Assuming the allowable stress as f and the specific weight of cable as γ, the weights of the cables are evaluated as

$$Q_1 = \left(\frac{F_1 L_1}{f\cos\alpha_1}\right) = \left[\frac{0.047(g+p+q)L^2\gamma}{f\sin\alpha_1\cos\alpha_1}\right]$$

$$Q_2 = \left[\frac{0.0696(g+p+q)L^2\gamma}{f\sin\alpha_2\cos\alpha_2}\right]$$

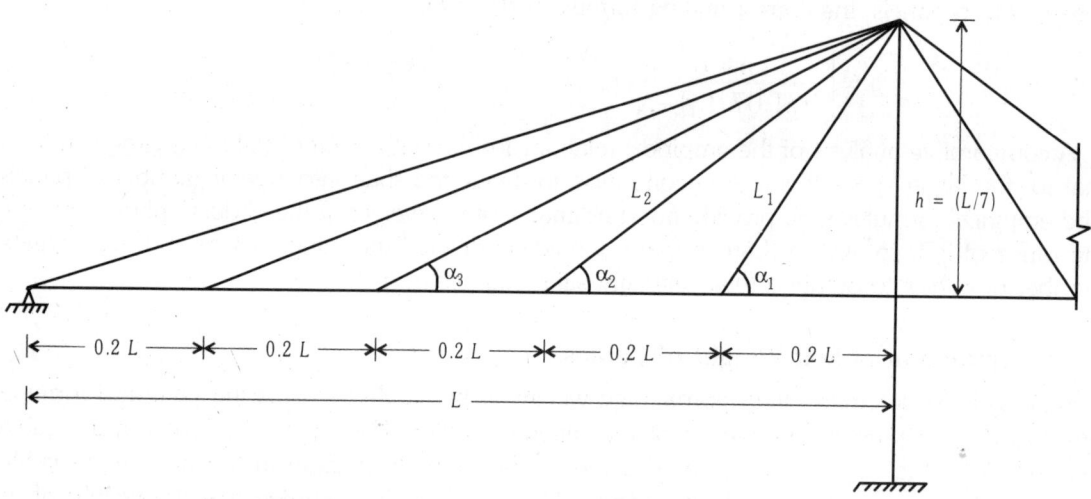

Fig. 10.17: Determination of Weight of Stay Cables

Assuming the height of the tower as $h = (L/7) = 0.143L$

$$\tan\alpha_1 = 0.713 \text{ and } \tan\alpha_2 = 0.357$$
$$\alpha_1 = 35°30' \text{ and } \alpha_2 = 19°40'.$$

and $\sin\alpha_1\cos\alpha_1 = 0.463$

$\sin\alpha_2\cos\alpha_2 = 0.312$

Substituting these values, we have

$$Q_1 = \left[\frac{0.102(g+p+q)\gamma L^2}{f}\right]$$

$$Q_2 = \left[\frac{0.223(g+p+q)\gamma L^2}{f}\right]$$

After distributing the weight of the four cables uniformly along the span, the theoretical weight per unit length of the span is obtained as

$$g_c = \left[\frac{0.65(g+p+q)\gamma L}{f}\right]$$

The backstay cable force is

$$F_0 = \left[\frac{F_1 \cos\alpha_1}{\cos\alpha_0} + \frac{F_2 \cos\alpha_2}{\cos\alpha_0}\right]$$

$$= \left[\frac{0.237(g+p+q)L^2}{\tan\alpha_1 \cos\alpha_0} + \frac{0.174(g+p+q)L^2}{\tan\alpha_2 \cos\alpha_0}\right]$$

By assuming $\alpha_0 = 30°$, the cable force is
$$F_0 = 0.948(g+p+q)L$$

Then considering the length of the two back stay cables as equal to $0.8L$, their weight can be expressed as,

$$Q_0 = \left[\frac{F_0(0.8L)}{f}\right] = \left[\frac{0.78(g+p+q)L^2}{f}\right]$$

And the weight per unit length is

$$g_0 = \left[\frac{0.78(g+p+q)L}{f}\right]$$

Hence the total weight of all cable stays is expressed as

$$g_{tot} = (g_c + g_0) = \left[\frac{1.43(g+p+q)}{f}\right]$$

The number of cables significantly influences the anchorage system and consequently the reinforcement in the stiffening girder to transfer moment, shear and axial forces. A relatively deep girder is required to span large distances between the stay cable attachments. A large number of cable stays supporting a continuous elastic medium simplifies the anchorage and distribution of forces to the girder and permits the use of shallower depth girders. Although more stays are used, the additional cost is more than offset by simpler connection details for the smaller cable and lesser force in cable stays. The erection work is also simplified since the deck structure can be constructed by cantilever method from stay connection point without any auxiliary means.

6. Degree of Redundancy

The degree of statical indeterminacy of the cable stayed bridge system is determined by the formula
$$I = (C + 2S - H - 3)$$
where C = Total number of cables
  S = Total number of stiffening girder systems
  H = Number of moveable connections or hinges considering even the moveable supports of the cable on towers.

For example, in the Akkar Bridge in Sikkim shown in Fig. 10.18, the degree of redundancy is computed using the following data:
  Number of cables = C = 34
  Number of supports = S = 3
  Number of moveable connections = H = 3

Prestressed Concrete Bridges

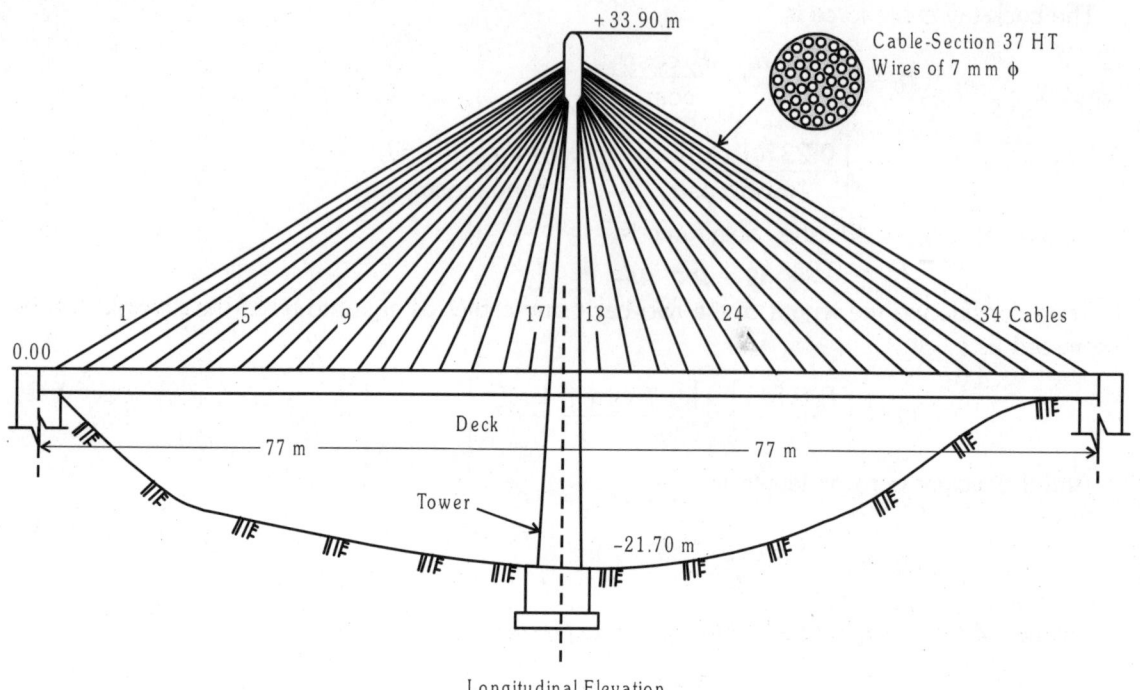

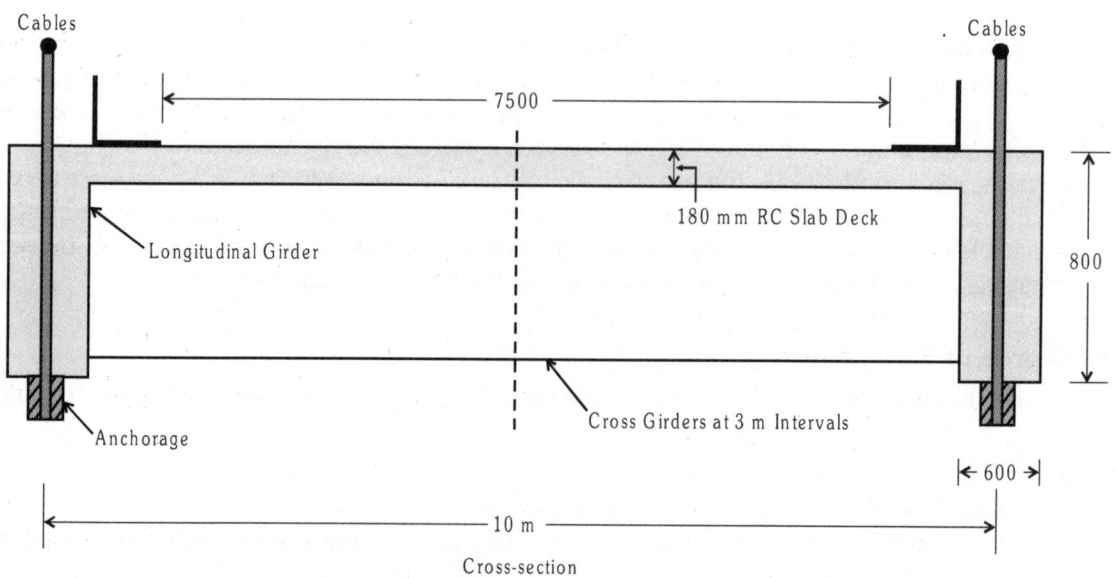

Fig. 10.18: Akkar Bridge (Sikkim)

Hence the degree of redundancy is computed as
$$I = (C + 2S - H - 3)$$
$$= [34 + (2 \times 3) - 3 - 3]$$
$$= 34.$$

Hence 34 equations have to be formulated by selecting suitable redundants and the equations are solved using a digital computer.

10.10 STRUCTURAL ANCHORAGES

The axial force in the stiffening girder depends upon the method of anchoring the cables and the provision of expansion joints and their location in the structure. Basically three different types of anchorage systems are considered such as

1. Self anchored system
2. Fully anchored system
3. Partially anchored system

1. **Self Anchored System:** Referring to Fig. 10.19 (a), in the self anchored system, there is no restraint at the supports to the horizontal components of the cable force. In this case the axial force distribution in the girder will vary from zero at the centre of main span to maximum compression near the towers.

2. **Fully Anchored System:** In this type, no provision is made for the movement at the supports but expansion are provided at the towers. The axial force distribution changes as shown in Fig. 10.19 (b) with zero force at towers increasing to a maximum value at the centre of span.

3. **Partially Anchored System:** In the partially anchored system, the axial forces are considerably reduced using a combination of the above two systems by providing horizontal restraint at the abutments with no expansion joints or expansion joints provided only in the end spans as shown in Fig. 10.19 (c).

The anchorages must be designed to allow adjustment of length and replacing a cable damaged by an accident without interrupting traffic. They must be further designed to prevent bending stresses in the wires or strands at the socket due to change of sag or due to slight oscillations. The anchorage should further comprise dampers to prevent resonance oscillations of the cables caused by wind turbulence. At the tower head, cables running over a saddle like in a suspension bridge should be avoided because their replacement would be rather difficult. It is preferable to anchor the cable at each side individually and design suitable devices to carry the horizontal component of the cable forces through the tower. At the top of the towers, the box must be wide enough to allow access and handling of the jacks.

10.11 DYNAMIC BEHAVIOUR AND AERODYNAMIC STABILITY

In the early stages when the first cable stayed Stromsund Bridge was built in Sweeden in 1955, the problem of aerodynamic stability in bridge design did receive considerable study. However, that study did not lead to explicit design rules and formulae. The aerodynamic phase of the problem is a real challenge to the bridge engineers. Catastrophic collapse of the Tacoma Narrows Bridge in 1940 prompted bridge engineers to plan comprehensive research investigations on aerodynamic stability of suspension and cable stayed bridge systems. The problem of aerodynamic stability is more important in the case of lighter bridges covering long spans such as the suspension and cable stayed bridges.

Prestressed Concrete Bridges

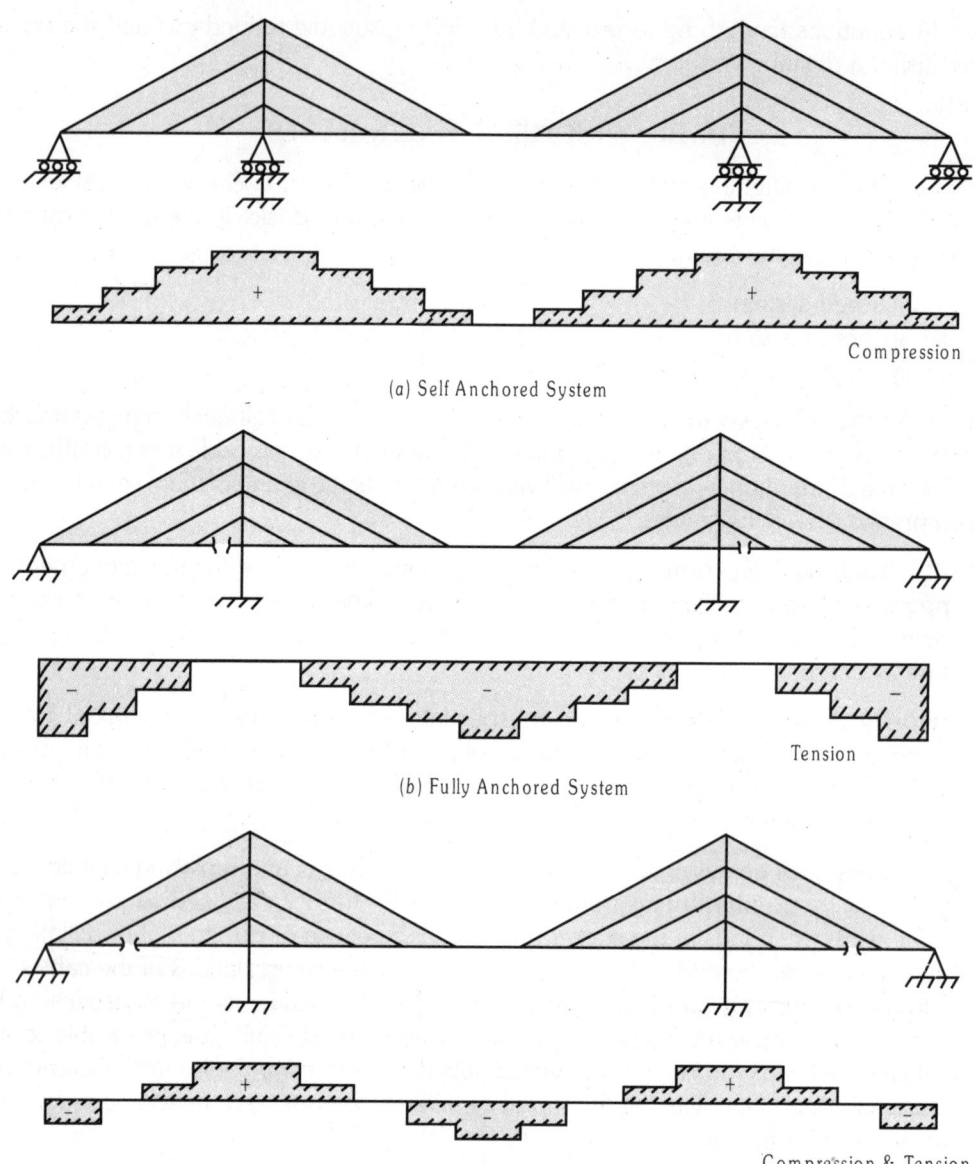

Fig. 10.19: Structural Anchorage Systems

According to Leonhardt, the cable stayed bridge with concrete decks and highly stressed cables exhibits superior dynamic behaviour under wind loads. The deflections under live loads are extremely small because the effective depth of the large cantilever truss formed by the cables is much larger than for beam girders. The main advantage of the multicable system being that the increase of amplitude due to resonance oscillation is prevented by system damping caused by the interference of the multicable system. Measurements made at the Tjorn bridge in Sweeden indicated that the damping increases with increasing amplitudes and the logarithmic decrement reaches a value of well above 0.10. This damping is very favourable for the stability of the structure under wind loads. Current theories which calculate critical wind speeds do not adequately represent the actual behaviour and have only limited validity. The same is true for wind tunnel tests with sectional models in which only a constant damping factor is applied. More tests should be made on proptotype bridges to improve the theories based on observed facts.

Based on the present knowledge, Leonhardt has suggested the following geometrical relations for obtaining wind stability in the case of concrete bridge decks supported with cables in two planes along the edges.

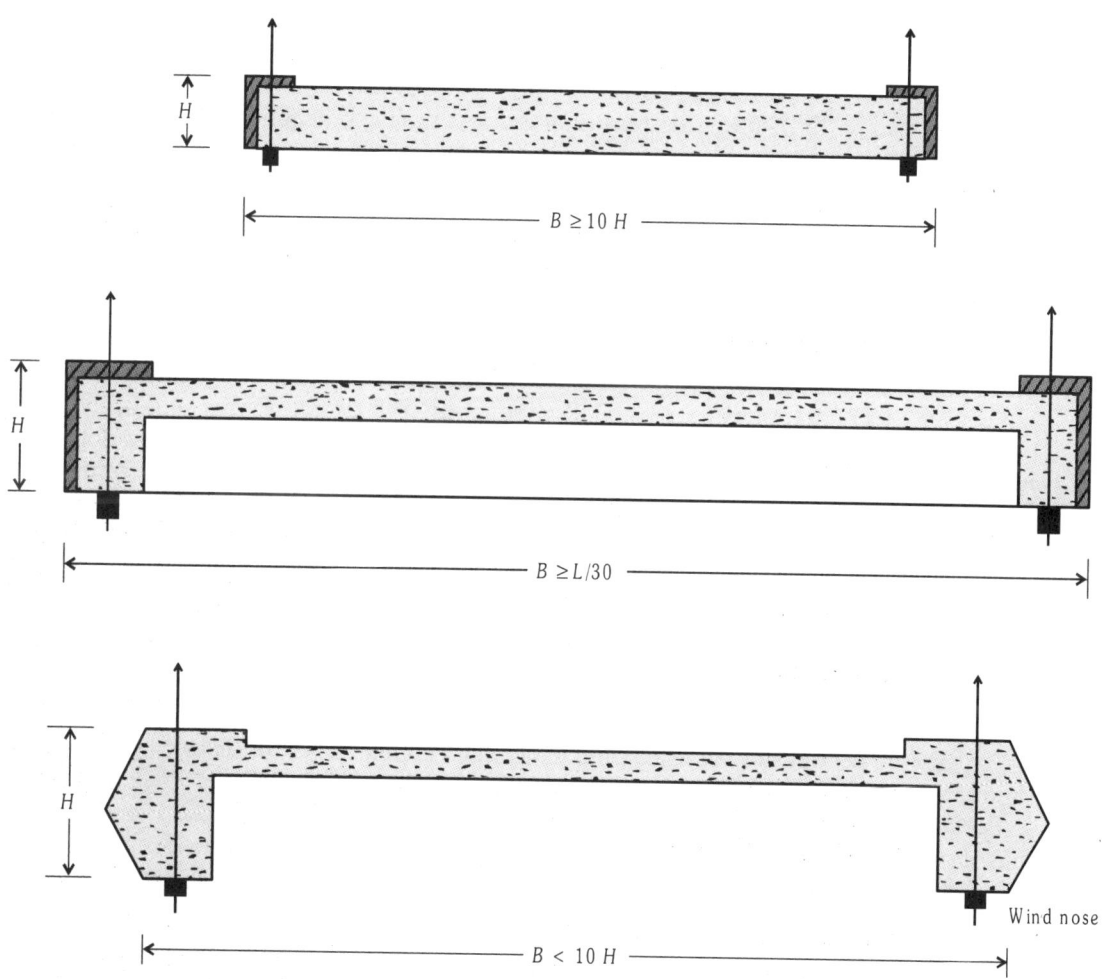

Fig. 10.20: Bridge Deck Dimensions to Ensure Aerodynamic Stability

Referring to Fig. 10.20 and using the following notations,
- L = Span of the bridge deck
- B = Width of bridge deck
- H = Depth of stiffening girder

The bridge will be safe against wind loads if the following geometrical relations are satisfied:
(a) $B \geq 10H$
(b) For $B < 10H$, a wind nose should be provided
(c) $B \geq (1/30)L$, which indicates that the width of the bridge should not be too small in relation to the main span. If this ratio gets smaller, then A-shaped towers and wind shaping of the cross section must be used. The A-shaped tower provides a triangular shape of the cable planes and deck, which increases the torsional rigidity. Bridges supported with cables in one plane along the centre line have negligible damping under torsional oscillations.

10.12 CONSTRUCTION METHODS

In the case of cable stayed bridge decks, the method of construction significantly affects the nature of stresses developed in the super structure and hence suitable precautions have to be taken during the construction procedure.

The prominent methods[12, 13] generally used for the erection of cable stayed bridges are:

1. **The Staging Method:** This method is adopted when low clearance is required below the deck and supporting form work does not interfere with traffic. This method facilitates rapid construction by maintaining correct geometry of the structure with relatively low cost. This method has been used in the construction of the Rhine river bridge at Maxau, Japan.

2. **The Push-out Method:** In this method, large precast sections of the bridge deck are pushed out over the piers on rollers or sliding teflon bearings. The deck units are pushed out from both abutments towards the centre or from one abutment all the way to the other abutment. Assembling the components in an erection bay and progressively pushing the components out into span simplifies construction and reduces costs. This technique was used in the construction of Julicher-Strasse bridge in Germany and Paris Massena bridge in France.

3. **The Cantilever Method:** The cantilever method of erection is the latest and most popular method of construction of cable stayed bridge decks. The erection proceeds from either side of the pylon in the form of two free cantilevers which balance each other. The units may also be supported directly by the stay cables depending upon the stay spacing and size of each precast unit. For bridges with box girders, prefabricated segments can be used with match cast joints but using a paste in the joint which compensates for differential shortening during the curing and hardening period.

   Bridges with a deck of composite beams allow the simplest and quickest erection. The grid of steel cross girders and light steel edge girders, including the cable anchors, are installed with light derricks and then the prefabricated concrete slabs are placed leaving gaps for overlapping reinforcement and shear connectors which are closed using cast *in situ* concrete. This method was adopted for the construction of the Annacis bridge in Vancouver, British Columbia.

   The main advantage of the cantilever method of construction is that the traffic below the bridge is not hindered during erection. This technique has been used in the construction of Pascokennewick intercity bridge, U.S.A, Kniebrucke bridge, Germany and the Stromsund bridge, Sweeden.

   During the last decade, construction methods have shown major progress towards simplification and reducing erection equipment. The construction procedure must however be well planned using sequential computations for the alignment, forces, exact lengths and angles considering temperature and creep influences which depend on seasonal, climatic and even daily weather conditions. The collective experience and knowledge gained during the past thirty years will help in evolving innovative applications and methods in the construction of cable stayed bridges in the future in various countries throughout the world to serve the needs of the human society.

10.13 ECONOMIC STUDIES

Exhaustive investigations by Fritz Leonhardt conclusively indicates that cable stayed bridges are structurally efficient, aesthetically superior and cost effective for low, medium and long spans, ranging from 40 m to 1800 m. pedestrian bridges with only 40 m span comprising a prestressed concrete deck with a depth of 250 to 300 mm supported by a few cable stays have been successfully built in Germany. Highway bridges can be built of prestressed concrete with spans up to 700 m and rail road bridges up to about 400 m. If composite action between the steel girders and a concrete deck slab is utilized, then spans in the range of 600 m and 1000 m for railway and highway bridges respectively can safely be used. For the crossing of the Messina Straits in Italy, Leonhardt has designed and constructed an all steel cable stayed bridge with a main span of 1800 m for six lanes of road traffic and two railway tracks without encountering any structural or construction difficulties.

Recent studies have indicated that cable stayed bridges are much more economical, stiffer and aerodynamically superior when compared with a suspension bridge. Figure 10.21 shows the comparison of the quantity of cable steel for suspension and cable stayed bridges in the span range of 500 to 1800 m. For a bridge with 1800 m span and 38 m width, a suspension bridge requires 46000 t of steel, whereas a cable stayed bridge needs only 19200 t of steel indicating more than 50 per cent savings in the quantity of steel required in the cable stayed bridge.

A comparative analysis indicates that a suspension bridge requires a stiffening girder with a flexural stiffness which must be about ten times larger than that required for a cable stayed bridge covering the same span. The suspension bridge needs additional heavy anchor blocks which can be prohibitively costly if the navigation clearance is high and foundation conditions are poor. The total cost of such a suspension bridge can easily be 20 to 30 per cent more than the cost of cable stayed bridge.

The second Hooghly bridge (Vidyasagr Sethu) at Kolkata is an excellent example of a cable stayed bridge comprising a main span of 457.2 m and two end spans of 182.8 m each. The deck is made up of a concrete slab 230 mm thick, with two outer steel I-girders 28.10 m apart and a central I-girder. The deck is suspended by cable stays comprising parallel wire cables of B.B.R.-HIAM type with their own anchorage system. The bridge provides for two 3-lane carriageways 12.3 m each and 2.5 m footways. The cable stayed bridge costing 600 million rupees was found to be cost effective in comparison with other types.

Economic studies of bridges constructed in Russia clearly indicate the cost effectiveness of cable stayed bridges. In this study, five types of concrete and four types of steel bridges were included. The concrete bridges were cable stayed, arch cantilever, rigid frame, suspension and continuous box girder type. Economic evaluation included the cost of piers, super structure and erection work. The analysis indicates that the volume of concrete per unit area of the bridge deck was the least in the case of cable stayed bridges in the span range of 60 to 300 m. Cost studies of the various types of bridges in U.S.A., also indicates that the cost per unit area of bridge deck is the least for cable stayed system in the span range of 60 to 400 m.

The survey indicates that cable stayed bridges of prestressed concrete and steel have wide applications in the future in countries throughout the world in the domain of highways.

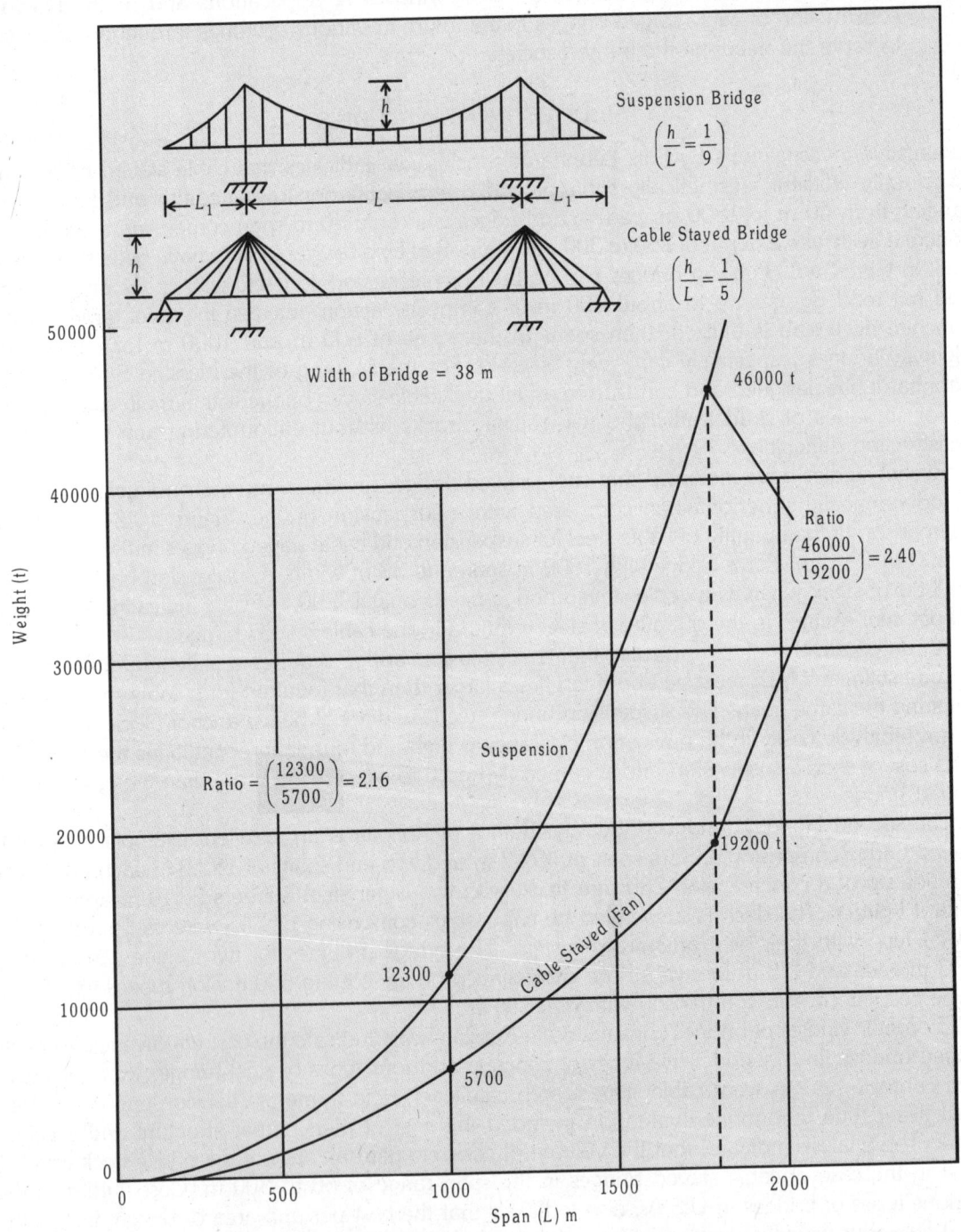

Fig. 10.21: Comparison of Quantity of Cable Steel for Suspension and Cable Stayed Bridges

Prestressed Concrete Cable Stayed Bridges

10.14 DESIGN EXAMPLE OF PRESTRESSED CONCRETE CABLE STAYED BRIDGE

Design a suitable prestressed concrete cable stayed bridge deck for crossing a deep valley of span length 200 m to suit the following data:

1. **Data**
   Effective span of bridge deck = 200 m
   Width of Road way = 7.5 m
   Width of Foot paths = 2.5 m on either side
   Wearing coat thickness = 80 mm
   Type of loading: IRC Class AA tracked vehicle
   The deck is proposed to be made up of a prestressed concrete solid slab with longitudinal stiffening girders supported by stay cables.
   Spacing of stay cables = 6 m intervals
   Type of concrete: M-60 Grade concrete
   Type of steel: Fe-415 HYSD bars for use as supplementary reinforcement
   Standard Freyssinet prestressing cables are available for use as stay cables and for prestressing the deck slab and stiffening girders.
   Design the deck slab, longitudinal girders and stay cables.
   The structural concrete components of the bridge should be designed as Class-1 type without developing any tensile stresses under service loads.
   Sketch the details of reinforcements and cables in the deck slab and girders and the elevation of the bridge showing the stay cable configuration
   The design should conform to the National Codes, IRC: 6, IRC: 18 and IRC: 21.

2. **Permissible Stresses**
   The permissible compressive stress in concrete at transfer and service loads as recommended in IRC: 18 are as follows:
   For M-60 Grade concrete,
   $$f_{ck} = 60 \text{ N/mm}^2$$
   $$f_{ci} = 45 \text{ N/mm}^2$$
   $$f_{ct} = 0.45 f_{ci} = (0.45 \times 45) = 20 \text{ N/mm}^2$$
   $$f_{cw} = 0.33 f_{ck} = (0.33 \times 60) = 20 \text{ N/mm}^2$$
   $$f_{tt} = 0 \text{ (Class-1 type member)}$$
   $$E_c = 5000 \sqrt{f_{ck}} = 5000 \sqrt{60} = 38730 \text{ N/mm}^2 = 38.73 \text{ kN/mm}^2$$

3. **Selection of Dimensions of Bridge Deck**
   The overall span of the bridge being 200 m, it is proposed to have a single tower at the centre of span with the deck comprising a prestressed concrete solid slab supported on two longitudinal stiffening girders which are suspended by stay cables in two planes at intervals of 6 m along the girder.
   Width of two lane carriage way = 7.5 m
   Foot paths with kerb on either side = 2.5 m
   Effective span of the deck slab = B = (7.5 + 2.5) = 10 m
   Thickness of slab assumed as 40 mm/m of span = (40 × 10) = 400 mm
   The depth of longitudinal girders supporting the slab is selected based on the wind stability criteria proposed by Leonhardt[12].

The stiffening girder dimensions should be such that
$$B \geq 10H$$
where B = width of the girder
H = depth of the girder
L = span of girder = 200 m
In this example, B = 10 m

$$\therefore \quad H = \left(\frac{B}{10}\right) = \left(\frac{10}{10}\right) = 1 \text{ m} = 1000 \text{ mm}$$

Also $\quad B \geq \left(\dfrac{L}{30}\right) \geq \left[\dfrac{200}{30}\right] \geq 6.66 \text{ m}$

Both the criteria are satisfied hence adopt the depth of stiffening girder = H = 1000 mm
Width of girder = $0.5H$ to $0.6H$
$\qquad\qquad\quad = (0.5 \times 1000)$ to (0.6×1000)
$\qquad\qquad\quad = 500$ mm to 600 mm

Adopt a stiffening girder of size 600 mm wide by 1000 mm deep.
The cross-section of the bridge deck is shown in Fig. 10.22.

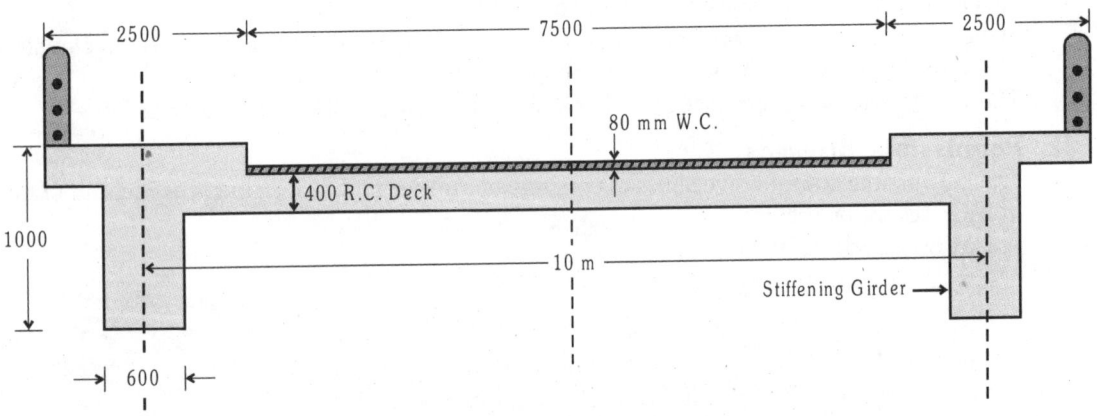

Fig. 10.22: Cross-section of Bridge Deck

4. Sectional Properties of Slab and Girder

(a) Deck Slab

Thickness of slab = 400 mm
Cross-sectional area per metre width = $A = (400 \times 1000) = (4 \times 10^5)$ mm^2

Sectional modulus = $Z = Z_t = Z_b = \left(\dfrac{1000 \times 400^2}{6}\right) = (26.66 \times 10^6)$ mm^3

(b) Stiffening Girder

Width of stiffening girder = 600 mm
Overall depth of girder = 1000 mm

Cross-sectional area = $A = (600 \times 1000) = (6 \times 10^5)$ mm²

Section Modulus = $Z = \left(\dfrac{600 \times 1000^2}{6}\right) = 10^8$ mm³

5. Design of Deck Slab

(a) Dead Load Bending Moments

The deck slab is rigidly connected to the edge stiffening girder.
Effective span = L = 10 m
Thickness of slab = 400 mm
Self weight of slab = (0.4×24) .... = 9.60 kN/m²
Self weight of W.C. = (0.08×22) ..... = 1.76
Foot paths, kerbs and finishes (L.S.) ...... = 0.64
Total dead load ....(g) = 12.00 kN/m²

Dead load bending moment at support = $M_g = \left(\dfrac{gL^2}{12}\right) = \left(\dfrac{12 \times 10^2}{12}\right) = 100$ kN·m

Dead load bending moment at centre = $M_g = \left(\dfrac{gL^2}{24}\right) = \left(\dfrac{12 \times 10^2}{24}\right) = 50$ kN·m

(b) Live Load Bending Moments

The slab is monolithically cast with the edge girders. Hence it is considered as fixed at supports over a span of 10 m. The IRC Class AA tracked load is positioned on the span as shown in Fig. 10.23 to yield maximum moments.

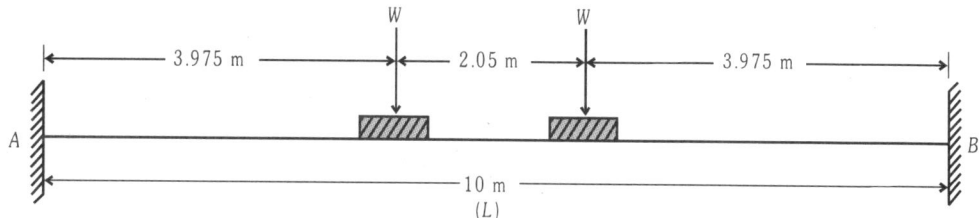

Fig. 10.23: Position of IRC Class AA Loads for Maximum Moments in Deck Slab

Live load per metre length = $W = (350/3.6) = 97.22$ kN
Maximum negative bending moment at support is obtained as

$$M_A = \left(\dfrac{Wab}{L}\right) = \left(\dfrac{97.22 \times 3.975 \times 6.025}{10}\right) = 233 \text{ kN.m}$$

Design Negative Bending Moment at support with impact factor is computed as
$M_q = (1.1 \times 233) = 256$ kN.m
Maximum positive bending moment at centre of span with impact factor is calculated as
$M_q = 1.1 (386.4 - 233) = 169$ kN·m

(c) Dead and Live Load Shear Forces
Dead load on slab = g = 12 kN/m²

Effective span = 10 m

Dead load shear force = $V_g = \left(\dfrac{12 \times 10}{2}\right) = 60$ kN

Maximum live load shear force occurs when the IRC Class AA tracked vehicle loads are positioned as shown in Fig. 10.24

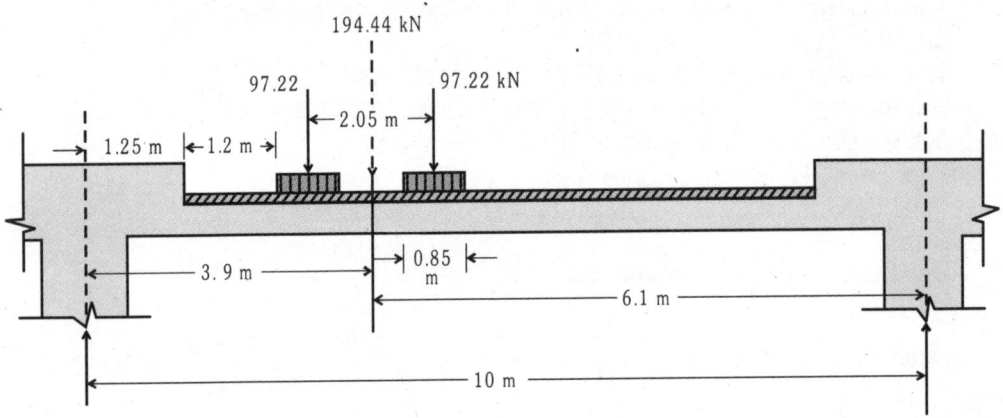

Fig. 10.24: Position of IRC Class AA Loads for Maximum Shear Force at Support

Maximum live load shear force with impact is computed as

$$V_q = 1.1\left[\dfrac{194.44 \times 6.1}{10}\right] = 130.5 \text{ kN}$$

Design ultimate shear force = $V_u = (1.5\,V_g + 2.5\,V_q)$
$= [(1.5 \times 60) + (2.5 \times 130.5)] = 416.25$ kN

(d) Check for Minimum Section Modulus

Dead load moment = $M_g = 100$ kN·m
Live load moment = $M_q = 256$ kN·m
$Z = Z_t = Z_b = (26.66 \times 10^6)$ mm^3
$f_{ct} = f_{cw} = 20$ N/mm^2 and $f_{tw} = 0$
Loss ratio = $\eta = 0.8$
$f_{br} = [\eta f_{ct} - f_{tw}] = [(0.8 \times 20) - 0] = 16$ N/mm^2

$$Z_b \geq \left[\dfrac{M_q + (1-\eta)M_g}{f_{br}}\right] = \left[\dfrac{(256 \times 10^6) + (1-0.8)\,100 \times 10^6}{16}\right]$$

$\geq (17.25 \times 10^6)$ mm^3 < (26.66×10^6) mm^3, (Section modulus provided)

Hence the section selected for the slab is adequate to resist the service loads safely without exceeding the permissible stresses.

(e) Minimum Prestressing Force

The minimum prestressing force required is computed using the relation

$$P = \left[\dfrac{A(f_{inf}Z_b + f_{sup}Z_t)}{Z_t + Z_b}\right]$$

Prestressed Concrete Cable Stayed Bridges

$$f_{sup} = \left[f_{tt} - \frac{M_q}{Z_t}\right] = \left[0 - \frac{(100 \times 10^6)}{(26.66 \times 10^6)}\right] = -3.75 \text{ N/mm}^2$$

$$f_{inf} = \left[\frac{f_{tw}}{\eta} + \frac{M_g + M_q}{\eta Z_b}\right] = \left[0 + \frac{(100 + 256)10^6}{(0.8 \times 26.66 \times 10^6)}\right] = 16.69 \text{ N/mm}^2$$

$$P = \left[\frac{(4 \times 10^5)(26.66 \times 10^6)(16.69 - 3.75)}{(2 \times 26.66 \times 10^6)}\right] = (2588 \times 10^3) \text{ N} = 2588 \text{ kN}$$

Using Freyssinet cables containing 12 wires of 8 mm diameter stressed to 1100 N/mm^2,

Force in each cable $= \left[\dfrac{12 \times 50 \times 1100}{1000}\right] = 660$ kN

Spacing of cables $= \left[\dfrac{1000 \times 660}{2588}\right] = 255$ mm

Provide cables at a spacing of 250 mm c/c.

(f) Eccentricity of Cables

The eccentricity of cables at support sections is obtained from the relation

$$e = \left[\frac{Z_t Z_b (f_{inf} - f_{sup})}{A(f_{sup} Z_t + f_{inf} Z_b)}\right] = \left[\frac{(26.66 \times 10^6)^2 (16.69 + 3.75)}{(4 \times 10^5)(26.66 \times 10^6)(16.69 - 3.75)}\right]$$

$= 105$ mm (towards the top of slab)

The positive bending moment at centre of span is of smaller magnitude and hence the eccentricity required at the centre of span is computed by limiting the tensile stress at top fibre to zero under the loading condition of prestress together with the self weight stress. At the centre of span section, we have $P = 2588$ kN and $M_g = 50$ kN·m. Using the stress relation,

$$\left[\frac{P}{A} - \frac{Pe}{Z_t} + \frac{M_g}{Z_t}\right] = 0$$

$$\left[\frac{2588 \times 10^3}{4 \times 10^5} - \frac{2588 \times 10^3 \times e}{26.66 \times 10^6} + \frac{50 \times 10^6}{26.66 \times 10^6}\right] = 0$$

Solving, $e = 86$ mm.

The cables are arranged in a parabolic profile with an eccentricity of 105 mm towards the top of slab at supports varying to an eccentricity of 86 mm towards the the soffit of slab at centre of span.

6. Check for Stresses in Deck Slab Under Service Loads

(a) Stresses at Support Section

$P = (2588 \times 10^3)$ N
$e = -105$ mm (towards top of slab)
$A = (4 \times 10^5)$ mm^2
$Z = (26.66 \times 10^6)$ mm^3

$M_g = 100$ kN·m
$M_q = 256$ kN·m
$\eta = 0.8$

$$\left(\frac{P}{A}\right) = \left(\frac{2588 \times 10^3}{4 \times 10^5}\right) = 6.47 \text{ N/mm}^2$$

$$\left(\frac{Pe}{Z}\right) = \left(\frac{2588 \times 10^3 \times 105}{26.66 \times 10^6}\right) = 10.19 \text{ N/mm}^2$$

$$\left(\frac{M_g}{Z}\right) = \left(\frac{100 \times 10^6}{26.66 \times 10^6}\right) = 3.75 \text{ N/mm}^2$$

$$\left(\frac{M_q}{Z}\right) = \left(\frac{256 \times 10^6}{26.66 \times 10^6}\right) = 9.60 \text{ N/mm}^2$$

Stress at transfer
   Stress at top = (6.47 + 10.19 − 3.75) = 12.91 N/mm²
   Stress at bottom = (6.47 − 10.19 + 3.75) = 0.03 N/mm²
Stress at working loads
   Stress at top = [0.8 (6.47 + 10.19) − 3.75 − 9.60] = −0.022 N/mm²
   Stress at bottom = [0.8 (6.47 − 10.19) + 3.75 + 9.60] = 10.37 N/mm²

(b) Stresses at Centre of Span Section

$P = (2588 \times 10^3)$ N $M_g = 50$ kN·m
$e = 86$ mm (towards soffit) $M_q = 169$ kN·m
$A = (4 \times 10^5)$ mm² $\eta = 0.8$
$Z = (26.66 \times 10^6)$ mm³

$$\left(\frac{P}{A}\right) = \left(\frac{2588 \times 10^3}{4 \times 10^5}\right) = 6.47 \text{ N/mm}^2$$

$$\left(\frac{Pe}{Z}\right) = \left(\frac{2588 \times 10^3 \times 86}{26.66 \times 10^6}\right) = 8.34 \text{ N/mm}^2$$

$$\left(\frac{M_g}{Z}\right) = \left(\frac{50 \times 10^6}{26.66 \times 10^6}\right) = 1.88 \text{ N/mm}^2$$

$$\left(\frac{M_q}{Z}\right) = \left(\frac{169 \times 10^6}{26.66 \times 10^6}\right) = 6.34 \text{ N/mm}^2$$

Stress at transfer
   Stress at top = (6.47 + 8.34 + 1.88) = 0.01 N/mm²
   Stress at bottom = (6.47 + 8.34 − 1.88) = 12.93 N/mm²
Stress at working loads
   Stress at top = [0.8 (6.47 − 8.34) + 1.88 + 6.34] = 6.72 N/mm²
   Stress at bottom = [0.8 (6.47 + 8.34) − 1.88 − 6.34] = 3.63 N/mm²
The stresses in the slab at various limit states are well within the safe permissible limits.

7. Design of Stay Cables

(a) Dead Loads
Total dead load of slab = 12 kN/m²
Weight of longitudinal girder = (1 × 0.6 × 24) = 32 kN/m
Weight of cross girders = (0.3 × 0.6 × 24) = 4.32 kN/m

(b) Live Loads
For maximum reaction, the critical loading position of IRC Class AA tracked vehicle loads are shown in Fig. 10.25.

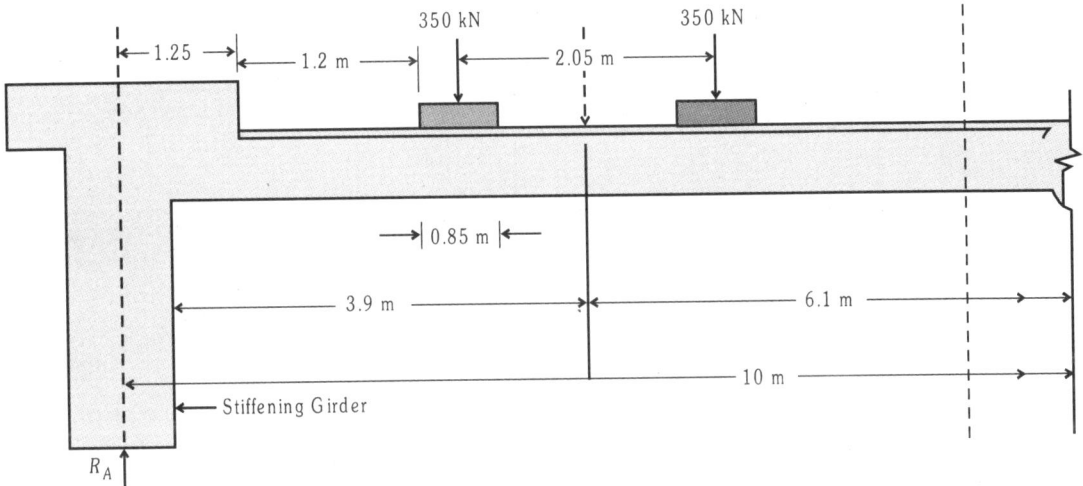

Fig. 10.25: Maximum Live Load Reaction in Stiffening Girder

Allowing for an impact factor of 10 per cent for live loads,

$$\text{Reaction } R_A = 1.1\left(\frac{700 \times 6.10}{10}\right) = 470 \text{ kN}$$

(c) Total Weight of deck for 6 m length and 10 m wide
Weight from deck slab .... = (6 × 10 × 12)   = 720 kN
Weight of longitudinal girders .... = (2 × 14.4 × 6) = 172.8 kN
Weight of cross girder .... = (4.32 × 10)   = 43.2 KN
Total load .... = 936.0 kN
Force transmitted on each cable due to deck loads = (0.5 × 936)   = 468 kN

(d) Total Reaction on Cable
Due to live loads .... = 470 kN
Due to dead loads .... = 468 kN
Add for foot paths, railings, etc. = 62 kN
Total reaction (R) .... = 1000 kN

(e) Design of Cable
Referring to Fig. 10.26,
The angle subtended by the cable to the horizontal = α = 46° and sinα = 0.72

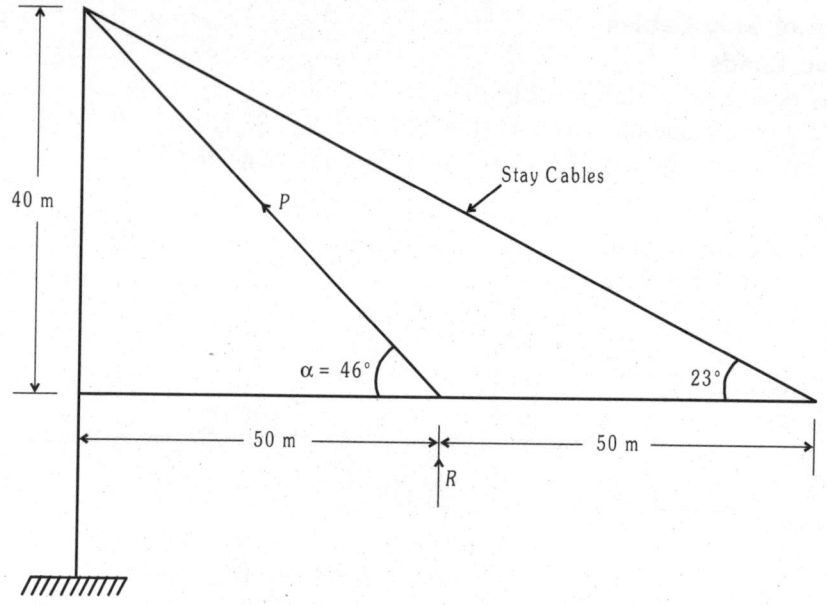

Fig. 10.26: Computation of Force in Cable Stay

The tensile force developed in the cable is computed as

$$P = \left(\frac{R}{\sin\alpha}\right) = \left(\frac{1000}{0.72}\right) = 1389 \text{ kN}$$

Using 15.2 mm diameter high tensile strands initially stressed to 1500 N/mm²

$$\text{Force in each strand} = \left(\frac{140 \times 1500}{1000}\right) = 210 \text{ kN}$$

$$\text{Number of strands in the cable} = \left(\frac{1389}{210}\right) = 7$$

Adopt 7K-15 Freyssinet cable stays at intervals of 6 m.

(f) Design of Stiffening Girder

The maximum bending moment developed in the stiffening girder including the effect of temperature (Refer section 10.9) is expressed as,

$$M = 0.05(\psi g + p + q)L^2$$

where L = span length of stiffening girder (panel length)
g = weight of the stiffening girder per unit length
p = uniformly distributed weight of deck per unit length of span
q = uniformly distributed live load carried by a single girder per unit length
ψ = the construction coefficient of the stiffening girder = 1.4

In the present example, the corresponding values are
L = 6 m
p = 60 kN/m
q = 130 kN/m

$$M = 0.05(1.4 \times 14.4 + 60 + 130)6^2$$
$$= 378 \text{ kN·m}$$

The cross-sectional dimensions of the stiffening girder assumed is 600 mm wide by 1000 mm deep.

Area of cross-section $A = (600 \times 1000) = (6 \times 10^5)$ mm^2
Breadth of section = b = 600 mm
Overall depth = h = 1000 mm
Effective depth = d = 940 mm (Cover to main steel = 60 mm)

Using M-60 Grade concrete and Fe-415 HYSD bars, the limiting moment of resistance of the section is computed as,

$$M_{u,\lim} = 0.138 f_{ck} bd^2$$
$$= (0.138 \times 60 \times 600 \times 940^2)$$
$$= (4389 \times 10^6) \text{ N·mm}$$
$$= 4389 \text{ kN·m} > 378 \text{ kN·m}$$

According to IRC: 21-2000, the minimum amount of reinforcement in the section should be not less than 0.2 per cent of the gross cross-section. Providing 0.3 per cent of reinforcement, we have

$$A_{st(min)} = \left[\frac{0.3}{100} \times 600 \times 1000\right] = 1800 \text{ mm}^2$$

Provide 4 bars of 25 mm diameter at top and bottom of the section (A_{st} =1963 mm^2) Provide 10 mm diameter 4-legged stirrups at 300 mm centres. Provide 10 mm diameter bars as side face reinforcement in web at 300 mm intervals.

Figures 10.27, 10.28, 10.29 and 10.30 show the longitudinal elevation and cross-section of the bridge deck.

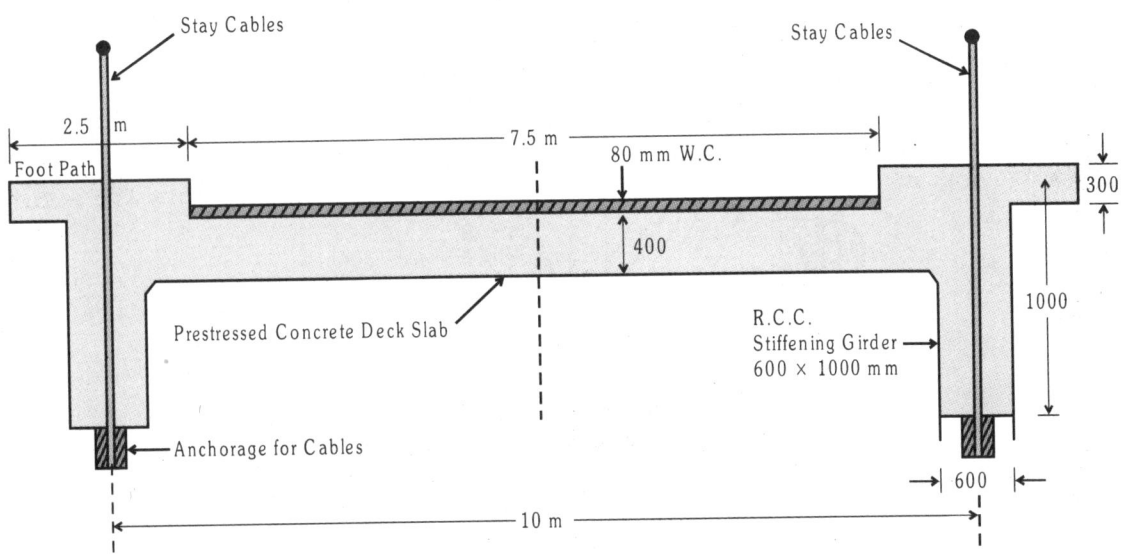

Fig. 10.28: Cross-section of Cable Stayed Bridge Deck

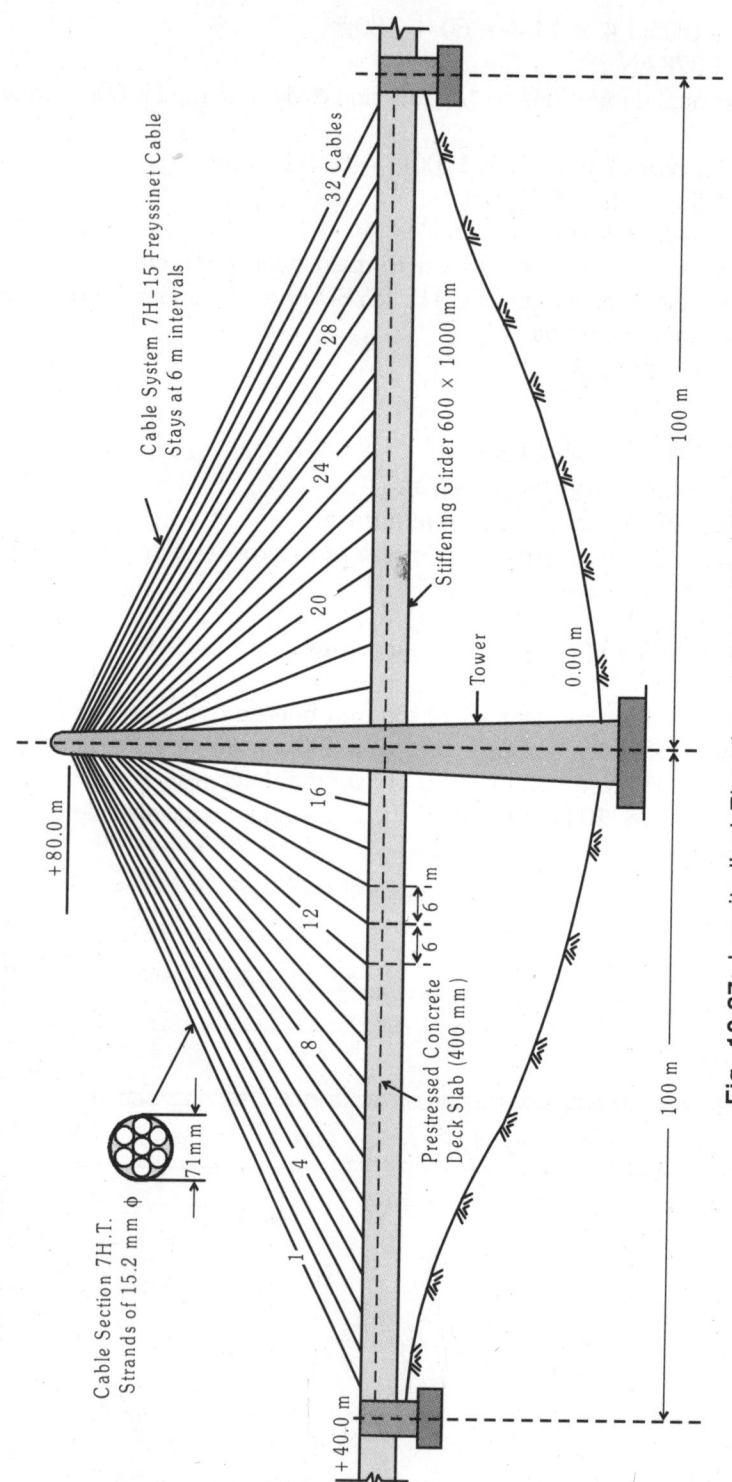

Fig. 10.27: Longitudinal Elevation of Cable Stayed Bridge

Prestressed Concrete Cable Stayed Bridges

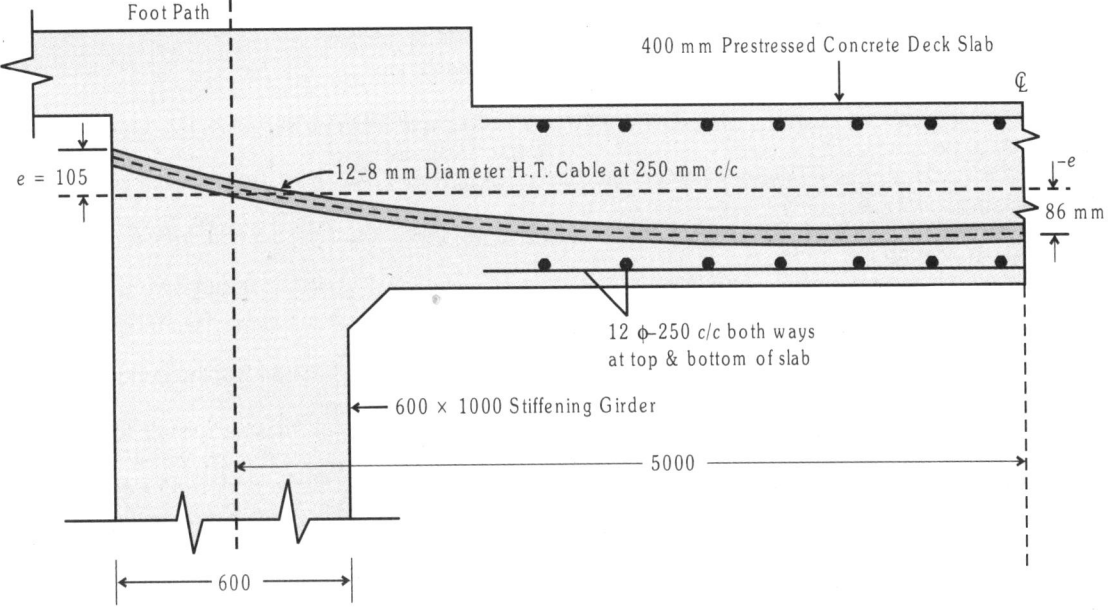

Fig. 10.29: Longitudinal Section of Prestresed Concrete Deck Slab

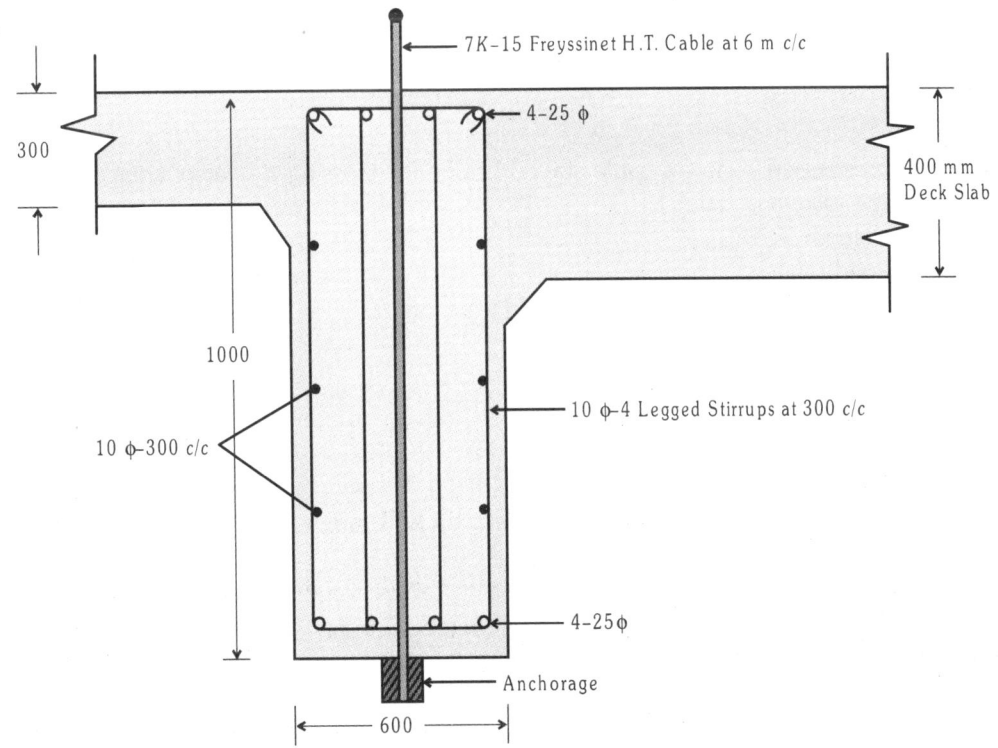

Fig. 10.30: Cross-section of Stiffening Girder

REFERENCES

1. RAINA, V. K., *Concrete Bridge Practice, Analysis, Design and Economics*. Tata McGraw Hill Publishing Co. Ltd, New Delhi, 1991, pp. 509-530.
2. KRISHNA RAJU, N., *Design of Bridges*, Third Edition. Oxford and IBH Publishing Co, Pvt, Ltd, New Delhi, 1998, p. 366.
3. LEONHARDT, F., *Latest Developments of Cable Stayed Bridges for Long Spans*. Bygningsstatiske Meddelsor, Copenhagen, 1974.
4. LEONHARDT, F. and W. ZELLNER, Cable Stayed Bridges, IABSE Surveys, S-13/80, *IABSE Periodical 2/1980*, May 1980.
5. MARSHALL, W. T. and H. M. NELSON, *Structures*. Pitman, London, 1970, pp. 1-442.
6. KRISHNA RAJU, N. and GURURAJA, D. R., *Advanced Mechanics of Solids and Structures*. Narosa Publishing House, New Delhi, 1997.
7. TROITSKY, M. S., *Cable Stayed Bridges, Theory and Design*. Crossby Lockwood Staples, London, 1977, p. 383.
8. SMITH, B. S., The Single Plane Cable Stayed Girder Bridge: A Method of Analysis Suitable for Computer Use. Proceedings of the Institution of Civil Engineers, London, Vol. 37, July 1967, pp. 183-194.
9. SMITH, B. S., A Linear Method of Analysis for Double Plane Cable Stayed Girder Bridges. Proceedings of the Institution of Civil Engineers, London, Vol. 39, January 1968, pp. 85-94.
10. TANG, M. C., Analysis of cable stayed girder bridges. *Journal of Structural Division*. Proceedings of the American Society of Civil Engineers, ST-5, 1481-1496, May 1970.
11. SCHREIER, G., North bridge at Dusseldorf, analysis, design, fabrication and erection of the bridge spanning the river. *Acier Sthal Steel*, No. 9, 1958 pp. 369-385.
12. LEONHARDT, F., Cable stayed bridges with prestressed concrete. *Journal of the Prestressed Concrete Institute*, Sept/Oct-1987 (Special report), pp. 52-79.
13. KRISHNA RAJU, N., *Prestressed Concrete* (Fourth Edition). Tata McGraw Hill Publishers, New Delhi, 2007, pp. 709-713.

EXERCISES

1. Design a prestressed concrete cable stayed bridge for crossing a river of span length 300 m satisfying the following data:

   Total span length = 150 m
   Width of Road way = 7.5 m
   Width of foot paths = 1.5 m on either side
   No. of towers: One at the centre
   Spacing of stay cables = 5 m
   Thickness of wearing coat = 80 mm
   Spacing of cross girders = 5 m
   Number of longitudinal girders = 2
   Dimensions of longitudinal girders = 500 mm by 800 mm
   Grade of concrete = M-60
   Type of cables: Standard Freyssinet prestressing cables comprising wires or strands
   Type of supplementary reinforcement: Fe-415 HYSD bars
   Loading: IRC Class AA tracked vehicle loading

   Design the deck slab, longitudinal girders and stay cables and sketch the details of reinforcements in the slab and girders. Sketch the longitudinal elevation of the bridge showing the stay cable arrangement along the span length. The design should conform to the National Codes IRC: 6-2000, IRC: 18-2000 and IRC: 21-2000.

2. A prestressed concrete cable stayed bridge has been proposed for a National Highway crossing to suit the following data:

Main span length between the pylons = 400 m
Number of pylons = 2
Width of road way = Four lane highway with a central median of 1.5 m
Number of cable planes = 2
Number of stiffening girders = 2
Spacing of cross girders = 6 m
Thickness of wearing coat = 80 mm
Foot paths: 1.5 m on either side
Type of loading: IRC Class AA or A whichever gives the worst effect
Spacing of cables = 6 m
Grade of concrete = M-60
Type of high tensile steel: Freyssinet standard wires and strand cables are available for use in girders and for stay cables

Design the deck slab, longitudinal and cross girders and stay cables and sketch the details of reinforcements in deck slab and girders and the elevation of the bridge showing the arrangement of stay cables. The design should comply with the recommendations in National Codes IRC: 6. IRC: 21 and IRC: 18-2000.

11

Planning and Economical Aspects of Prestressed Concrete Bridges

11.1 INTRODUCTION

Conceptual planning, critical analysis and comprehensive design are the logical steps to be followed before embarking on the construction activity of any prestressed concrete bridge project. Bridge design and construction have progressed significantly with the development of new and revolutionary materials and the advent of computers paving the way for exhaustive analysis. In fact, large number of today's existing bridges have been built only during the past fifty years or so and the rate of design and construction activity has been the highest particularly during the last two decades. The domain of construction activity comprises several known and unknown features such as planning and scheduling of the construction process to a time bound frame, management of materials and labour, mobilization of suitable cost effective techniques, treacherous foundation problems, adverse weather conditions, constant interaction with the design engineer, architect, site engineer, construction workers and the most important of all, the ability to take sound and daring decisions at times of crisis.

According to Raina[1], "Engineering is not just solving problems, nor is it a matter of blind adherence to graphs, design charts and formulae. It is more meaningfull to have an approximate solution to an exact problem rather than an exact solution to an approximate problem. Practical engineers must be more conceptual than mere perceptual, more creative than mere analytical and more visual than mere mathematical. Construction engineers should have wide experience involving several types of structures rather than isolated narrow specialization. Expertise and original skills are attained from relentless understanding and practice rather than mere theoretical knowledge. Good and sound judgment is attained from wide practical experience and often experience comes from bad judgment and wrong decisions.

11.2 STRUCTURAL FORMS FOR BRIDGES

Prestressed concrete is ideally suited for the super structure of medium and long span bridges. In fact, the development of the revolutionary material "Prestressed Concrete" by Eugene Freyssinet[2] facilitated the rapid construction of innumerable number of bridges in Europe destroyed in World War II. French and German engineers contributed immensely through research and practice for the widespread use of this material for various types of structures.

Planning and Economical Aspects of Prestressed Concrete Bridges

The structural configurations generally adopted for bridges are listed as follows:

1. Solid slabs (10-15 m)
2. Voided or hollow slabs (15-25 m)
3. Rigid frame bridges (15-30 m)
4. Balanced cantilever type bridges (20-30 m)
5. Tee beam and slab (20-40 m)
6. Continuous girders of variable depth (30-40 m)
7. Twin cell box girders (30-70 m)
8. Multi cell box girders (40-80 m)
9. Cable stayed bridges (100-500 m).

Prestressed concrete has more or less replaced reinforced concrete as the most suitable material for bridge construction due to its inherent advantages of high strength coupled with durability, fatigue resistance under repetitive loads, energy absorption under dynamic loads, freedom from cracks, easy mouldability to desired shape, economy and ease of maintenance

11.3 COST CONSIDERATIONS OF DIFFERENT TYPES OF BRIDGE DECKS

It is a well established fact that there is no single unique form of design which will always be the most economical in a particular set of conditions. A comparative analysis of several types of designs using the available materials together with the cost of construction utilizing the locally available labour will lead to an economical design.

However, in general, the quantities of concrete and steel expressed per unit area of bridge deck can be considered as indicative of economy although these figures are not the only ones which govern the overall cost of the bridge.

The various factors which influence the cost of a bridge are,

1. The length of individual spans
2. The type of cross-section of deck
3. The number of longitudinal girders
4. The total width of bridge deck
5. The depth and type of foundations, excavation, etc.
6. The cost of materials and labour
7. The cost of form work
8. The type of construction, such as, cast in situ or precast
9. The method of erection of precast segments
10. The time constraint for completion of the project.

Based on comprehensive analysis and extensive practical experience, Raina[3], has compiled the approximate quantities of concrete and steel required per unit area of deck with different types of bridge configuration, for a specified loading and design criteria as shown in Fig. 11.1. For spans less than 35 m, simply supported beams are the cheapest form of construction. As the span increases, the cost of simply supported beams increase rapidly and it is prohibitively costly for spans exceeding 60 m.

Continuous girder decks with unequal spans are more economical than equal spans in the span range of 50 to 100 m. Portal frames also use the same quantity of materials as those of continuous girder types with unequal spans. For spans less than 90 m, portal frames are slightly more expensive than continuous girder decks of unequal spans.

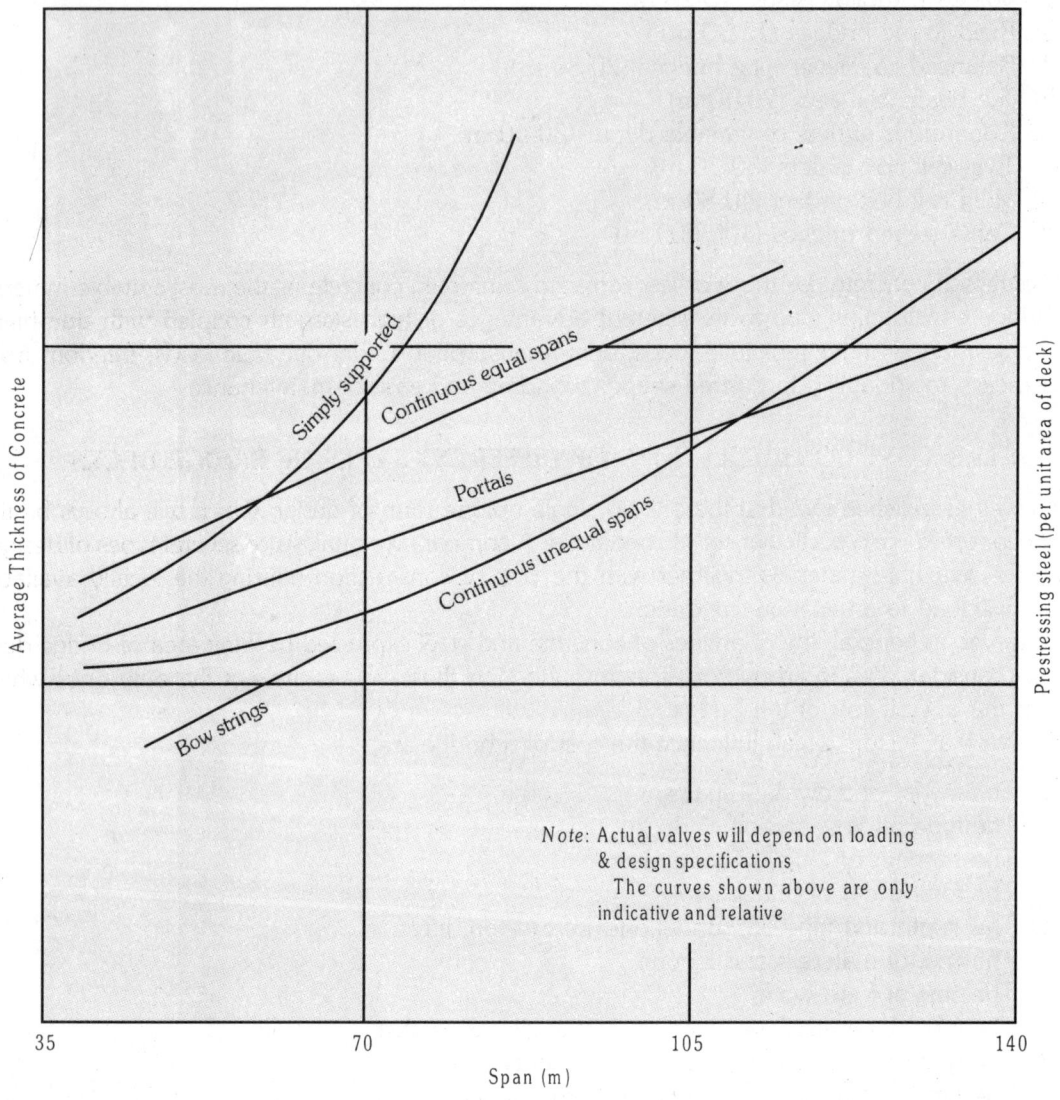

Fig. 11.1: Approximate Quantities of Concrete and Steel in Various Types of Bridge Decks

However, for very long spans especially in deep ravines, reinforced concrete arched bridges are more economical. Concrete arches are seldom prestressed because prestressing adds very little to the natural advantage of thrust in arch design. Bow string girders are economical, considering the material quantities for spans up to 50 m. The dimensions of the tie beam of the bow string girder bridge which develops tension, can be considerably reduced by axial prestressing resulting in overall economy. Although the bow string girder bridge is aesthetically superior to other types of girder bridges, the cost of form work being significantly higher, this type is rarely preferred.

The cross-sections of various types of prestressed concrete bridge deck configurations are shown in Fig. 11.2. Cast *in situ* post tensioned voided slabs are preferred for spans up to 40 m, either in simple or continuous spans. Span/depth ratios as high as 40 (plus or minus) have been used resulting in high torsional resistance and making it highly suitable for curved alignment with the bridge deck supported by single central column for arterial highways.

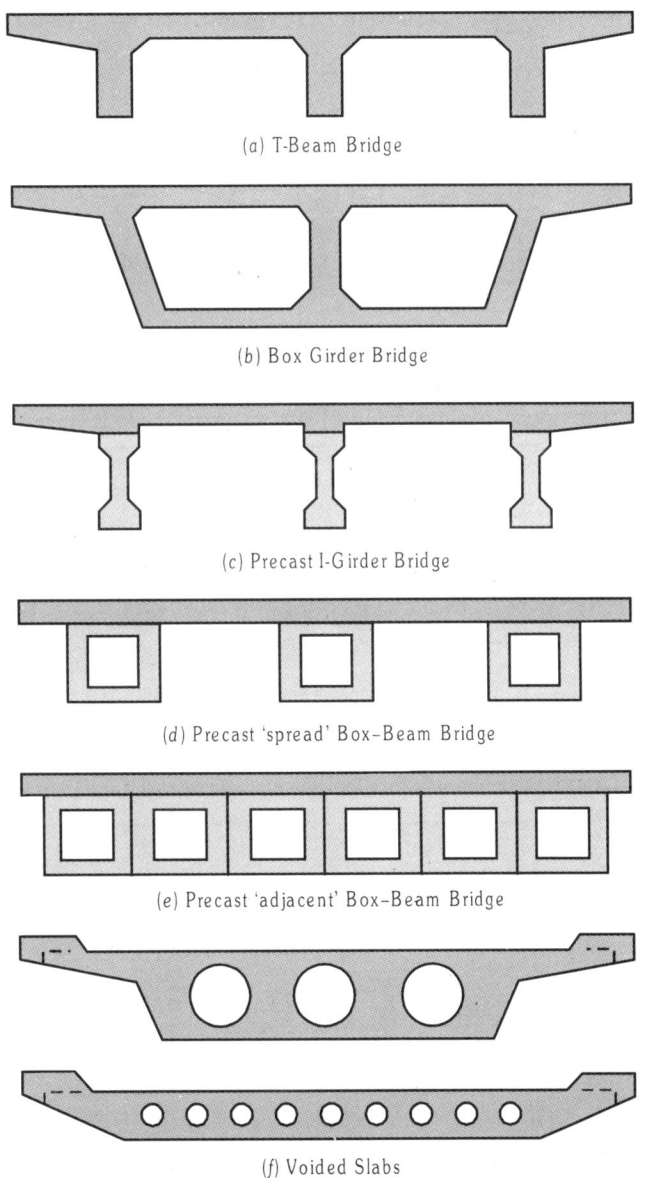

(a) T-Beam Bridge

(b) Box Girder Bridge

(c) Precast I-Girder Bridge

(d) Precast 'spread' Box-Beam Bridge

(e) Precast 'adjacent' Box-Beam Bridge

(f) Voided Slabs

Fig. 11.2: Typical Cross-sections of Prestressed Concrete Bridge Decks

Precast pretensioned or posttensioned voided slabs are economical for spans of 10-25 m with span/depth ratios of 25 to 30. The precast units can be placed using cranes resulting in minimum construction time and hence are ideally suited for busy city road flyovers. Precast tee, I and box shaped girders with cast *in situ* slab are suitable for spans up to 50 m. The span/depth ratios vary from 18 to 20 for simply supported spans to 25 to 30 for continuous spans. Raina[3] has presented graphically the variation in the quantities of concrete, prestressing steel and reinforcement with the average thickness of concrete deck for different types of structural configurations generally used in Fig. 11.3. In general, box sections are the most economical types requiring the least concrete content, prestress and reinforcement compared to a voided slab for a given depth of bridge deck.

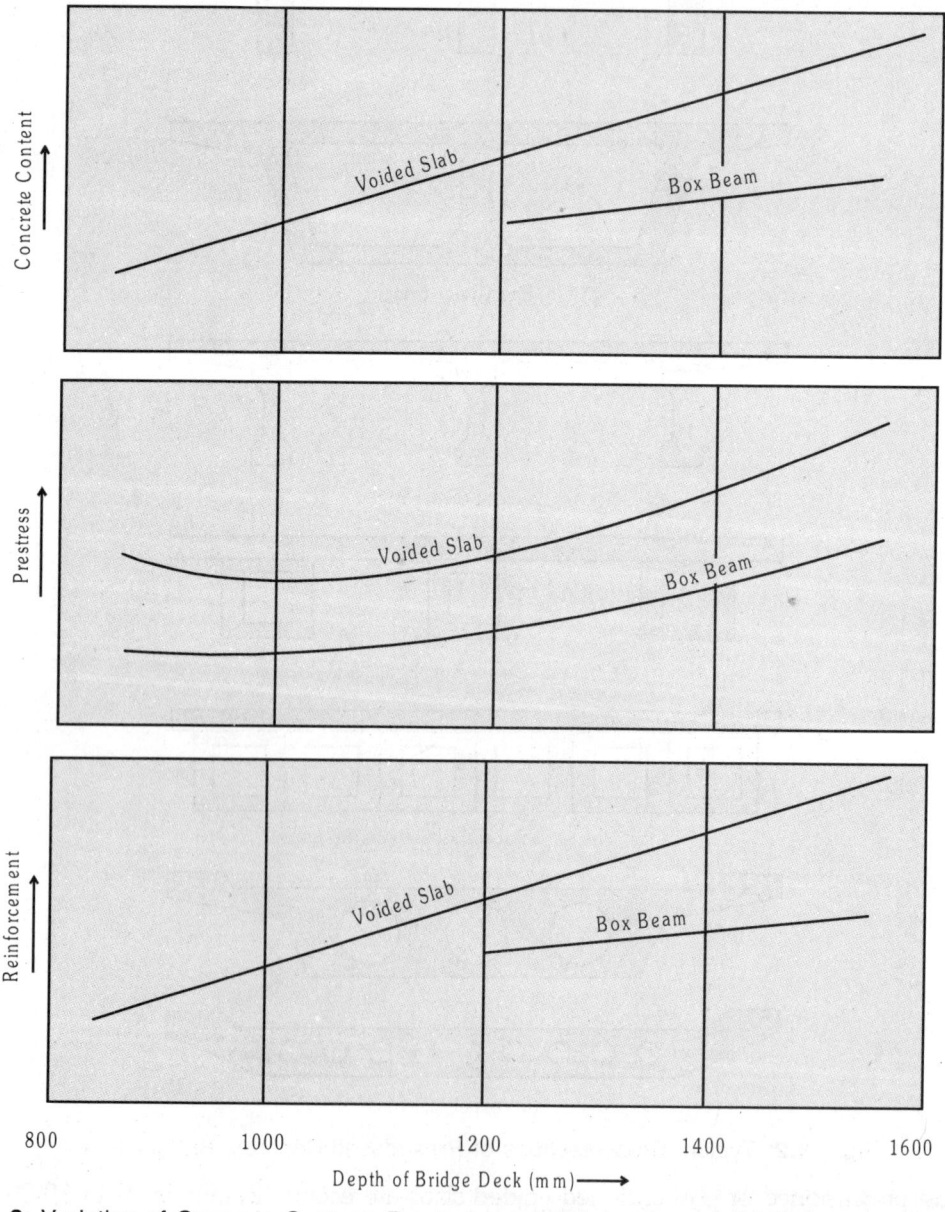

Fig. 11.3: Variation of Concrete Content, Prestress and Reinforcement with Depth of Bridge Deck

Planning and Economical Aspects of Prestressed Concrete Bridges

The variation of cost per unit area with the depth of bridge deck using prestressed voided slab and box beams is examined in Fig. 11.4. Also Fig. 11.5 shows the variation of cost of these two bridge configurations as a function of span. In general for longer spans, box sections are more economical in overall costs while voided slabs are preferable for short spans like culverts.

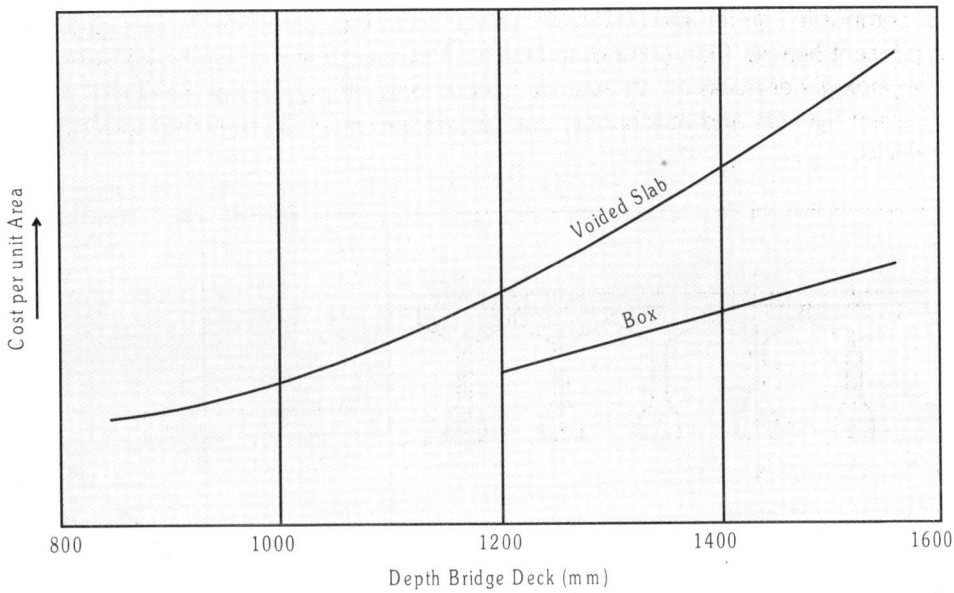

Fig. 11.4: Variation of Cost per unit Area of Deck with Depth of Bridge Deck

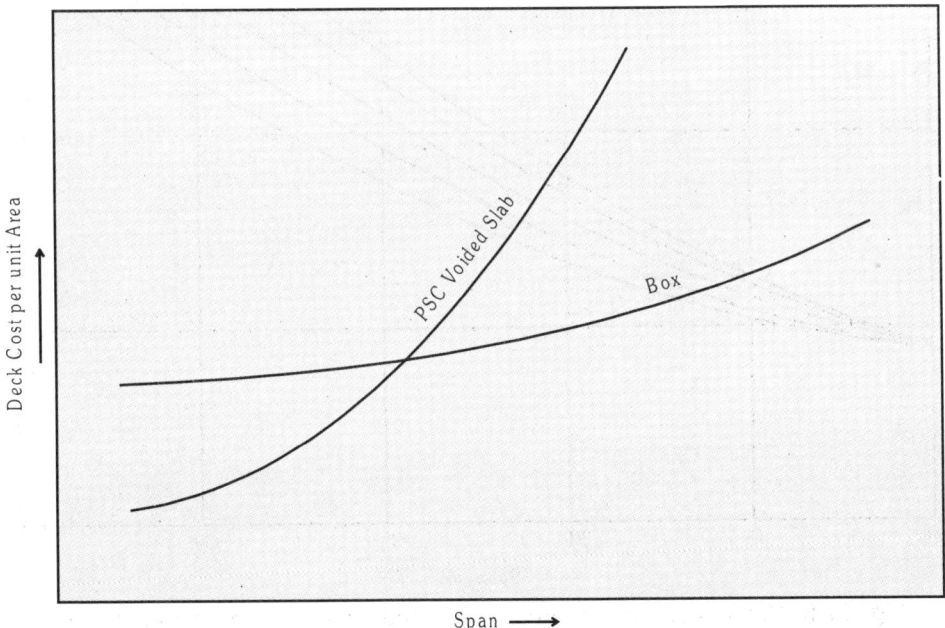

Fig. 11.5: Variation of Bridge Deck Cost with Span and Type of Cross-section

Investigations by Sarkar *et al.*[4] in 1969 based on rigorous computational analysis have examined the variation of the cost of bridge deck in relation to span and the number of longitudinal girders in a prestressed concrete tee girder bridge. Although the cost was more or less the same when the number of girders varied from 5 to 9 for spans up to 15 m, the cost increased with the increasing number of girders for spans in the range of 15 to 35 m as shown in Fig. 11.6. A comparative analysis of composite I girder and box girder bridge decks indicated that the box girder decks were about 23 per cent heavier than composite I decks. However if the minimum construction depth is the criterion, box girder decks on an average needed only 55 per cent of the depth of a composite I deck. In urban flyovers and autobahns, box girders are generally preferred to restrict the depth of bridge decks.

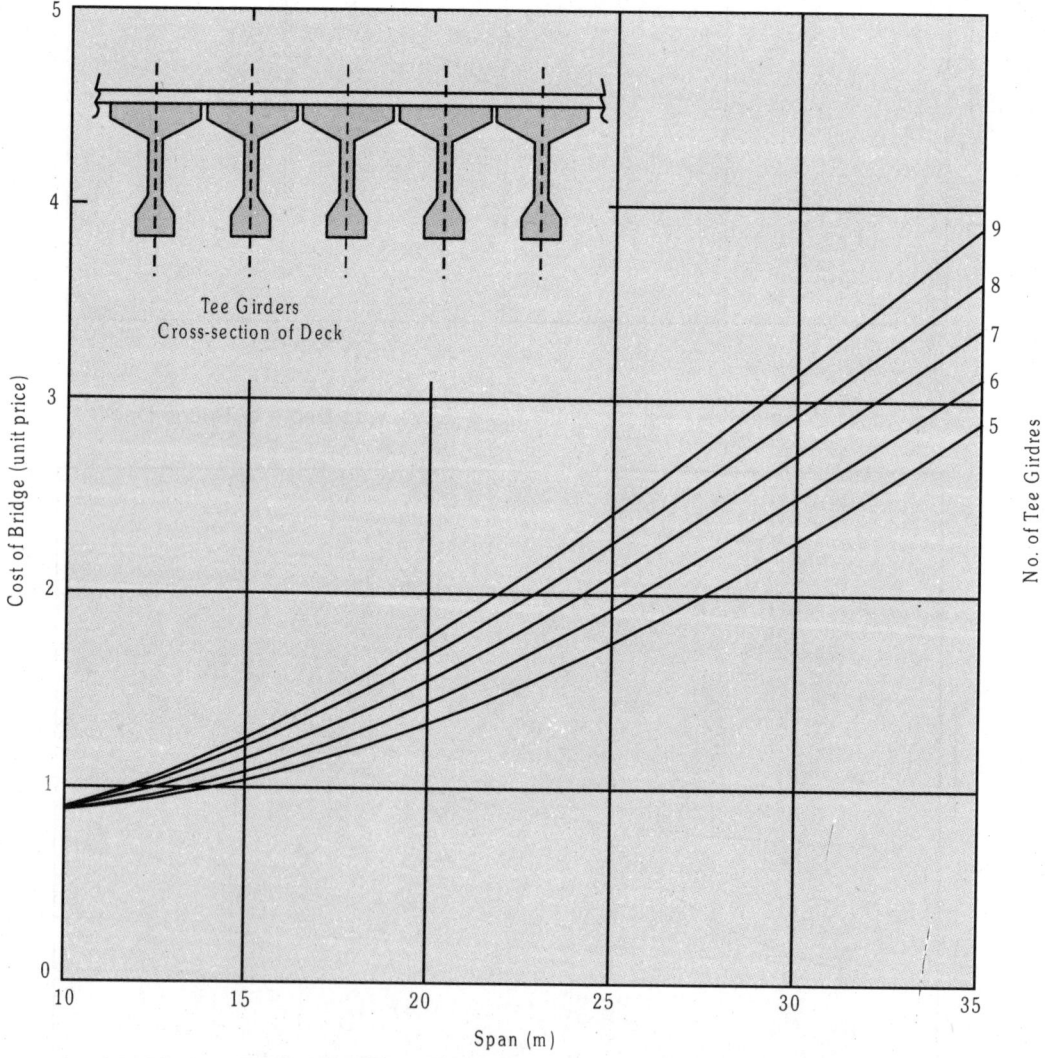

Fig. 11.6: Variation of Cost of Bridge with Span and Number of Tee Girders

A critical survey of the various types of bridges built during the last three decades, clearly indicates the economical advantages of cable stayed bridges especially for long spans. Prof. Leonhardt[5] designed the first cable stayed bridge across the river Rhine in Dusseldorf in 1952. During the last two decades, hundreds of cable stayed bridges have been built in different countries with spans in the range of 100 to 624 m as outlined in Table 10.1.

Cable stayed bridges are preferred to conventional suspension bridges for long spans mainly due to the reduction in bending moments in the stiffening girder resulting in smaller section of the girders leading to considerable savings in overall costs. Economic analysis by Fritz Leonhardt[6] indicates that cable stayed bridges are structurally efficient and cost effective for low, medium and long spans varying from 40 to 1800 m. Highway bridges can be built of prestressed concrete with spans up to 700 m and rail road bridges up to a span range of 400 m.

Comparative studies have indicated that cable stayed bridges are much more economical, stiffer and aerodynamically safer when compared with suspension bridges. Economic analysis conducted by Leonhardt has clearly shown that for a bridge of 1800 m main span and 38 m width, a suspension bridge requires 46000 t of steel whereas a cable stayed bridge needs only 19200 t indicating more than 50 per cent savings in the quantity of steel required in the cable stayed bridge (Refer Fig. 10.21).

The second Hooghly Bridge (Vidyasagar Sethu) at Kolkata is an excellent example of cable stayed bridge comprising a main span of 457 m and two side spans of 182 m each. The deck is made up of concrete slab 230 mm thick, two outer steel I-girders 28.10 m apart and a central I-girder. The deck is suspended by cable stays comprising parallel wire cables of B.B.R.-HIAM type with their proprietary anchorage system. The bridge provides for two 3-lane carriage ways 12.3 m each with 2.5 m foot paths. The cable stayed bridge costing 600 million rupees was found to be cost effective in comparison with other types.

The Bassien Creek Bridge shown in Fig. 11.7 and constructed in 1970, is an excellent example of long span prestressed concrete bridge built in the state of Maharashtra, using precast segments of variable depth prestressed to form continuous spans. The bridge is made up of two cantilever continuous spans of 114.6 m and four spans of 57.3 m built using precast girders together with two end spans of 48.5 m with a total length of 556 m. One of the notable contributions of India

Fig. 11.7: Bassien Creek Bridge, Maharashtra

to the technology of bridge engineering is the prestressed concrete submersible bridge which remains closed for traffic under high flood conditions for a few hours. The Krishna Bridge at Deodurg shown in Fig. 11.8 belongs to this type. The bridge is 549 m long and has a box decking of aerofoil design having three compartments. The deck is continuous over a length of 183 m, comprising 6 spans of 30.58 m each and supported on Teflon bearings. The bridge consists of 1.2 m long precast segments erected on staging and prestressed after joining the segments with epoxy resin. The bridge located in Karnataka state was built by Gammon India Ltd.

The world's tallest and longest bridge is a cable stayed bridge located outside the French town of Millau extending over a length of 2.46 km. The French bridge is considered to be an engineering marvel since some of the bridge pillars rise gracefully to a height more than 300 m. The bridge was opened for traffic in December 2004. Cost analysis of various types of bridges in U.S.A. also indicate that the cost per unit area of a bridge deck is the least for cable stayed bridges in the span range of 60 to 400 m. During the last four decades, hundreds of prestressed concrete bridges have been built to suit the local conditions. The reader may refer to the introductory article by Subba Rao[7,8] and Krishna Raju[9,10] for a critical survey of prestressed concrete bridges in India.

11.4 ECONOMIC EVALUATION

The form of structural components of a bridge chosen should be functional and well proportioned for the conditions imposed by the layout and location of the structure.

In coastal areas exposed to aggressive environmental conditions, concrete bridges should be preferred to steel bridges to minimize maintenance costs. According to Raina[1], the appropriateness of various forms of construction to differing situations and ranges of span is a judgment usually attributed to professional experience of bridge engineers. It is certainly a waste of time and effort to indulge in lengthy computations on optimization for component parts of a bridge system within a solution which in itself inappropriate. Generally, most of the engineering design situations involve myriad practical constraints on the proportions and every bridge is unique with its own practical constraints.

It is important to note that there are no permanent or guiding rules about the comparative costs of differing solutions for construction, because changes take place in the relative cost and availability of labour and materials with the time factor and developments in constructional techniques. Also there are inevitable difficulties in forecasting the cost of bridge projects during times of economic uncertainty or rapid inflation. Economy of a particular bridge system should be examined by considering the various costs of materials of construction such as, concrete, reinforcing and prestressing steel, etc. expressed per unit area of bridge deck together with the cost of construction. Production of very high strength and high performance concrete together with revolutionary developments in prestressing tendons and innovative construction techniques have paved the way for selection of prestressed concrete as the choicest material for construction of bridges.

Planning and Economical Aspects of Prestressed Concrete Bridges

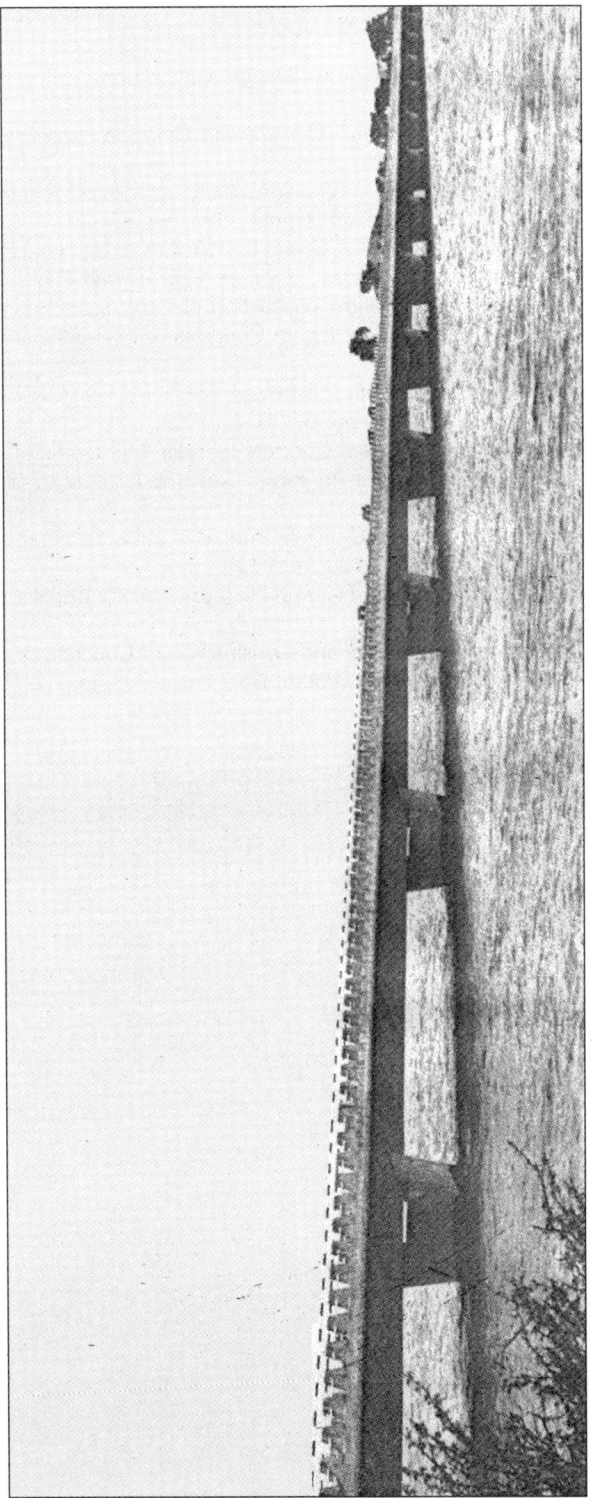

Fig. 11.8: Krishna Bridge at Deodurg

REFERENCES

1. RAINA. V. K., *Concrete Bridge Practice, Analysis, Design and Economics*. Tata McGraw Hill Publishing Co. Ltd., New Delhi, 1991, pp. viii.
2. FREYSSINET, E., "The Birth of Prestressing". *Cement and Concrete Association Translation*, No. CJ. 59, London, 1956, p. 44.
3. RAINA. V. K., *Concrete Bridge Practice, Construction, Maintenance and Rehabilitation*. Tata McGraw Hill Publishing Co. Ltd., New Delhi, 1988, Second Reprint 1993, pp. 87-111.
4. SARKAR, S., KAPLA. M. S., PRASADA RAO. A. S. and CHHAUDA. J. N., Hand Book of Prestressed Concrete Bridges. Structural Engineering Research Centre, Roorkee, Uttaranchal, 1969, pp. 226-230.
5. LEONHARDT. F., New Trends in Design and Construction of Long Span Bridges and Viaducts (skew, flat slabs, torsion box), Preliminary Publication of Eighth Congress of International Association for Bridge and Structural Engineering, New York, 1968.
6. LEONHARDT. F., Cable Stayed Bridges with Prestressed Concrete. *Prestressed Concrete Institute Journal*, September-October-1987 (Special Report), pp. 52-79.
7. SUBBA RAO. T. N., Application of Prestressed Concrete in India, Outstanding Concrete Structures of India, Compiled by Maharashtra India Chapter of American Concrete Institute's Fall Convention at New York, 1984, pp. 1-5.
8. SUBBA RAO. T. N., Long Span Prestressed Concrete Bridges in India, Seminar on Problems of Prestressing, Preliminary Publication, Madras, 1970, pp. I-113 to I-130.
9. KRISHNA RAJU. N., Advances in Design and Construction of Concrete Bridges. *Construction India Manual*, Bombay, Vol. 134, 1992, pp. 50-53.
10. KRISHNA RAJU. N., Developments in Design and Construction of Concrete Bridges. *Civil Engineering and Construction, Special Publication on Bridges*, Vol. 6, No.7 July 1993, pp. 12-15.

12. Construction of Prestressed Concrete Bridges

12.1 INTRODUCTION

Construction domain of modern prestressed concrete bridges has been revolutionized by rapid developments in the field of materials of construction, types of form work, and innovative construction techniques together with the application of modern management principles during the last few decades. The present day approach in construction management comprises several diverse functionaries such as architects, structural analysts and designers, estimators, construction engineers, field supervisors, accountants, financial managers, corporate secretaries, tax planners working under professional managers.

The ten major basic construction management functions for the successful execution of a bridge project as identified by Raina[1] are listed below:

1. Preparation of bridge project, tendering and winning the contract
2. Negotiations on contractual obligations like time and financial aspects
3. Developing liaison with clients
4. Mobilizing financial resources and labour for timely completion of project
5. Project work planning
6. Supervision of project work
7. Monitoring and control of progress in project work
8. Maintenance of good labour relations
9. Maintenance of proper accounts of flow of funds and expenditure
10. Engineering aspects and completion of project in time

Modern management techniques like Critical Path Method (CPM) and Project Evaluation and Review Techniques (PERT) are widely used in the management of major project works. The major activities like work scheduling, management of materials, controlling the various activities of the project and updating the various tasks have become more simpler and they can be efficiently handled with less paper work with the advent of computers.

12.2 HIGH STRENGTH CONCRETE MIX CONSIDERATIONS

The two most important constituents of prestressed concrete which will influence the quality of bridge decks are (a) Concrete and (b) High tensile steel tendons. The quality of steel is more or less assured since it is manufactured in a factory under intensive supervision. Concrete is generally made at the site of bridge construction and hence it is essential to consider the minimum requirements for material and workmanship which will result in a bridge structure performing satisfactorily at various limit states[2, 3, 4].

The most important consideration in the construction of prestressed concrete bridges hinges on the design, production and control of high strength and high performance concrete with desirable properties. The following information is essential for making any controlled concrete mix.

1. **Grade of Concrete:** The grade of concrete specified in the Indian Standard Code IRC: 18-2000[5] varies from a minimum of M-35 with values of permissible stresses listed up to a maximum of M-60. The code also specifies that only 'Design Mix Concrete' should be used to ensure the desired strength and durability. The designed concrete mix should conform to the specifications prescribed in the Indian Standard Code IRC: 21-2000[6]. The reader may refer to Section 3.2 for the methods of design of high strength concrete mixes and Section 3.5 for permissible stresses in concrete.

2. **Cement Requirements:** The IRC: 21-2000 Code also prescribes the minimum cement content of 400 kg/m^3 for all exposure conditions for prestressed concrete bridges. The maximum cement content is limited to 540 kg/m^3 mainly to limit cracking due to drying shrinkage. The maximum water/cement ratio is limited to a value of 0.4 for all exposure conditions mainly for durability considerations. The various types of cements like Ordinary Portland, Portland Slag, Portland Pozzolana, Portland Blast Furnace Slag are recommended for use in bridge structures. In situations where sulphate attack is likely, Sulphate resisting Portland Cement should be preferred (Refer Table 5 of IRC: 21-2000).

3. **Types of Aggregates:** Coarse aggregate consisting of crushed granite or gravel with a nominal size of 20 to 40 mm is generally preferred. However the maximum size of aggregate should not exceed one quarter of the minimum thickness of the structural member or 10 mm less than the minimum lateral clear distance between individual reinforcements or 10 mm less than the minimum clear cover to any reinforcement. Also the nominal maximum size of the aggregate shall be 5 mm less than the spacing between the cables, strands or sheathings. Aggregates containing particular varieties of silica, which are susceptible to attack by alkalis present in cement resulting in expansive reaction should be avoided. Crushed granite aggregates are better than rounded gravel for high strength concrete.

4. **Workability of Fresh Concrete:** The water/cement ratios generally used in the production of high strength concrete usually varies in the narrow range of 0.3 to 0.4. Consequently the workability of concrete will be very low requiring admixtures to improve the flowability of fresh concrete. Fresh concrete deposited into the moulds or form work needs proper compaction using form and immersion vibrators to expel air bubbles and improve the density and strength of hardened concrete. Concrete which is more workable than necessary (with more water or more cement content and sand) for the actual job conditions will often result in concrete of lower strength, durability and abrasion resistance.

Construction of Prestressed Concrete Bridges

The aim should be to use the lowest percentage of fine aggregate which is consistent with the job conditions and which will permit placement of concrete without honeycombs and finishing it to a satisfactory surface. The use of super plasticizing admixtures is common in the case of concrete with low to extremely low workability. The plasticizers in conjunction with vibration will improve the mobility and finishability of concrete besides resulting in dense concrete.

5. **Water/Cement Ratio:** Water/cement ratio is the principal parameter controlling the strength and properties of hardened concrete and hence the primary step in proportioning a high strength concrete mix should be the selection of an appropriate water/cement ratio depending upon the type of cement used to achieve the desired durability and compressive strength of hardened concrete intended for the bridge work. Lower water/cement ratios generally result in the following advantages:

   1. Increased density with less voids
   2. Increased compressive, flexural and tensile strength
   3. Increased water tightness
   4. Increased resistance to weathering
   5. Superior dimensional stability
   6. Lower absorption of moisture due to less porosity
   7. Lesser shrinkage cracks
   8. Increased modulus of elasticity of concrete
   9. Reduced creep and long term deflections
   10. Increased fatigue resistance

   The reader may refer to a separate monograph[5] by the author for more information on the design of high strength and high performance concrete mixes using the Indian, British and American Concrete Institute methods.

12.3 BATCHING AND MIXING OF CONCRETE

Weigh batching and machine mixing are essential to produce concrete of uniform quality. Good specification requires that batching is done by weight rather than by volume. Weigh batching ensures greater accuracy, simplicity and flexibility. Specifications require that the main ingredients, cement, aggregates and water are measured to an accuracy of ±0.5%. Homogeneity in concrete improves if cement, fine aggregate, and coarse aggregate are simultaneously charged into the mixer. Chemical admixtures should be added into the mix in the form of solution and the liquid should be considered as part of the mixing water. Admixtures that cannot be added in solution may be measured by volume as directed by the manufacturer. Daily checking of admixture dispensers is essential to prevent over dosages leading to serious problems in both fresh and hardened concrete.

Concrete may be mixed using stationery mixers on job site or central mixers as used in ready mixed concrete plants. Under normal conditions, up to about 10% of the mixing water should be placed in the mixer drum before the solid materials are added. Thereafter water should be added uniformly with the solid materials leaving about 10% to be added after all the other materials are in the mixer drum. When mixed in a central mixing plant, the mixing time should be not less than 50 seconds nor more than 90 seconds. Ready mixed concrete is produced by any one of the following three methods of mixing:

1. Centrally mixed concrete is mixed completely in a stationery mixer and is delivered either in a truck agitator or a special non agitating truck.
2. Transit mixed concrete is mixed partially in a stationery mixer and then mixing is completed in a truck mixer.
3. Truck mixed concrete is mixed completely in a truck mixer during transportation. When truck mixer is used, a minimum of 70 to a maximum of 100 revolutions of the drum or blades at mixing speed are required for complete mixing.

12.4 PLACING OF CONCRETE IN FORMS

Bridge girders being massive due to their large depth, concrete should be placed in horizontal layers of not more than 300 mm thick. When less than a complete layer is placed in one operation, it should be terminated in a vertical bulk head. Each layer should be placed and compacted before the preceding batch undergoes the initial set to prevent injury to the green setting concrete and to avoid surfaces of separation between the adjacent layers. In the case of columns with taller lifts, suitable retarding agents should be used. In the case shallow beams, concrete should be deposited preferably for the full length and brought up evenly in horizontal layers. Concrete in slab panels should be placed in one continuous operation for each span.

In the case of continuous bridge decks, the floors slabs and girders should be placed in one continuous stretch unless otherwise specified, in which case a special shear anchorage should be provided to ensure monolithic action between the girders and floor slab. In the case of tee beam slab floors, concrete is preferably deposited up to the top of the beam ribs followed by concreting of the slab in one continuous operation. If the slab concrete is delayed, suitable shear keys should be formed by roughening the top of the beams before depositing the concrete in the slab.

12.5 COMPACTION OF CONCRETE BY VIBRATION

Bridge deck concrete deposited in slabs and beams should be compacted by mechanical vibration except when certain types of extrusion are used which consolidate the concrete by tamping. Depending upon the type of structural members, internal external or surface type vibrators are used for compaction of concrete. Normally vibrators with frequencies in the range of at least 3200 to 3600 cycles per minute are generally used to achieve good compaction.

The following techniques of vibration are recommended to achieve well compacted concrete.

1. Vibration should be well distributed so that the concrete reaches a state of plastic mass with uniform density.
2. Vibrators should be used for compaction only and not for moving concrete horizontally along the forms.
3. When vibrators are used for compacting concrete in slabs and beams, the spacing points of vibration should be such that the zones of influence overlap.
4. When concrete is deposited in vertical members like columns or walls, the vibrator should be inserted vertically and allowed to sink due to its own weight to the bottom of the layer and then slowly withdrawn. During vibration of succeeding layers, the vibrator should preferably penetrate the surface of the preceding layer by at least 150 mm. Good vibration should result in a surface without honey combing, aggregate or mortar pockets or excessive air bubbles.

Construction of Prestressed Concrete Bridges

12.6 RHEODYNAMIC CONCRETE

Revolutionary developments in the field of Polymer technology and Nano science has made inroads in the field of concrete technology. Rheodynamic concrete generally referred to as self compacting concrete (SCC) is able to flow under its own weight and completely fill the form work, even in the presence of dense reinforcements without the need for any vibration whilst maintaining homogeneity and resulting in concrete of high early strength and durability. Degussa-MBT Construction Chemicals (India)[6] have developed innovative type of admixtures using nano polymers which can be used to bring together functional groups aimed at targeted performances in concrete. Based on nano science, they have developed a range of nano polymers with the following applications:

1. **Zero Energy System:** A system of polymers with longer side chains and shorter main chains to facilitate high early strength in concrete without steam curing and with specific applications in precast reinforced and prestressed concrete unit manufacturing industry.

2. **Glenium Sky:** A custom made nano polymer which facilitates long haul concrete mix stability with development of high early strength coupled with high durability of hardened concrete suitable for prestressed concrete fly overs.

3. **Rheo Fit:** A nano polymer range which meets wider expectations such as aesthetics, economics, durability and performance of manufactured concrete products.

The application of nano technology in the production of nano polymers has revolutionized the concrete industry. Self compacting concrete is commonly used in Europe, Japan, North America, Singapore and Taiwan. In India, this technique has gained popularity with some of the major projects such as Delhi Metro Rail Corporation, Nuclear Power Corporation and Indian Space Research Organization.

12.7 EXPANSION JOINTS FOR BRIDGE DECKS

According to Lee[7], construction and expansion joints are invariably provided in all concrete bridge decks at locations of structural discontinuity between two elements. Construction joints should be planned in advance and preferably they should be located at points of minimum shear and they should be nearly perpendicular to the principal lines of stress. Keys should be made by embedding water soaked beveled timbers in soft concrete and they should be removed after the concrete has set. When the work is resumed the surface of the concrete previously placed should be thoroughly cleaned of dirt, scum, laitance, loosely projecting aggregates and other soft material using stiff wire brushes. The surface should then be thoroughly soaked with clear water for two or three hours before further concreting using a thin layer.

Construction joints are generally either vertical or horizontal. In road way slabs, construction joints shall be formed vertical and in true alignment. Shear keys in construction joints should be constructed as shown in working site plans. In the case of box girder webs, these shear keys are normally shown on the plans to the full width. Before resuming the day's concrete work, all joints should be roughened and prepared for the next pour of concrete in accordance with the standard specifications. An expansion joint implicitly also refers to a contraction joint and hence it is more rational to designate it as a movement joint. These joints become necessary due to the following reasons:

1. Differential shrinkage of concrete
2. Creep or inelastic deformation of concrete
3. Thermal expansion and contraction of the super structure and in some cases even the sub structure.
4. Elastic shortening of concrete due to prestress
5. Displacements of the structure under load or any other action.

The joint should be designed to allow free translation, deflection and rotation of the structure at the edges without damage or inconvenience to the traffic. The expansion joint should be strong enough to withstand the knocking of wheels of vehicles passing over the bridge deck. The modern trend is to adopt elastomeric (Neoprene) compression seals for expansion joints in bridge decks. Different types of standard compression seals are shown in Fig. 12.1. These elastomeric seals are made of polychloroprene otherwise known as neoprene. Single compression seal is used for small movements and newer type modular joint system using several seals are preferred to accommodate larger movements. Typical details of expansion joints for small and large movements are shown in Fig. 12.2 (a) and (b) respectively.

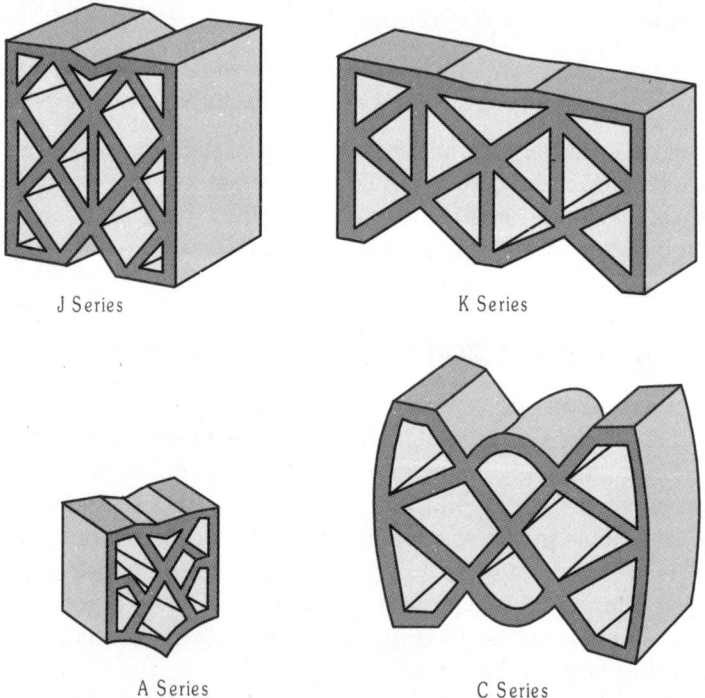

Fig. 12.1: Standard Compression Seals for Expansion Joints

There are many organizations which manufacture the proprietary expansion joints. According to Raina[8], the well known brands are: Maurer, Freyssinet International, Thormajoint, Transflex, Waboflex, Gutehoffnungshutte (GHH) and Demag. In selecting an appropriate joint the engineer must carefully weigh the manufacturer's claims on performance against the actually required functions of rotation and translation and the required directions of the incumbent movement, the fixing details, the initial costs and the maintenance problems. Strict adherence to quality controls

Construction of Prestressed Concrete Bridges

and actual performance record should be the main criteria for impartial selection of a particular brand of joint seal.

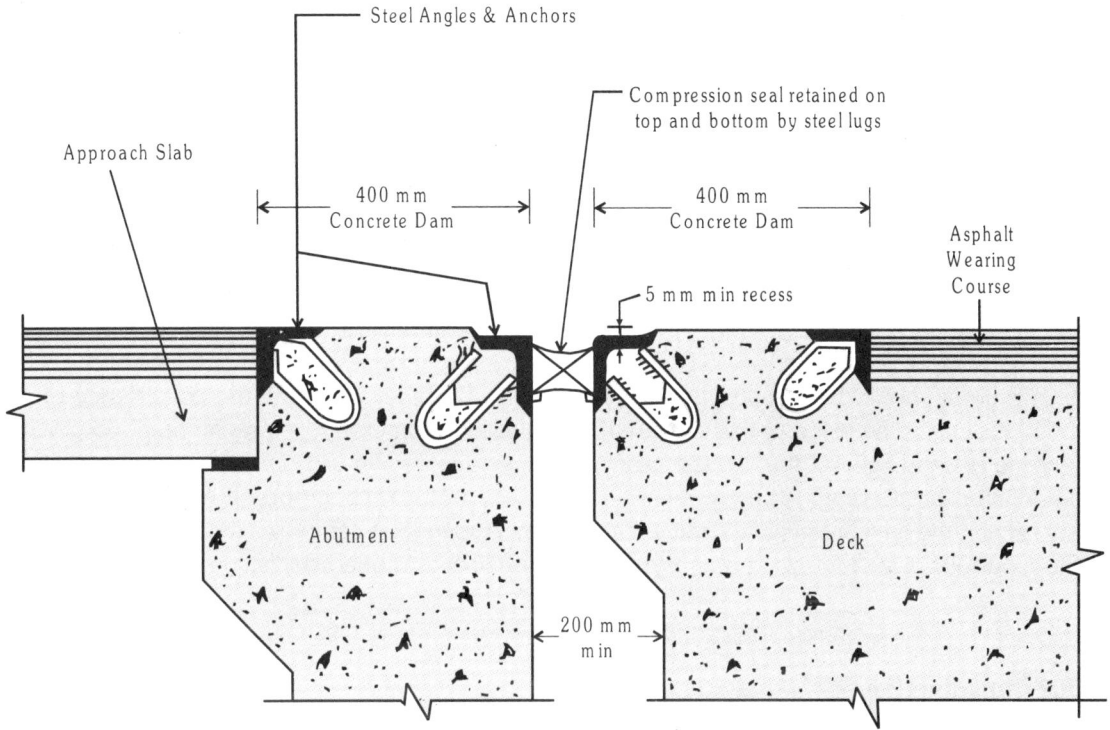

(a) Expansion Joint for Small Movements

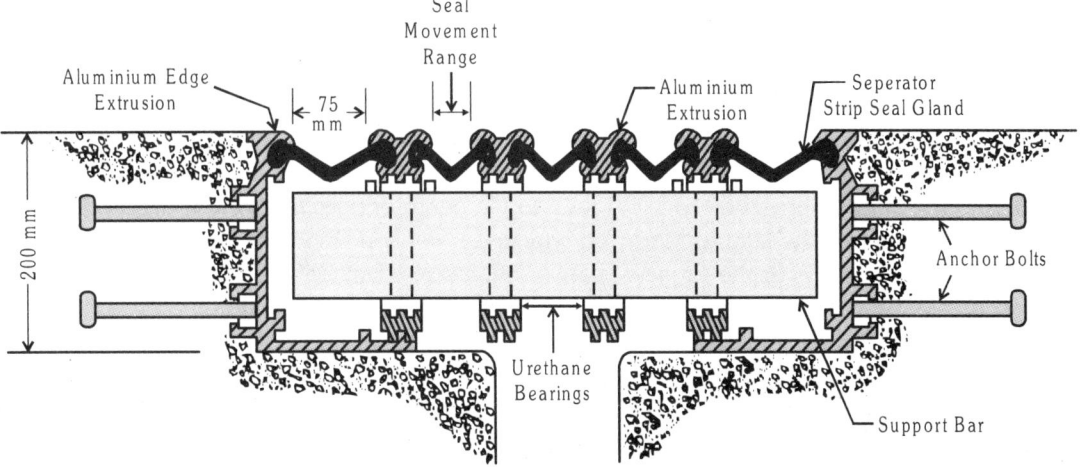

(b) Expansion Joint for Large Movements

Fig. 12.2: Expansion Joints for Bridges

12.8 ASSEMBLY OF PRESTRESSING STEEL AND GROUTING OF DUCTS

Prestressing tendons are generally made up of high strength wires, strands, bars or cables containing wires or strands. Long span highway bridge girders require large prestressing forces and hence they are generally prestressed using cables containing strands. The mechanical properties of various types of high tensile steels are compiled in Section 3.3. Prestressing steel should be carefully and accurately located in the exact position indicated in the design drawings with a permissible tolerance of ±5 mm. Coupling units used for joining of high tensile tendons should have an ultimate strength not less than the individual strengths of the wires or strands being joined. Welding is not permitted for joining of high tensile tendons.

The coils of high tensile steel, sheathing and anchorages should be stored at site in such a way as to provide them with adequate corrosion protection. After stressing operations, the tendons in the cables should be protected by pressure grouting using cement slurry. Grouting is commenced initially with a low pressure of up to 0.3 N/mm^2, increasing it until the grout emerges at the other end. The grout should be allowed to flow freely from the other end until the consistency of the grout at this end is the same as that of the grout at the injection end. It is a recommended practice to provide a stand pipe at the highest point of the tendon profile to hold all water displaced by sedimentation or bleeding. After grouting is completed, the projecting portion of the vents should be cut off and the face protected to prevent corrosion. Appendix 4 comprises the recommendations of the Code IRC: 18-2000 regarding the grouting of post-tensioned ducts.

12.9 LONG SPAN BRIDGE CONSTRUCTION TECHNIQUES

12.9.1 Introduction

Over the last several decades, rapid developments and innovations in the domain of prestressed concrete bridges have resulted in novel methods of construction along with the advances in materials of construction. Prestressed concrete being ideally suited for long spans, its application is much less for sub structures than for super structures in bridge decks. Although Freyssinet invented the most exciting building material and christened it as **"Prestressed Concrete"** in 1928, the big boom in its application came only after the Second World War. Prestressed concrete dominated the bridge construction during the post war period in Germany and Europe. Out of the 500 bridges built in war torn Germany during 1948-53, seventy per cent of them were built using prestressed concrete.

New construction techniques rapidly developed in America, Russia, Europe, Japan and even in India during the period from 1960 to 1970 resulting in beam bridges reaching spans up to 160 m. Morandi's bridge across the lake of Maricabo was under construction with spans of 235 m by the help of stay cables. In 1970, the longest span of beams reached 230 m in Japan and for cable stayed concrete bridges, spans up to 300 m were successfully implemented.

Cantilever construction technique developed by Ulrich Finisterwalder[9] revolutionized the prestressed concrete bridge construction by eliminating ground supporting form work for long span bridges. The span range for economical application of box beam bridges built by cantilever construction technique is believed to lie normally in the range of 50-200 m. However, Hamana bridge[8] in Japan has pushed this upper limit further with a span of 240 m and Urado bay bridge in Japan has a span of 270 m built by using the technique of cantilever construction.

A critical survey indicates that research efforts to expand the scope of application of prestressed concrete in bridges has led to growingly imaginative forms and new methods of construction over

Construction of Prestressed Concrete Bridges

the last few decades. Many innovative types of prestressed concrete bridges built in India and abroad carry a central message that the man made environment of bridge structures in the last four decades of the 20th century can be built not only economically but also with elegance and dignity. The idea of prestressing arose out of bridges in the last century and the 21st century will witness the development of innovative construction techniques in the field of prestressed concrete bridges.

12.9.2 Cantilever Construction

Cantilever construction technique is widely used universally for long span prestressed concrete bridge construction using *cast-in-situ* work or precast structural elements. Cantilever construction is a method of progressively constructing a cantilever in segments and stitching them to the segments already completed, by prestressing. The cantilever segments are constructed/erected from the pier outwards on either side, and stitched back simultaneously. This method eliminates the use of expensive form work and scaffolding especially for bridges in deep valleys and rivers with large depth of water resulting in faster rate of work progress coupled with overall economy. Basically there are two major types of cantilever construction classified as,

1. *Cast-in-situ* Construction
2. Construction using Precast Segments

1. **Cast-in-situ Construction:** In this method, the piers of the bridge are first constructed and the bridge work progresses simultaneously on either side of the pier. The bridge deck is *cast-in-situ* with units of 2.5 to 3 m length cantilevering symmetrically on both sides of the pier. The form work for *cast-in-situ* construction is supported by steel frame work attached to the completed part of the bridge deck and the form work moves from one completed section to the next part. The *in situ* construction is done by a pair of traveling gantries, each weighing about 40 t (for casting 2.5 to 3 m segments of a 2 lane deck). The gantry systems proceed in a systematic manner from section to section on either side of the pier after the prestressing of the segment last cast. The gantry system also supports a suspended scaffolding for constructional convenience and labour safety.

The *cast-in-situ* method involves a 10 day working cycle as summarized below:

(a) Shifting of traveling gantry system .... 1 day
(b) Completion of entire shuttering and placing of reinforcement bars and prestesing cables for the segment .... 3 days
(c) Casting the segment .... 1 day
(d) Curing .... 4 days
(e) Stressing operations and miscellaneous works ... 1 day
    Total cycle time .... 10 days

It is possible to reduce the working cycle to 6 days from ten by resorting to more organized and mechanical infrastructure. Figure 12.3 shows the *cast-in-situ* cantilever method of construction used for Boussen's bridge over the River Garonne in France having spans in the range of 49 to 96 m. The typical cross-section comprises multicell box girders of variable depth with prestressing cables running in the ribs and flanges. *Cast-in-situ* box girders of variable depth were used in the construction of Bassein Creek Bridge built by Gammon India Company at Bombay. The supporting form work used in the cantilever construction method is shown in Fig. 12.4.

Fig. 12.3: *Cast-in-Situ* Cantilever Construction of Boussen's Bridge in France

2. **Construction Using Precast Segments:** In the case of long span bridges, it is more advantageous and economical to adopt precast segmental construction in preference to the *cast-in-situ* type of construction. In this method, the bridge segments comprising single cell or multicell box girders are match cast in a casting yard using special forms and they are transported to the worksite. The precast segments are placed in position by means of a mobile launching girder or when access under the bridge is possible, with barges or trucks, by means of a crane or a mobile hoist located at the extremity of the cantilevers.

Fig. 12.4: *Cast-in-Situ* Construction of Segmental Box Sections of Bassien Creek Bridge

The main advantage of precast segmental construction is that they can be cast on ground near the bridge site well in advance and the quality of units will be better than those which are *cast-in-situ*. An added advantage of this method is that the segmental units can be cured to attain their full strength before they are transported to the work site. In the *cast-in-situ* method at least a week's time is required to move the form work to the next incremental length. In the precast segmental system, the fully cured units can be transported to the site by means of heavy duty trucks and lifted by cranes to join them to the previously erected units by using temporary stressing cables. The rate of progress achieved in the construction will be faster in the precast method than in the *cast-in-situ* method. In both the methods, a typical cross-section would be a box girder of constant or variable depth to suit the longitudinal profile of the bridge deck.

Figure 12.5 shows the bridge between Oleron Island and the continent in France built by the precast segment cantilever system of construction. This bridge is made up of 26 main spans of 79 m each built by using the single cell precast box girders. Mahatma Gandhi Sethu Bridge at Patna in Bihar was built by using single cell precast box girders using the segemental method of construction. The hoisting of the precast box girder unit by a crane is shown in Fig. 12.6. The erection of massive twin cell precast segments with cable ducts used for the Al Khalij Bridge in Riyadh, Saudi Arabia is shown in Fig. 12.7. The method of casting cellular units in a casting yard is shown in Fig. 12.8. The units are provided with preformed ducts to house the prestressing cables. Figure 12.9 shows the precast multicell box cellular units stacked for storage at the stocking yard. The transportation of the precast segments by a heavy duty truck is shown in Fig. 12.10.

A novel method of positioning the precast single cell box girder segments was used in the construction of Castejon Bridge in Spain. Figure 12.11 shows the transportation of the segment by rope way and a winch traveling between the steel towers positioned over the piers. In the case of large spans, several launching girders supported on three piers are used to lift the precast elements to the desired location. An excellent example of this type of erection can be seen in Fig. 12.12, where piers are separated by 80 m spans for the bridge between Rio and Niteroi across the Gunabara Bay in Brazil. The schematic diagram shows the precast elements towed to the work site by barges and lifted to join them to the cantilevers on either side of the pier.

Fig. 12.5: Precast Segmental Cantilever Construction of Bridge between Oleron Island and the Continent (France)

12.9.3 Choice of Spans and the Method of Construction

The choices of the method of construction between the *cast-in-situ* and precast segmental type depends upon the size of the bridge, span lengths, precasting facilities available at site and the equipment available for transportation of heavy precast girders along with suitable cranes and mechanical equipment. The cantilever method has been successfully in the span range of 50 to 200 m. For small spans of less than 50 m and for elevated roads or flyovers where scaffolding beneath the structure must be avoided, precast segmental construction is preferred. At present for spans greater than 70 m, prestressed concrete single or multicell box girders compete successfully with steel construction.

Construction of Prestressed Concrete Bridges

Fig. 12.6: Transportation and Hoisting of Single Cell Box Girders of Mahatma Gandhi Sethu at Patna

Fig. 12.7: Precast Segmental Construction of Al Khalij Bridge in Riyadh

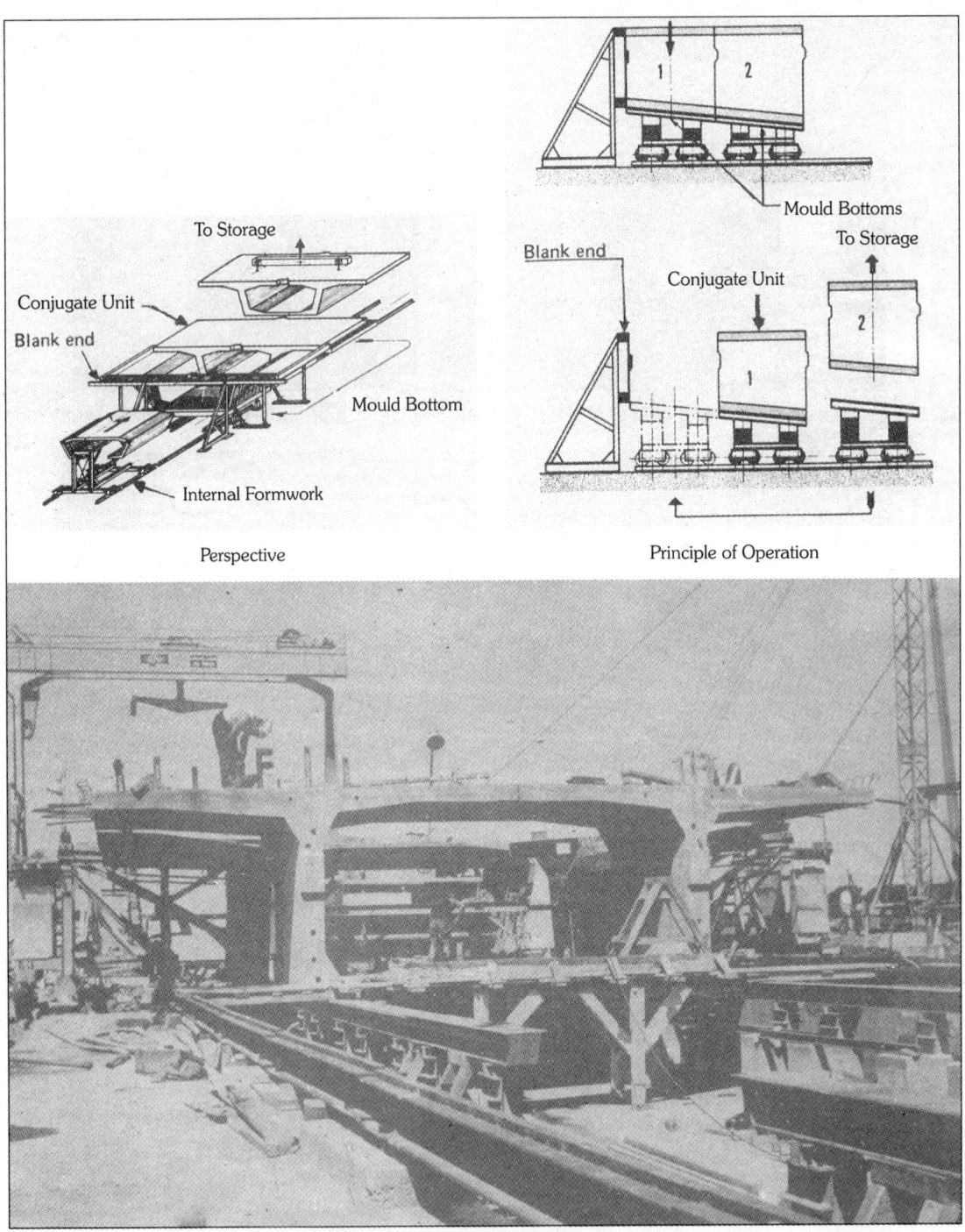

Fig. 12.8: Precasting Scheme of Cellular Box Girders

Construction of Prestressed Concrete Bridges

Fig. 12.9: Storage of Precast Cellular Multi Cell Box Units

Fig. 12.10: Transportation of Precast Cellular Box Units

Fig. 12.11: Erection of Precast Units using Cableway for Castejon Bridge in Spain

According to Raina[8], for average conditions of profile, piers and foundations, the most economical span range is from 70 to 110 m for cantilever construction. In this span range, variable depth girders are often used whereas constant depth elements are preferred for smaller spans. For very large spans exceeding 110 m, *cast-in-situ* method of construction is adopted since the precast segments are too heavy and the equipment for transportation and erection is prohibitively costly. Generally the choice between *cast-in-situ* and precast segmental construction is influenced mainly by the surface area of the bridge, precasting plant, erection facilities and construction time limitations. Hence for precast construction, a minimum approximate area of 5000 m^2 is recommended based on experience. Using standardized equipment such as floating or truck cranes and adopting smaller modular segments for short span, precast segmental construction has been used successfully for city flyovers[10] and via ducts.

12.9.4 Prestressing Cables

The salient points to be followed in using different types of cables and the prestressing system in cantilever construction are as follows:

1. Prestressing cables are threaded into the ducts after erection of each modular unit
2. The forces in the cables should be such that the congestion of cables at the top of cross-section near the piers are avoided, overcoming the possibility of excess stress concentration near the anchorages.
3. The voids in the duct should be large enough for proper grouting operations
4. The commonly used cables comprise 12 wires of 8 mm diameter using the Freyssinet system or 12 numbers of 13 mm strands and other similar cables of B.B.R. systems.

Construction of Prestressed Concrete Bridges

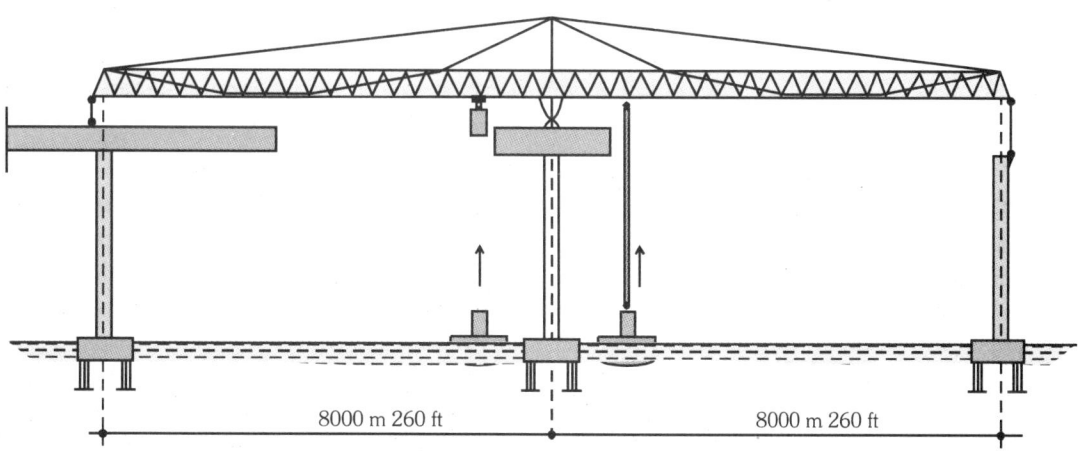

Fig. 12.12: Cantilever Construction of Bridge between Rio and Niteroi across Gunabara Bay in Brazil

As an alternative to steel construction, prestressed concrete cantilever system is economically cheaper and more durable since maintenance costs are higher in steel construction. Hence the cantilever construction technique is adopted world wide from the point of view of economy.

12.9.5 Epoxy Bonding of Segmental Box Girders

The precast segmental box units are usually bonded together at site by using epoxy resins. The units are held together under pressure by means of post tensioning the high tensile cables. The pressure applied uniformly on the mating surfaces should be not less than 0.25 N/mm^2 at any point. The thickness of the epoxy joint is normally in the range of 1 to 1.5 mm. Epoxy bonding agents comprising the resin and hardener should be mixed until it is of uniform colour using a mechanical mixer operating at no more than 600 rpm.

The epoxy bonding agents used should be insensitive to damp conditions during application and after curing. It should exhibit high bonding strength between the mating surfaces of the segmental units. It should be resistant to water penetration and should exhibit low creep characteristics. The bonding agent should have a tensile strength greater than the concrete in the precast units. The epoxy agent should function as a lubricant during the joining process of the precast elements, serve as a filler to accurately match the mating surfaces of the segments together with proper durability and serve as a water tight bond at the joint.

12.9.6 Innovative Construction Techniques

After the Second World War, the rapid reconstruction of highways in Germany, necessitated the widespread use of prestressed concrete in bridges which are aesthetically superior and economically cheaper than other types of material. Prominent Engineers like Dischinger, Finisterwalder and Birkenmaier[11] pioneered the development and use of cable stayed bridges with prestressed concrete decks. Further Leonhardt contributed immensely for the innovative construction techniques resulting in widespread use of cable stayed bridges throughout the world during the later half of the 20th century. Podolony and Muller[12] have also contributed to the design and construction techniques of prestressed concrete segmental bridges. Segmental prestressed concrete construction significantly affects the nature of stresses developed in the super structure and hence suitable innovative techniques for the construction of prestressed concrete bridges outlined elsewhere[13] are listed as follows:

1. **The Staging Method:** The staging technique of span by span assembly of precast segments is adopted when proper clearance is required below the deck and supporting form work does not interfere with the traffic in cases like urban flyovers. This method facilitates rapid construction by maintaining correct geometry of the structure with relatively low cost. This method has been used in the construction of Rhine River Bridge at Maxau, Japan.

    In constructing bridges across long via ducts using precast segmental units, span by span construction adopting the staging method is particularly advantageous resulting in speed of construction coupled with economy. The moveable form work may be supported from the ground. The traveler consists of a steel super structure which is moved from the completed portion of the structure to the next span to facilitate the casting or supporting of the precast units. A typical span by span assembly of precast segmental units using the staging method is shown in Fig. 12.13.

Construction of Prestressed Concrete Bridges

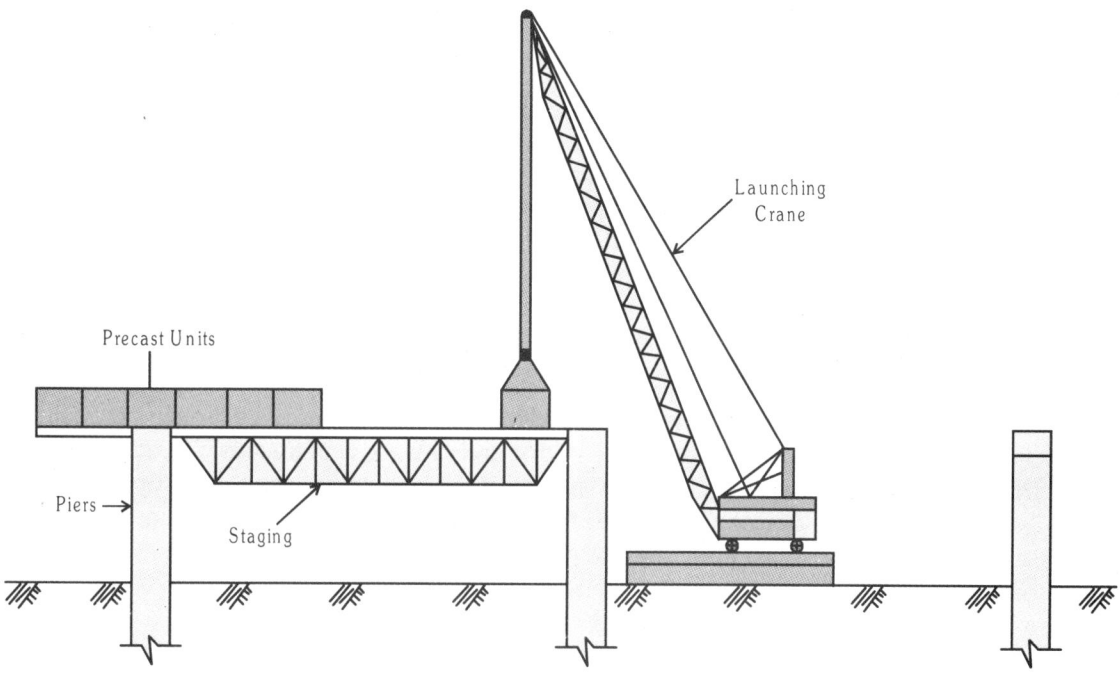

Fig. 12.13: Staging Method of Span by Span Assembly of Precast Segments

2. **Incremental Launching or Pushout Technique:** The incremental launching or push out technique was first developed by Fritz Leonhardt and Willi Baur and used in the construction of Rio Caroni Bridge in Venezuela built in 1962. The bridge deck segments were cast at site to cover lengths of 10 to 30 m in stationery forms located behind the abutments. Each unit was cast directly against the previous unit. After the concrete attained the desired strength, the new unit was joined to the previous unit by post tensioning. The assembly of units was pushed forward in a step wise manner to permit casting of the succeeding segments. Figure 12.14 shows the incremental launching technique adopted in building the bridge.

In this method, a work cycle of one week is normally required for casting and launching the segments. Low friction sliding bearings provided at the top of the piers with proper lateral guides facilitate the easy movement of the segmental units. Care should be taken to

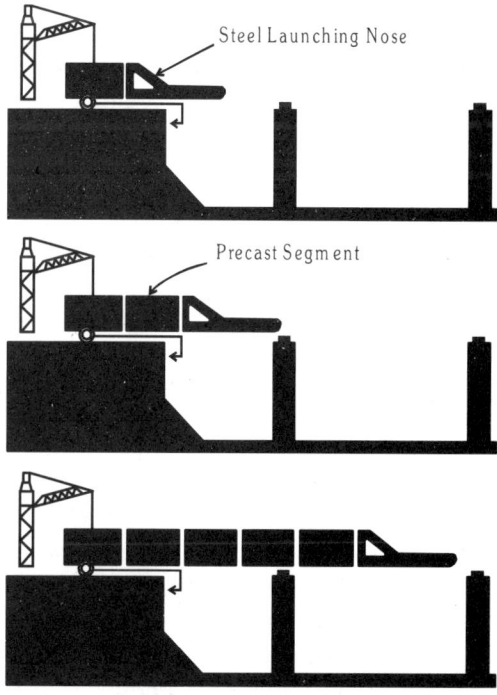

Fig. 12.14: Incremental Launching Technique

ensure the safety of stresses at critical sections of the super structure under its own weight during the various stages of launching. For this purpose, the first stage prestress is applied concentrically to the whole cross-section and in successive increments over the entire length of the super structure. Later, a fabricated structural steel launching nose is attached to the lead segment to reduce the large negative bending moments developed in the front portion (particularly just before the super structure reaches a new pier). If the spans are large, they are sub divided by means of temporary piers to control the magnitude of bending moments within safe limits. According to Raina[1], the incremental launching technique has been used to construct bridges of spans up to 60 m without the use of temporary false work bents. Also spans up to 100 m have been built using temporary supporting bents. In this technique the main girders must have constant depth generally varying from 1/12 to 1/16th of the longest span.

3. **Progressive Placement Construction Method:** In the progressive placement method, the construction starts at one end of the bridge and proceeds continuously to the other end. In contrast to the balanced cantilever construction method in which the super structure proceeds on both sides of the pier, in the progressive placement technique, the precast segmental units are placed from one end of the structure to the other in successive cantilevers on the same side of the various piers. At present this method has been found to be practicable and economical in the span range of 30 to 90 m.

   The main feature of this method comprises a moveable temporary stay arrangement to limit the stresses in the cantilever portion of the structure during the construction. The precast segmental units are transported over the completed portion of the deck to the tip of the cantilever span under construction, where they are positioned by a swivel crane moving over the deck. The segments are held in position by temporary stay cables passing through a tower located over the preceding pier as shown in Fig. 12.15. The stays are anchored to the top flange of the box girder segments so that the tension in the stays can be adjusted by light jacks. This method has been used for the construction of the Linn Cove Via duct in North Carolina, USA. The progressive placement method can also be used for *cast-in-situ* construction of bridge decks.

4. **Cantilever Construction Method:** The cantilever method of erection is the latest and the most economical and popular method for the construction of long span precast or *cast-in-situ* segmental bridges. This method is also ideally suited for the construction of cable stayed bridges decks. Normally the erection proceeds from both sides of the tower or pylon in the form of two free cantilevers which balance each other. The units may also be supported by the stay cables depending upon their spacing and size of precast segmental units. The primary advantage of the cantilever technique of construction is that the traffic below the bridge is not hindered during the erection process.

   The Second Vivekananda B.O.T., toll way bridge recently opened for traffic just north of Kolkata is an excellent example of multi span extradosed bridge, a hybrid structure combining elements of cable stayed post tensioned prestressed concrete box girders. The nine span extradosed bridge stretching across India's mighty Hooghly river is Asia's first multispan extradosed bridge and of only three extradosed bridges in Asia outside Japan according to Egeman Ayna, principal engineer of the International Bridge Technologies (IBT) who are the design consultants for the innovative bridge project. The bridge with a total length of

880 m comprises seven spans of 110 m and two 55 m long spans. The bridge deck comprising post tensioned prestressed concrete box girders is supported by a single plane central cable stay system instead of two planes of cable stays.

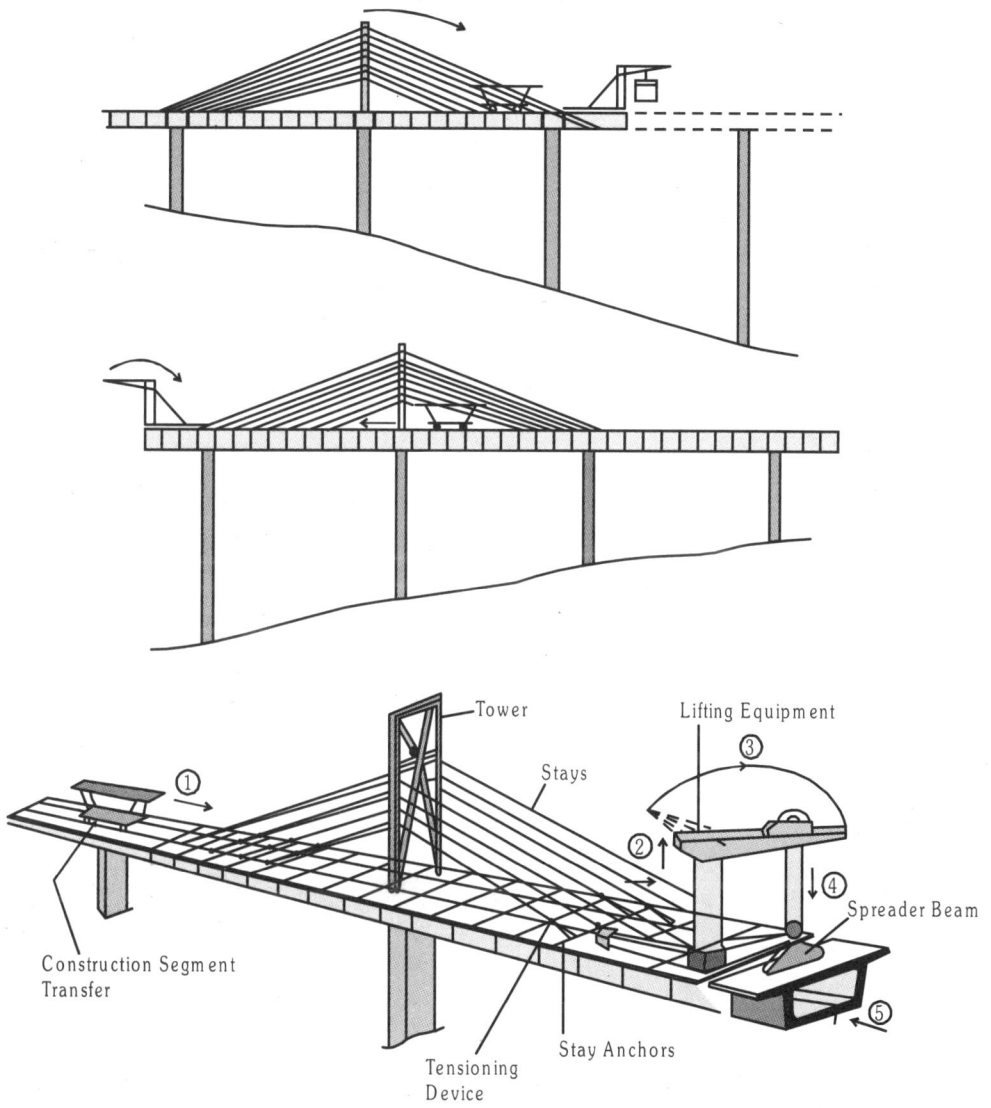

Fig. 12.15: Progressive placement method of Construction

According to Ayna[14], a typical box girder bridge would have had a depth of 6 m from span considerations. However the second Vivekananda Bridge lowers that profile by approximately 2.5 m. Also the bridges constant depth profile is a departure from the variable depth seen in other extradosed bridges. The bridge girders are prestressed using eight cables composed of 63 to 73 strands each of 15 mm diameter, extending from both sides of the 2 m wide piers as shown in Figs. 12.16 and 12.17.

Fig. 12.16: Second Vivekananda Extradosed P.S.C. Bridge across Hooghley River Near Kolkata

Fig. 12.17: Second Vivekananda Extradosed P.S.C. Bridge across Hooghley River Near Kolkata

The construction techniques developed over the last few decades have shown major progress towards simplification and reduction of erection equipment. The construction procedure must however be well planned using sequential computations for the alignment, forces, exact lengths and angles considering temperature and creep influences which depend on seasonal, climatic and daily environmental conditions. The collective experience and knowledge gained through research and practice during the last fifty years will immensely help in evolving innovative applications and methods in the construction of prestressed concrete bridges.

REFERENCES

1. RAINA, V. K., *Concrete Bridge Practice, Construction, Maintenance and Rehabilitation.* Tata McGraw Hill Publishing Co, Ltd, New Delhi, pp. 381-396.
2. ROWE, R. E., CRANSTON, W. B., BEST, B. C., New concepts in the design of structural concrete. *Structural Engineer,* Vol. 43, 1965, pp. 399-403.
3. KRISHNA RAJU, N., Limit state design of prestressed concrete members, engineering design, Special Issue on Prestressed Concrete. *Journal of National Design and Research Forum,* Institution of Engineers, India, April 1979, pp. 16-21.
4. KRISHNA RAJU, N., Limit State Design for Structural Concrete, Proceedings of the Institution of Engineers, India, Vol. 51, Jan 1971, pp. 138-143.
5. KRISHNA RAJU, N., *Design of High Strength Concrete Mixes.* Fourth Edition, C.B.S. Publishers and Distributors, New Delhi, 2002, pp. 103-125.
6. Degussa Construction Chemicals (India) Pvt. Ltd. Product Promotional Pamphlet, C-68, MIDC, Thane, Belapur Road, Yurbhe, Nava Mumbai, July 2006.
7. LEE, D. J., *The Theory and Practice of Bearings and Expansion Joints for Bridges.* Cement and Concrete Association, U.K, 1971.
8. RAINA, V. K., *Concrete Bridge Practice, Analysis, Design and Economics.* Tata McGraw Hill Publishing Co, Ltd, New Delhi, 1991, pp. 445-473.
9. FINISTERWALDER, U., Prestreesed concrete bridge construction. *Journal of American Concrete Institute,* Sept. 1965.
10. NAIDU, M. P. and SANKARALINGAM, C., Precast Segmental Construction of Sirsi Flyover, Souvenir, IRC Diamond Jubilee Session, Highways Department, Tamilnadu, January 2000, pp. 97-108.
11. BIRKEMMEAIER, M., Recent Developments and Trends in Design and Construction of Cable Stayed bridges, proceedings of the 9th Congress, F.I.P., Stockholm, June, 1982.
12. PODOLONY, W. and MULLER, J., *Construction and Design of Prestressed Concrete Segmental Bridges.* John Wiley and Sons, New York, 1982.
13. KRISHNA RAJU, N., *Design of Bridges,* (Third Edition), Oxford and I.B.H. Publishers, New Delhi, 1998, pp. 396-399.
14. JESSICA BINNS, Extradosed Bridge Distinguishes Tollway Project in India. *Civil Engineering,* Vol. 75, No. 2, February 2005, p. 20.

13

MAINTENANCE AND REHABILITATION OF PRESTRESSED CONCRETE BRIDGES

13.1 INTRODUCTION

Effective maintenance and rehabilitation of prestresssed concrete bridges forms an important part during the life time of any bridge system. The main objective of maintenance management of prestressed concrete bridges is to preserve the structure in such a way that it will function satisfactorily at the various limit states[1,2,3] immediately after construction and also during the projected life span of the structure. Periodical inspections, identification of minor and major defects and local damages, deterioration and loss of durability of the bridge structure due to environmental and other local effects form the basic aspects of good maintenance practice. Basically prestressed concrete bridge decks possess superior dynamic and fatigue resistance in comparison with other types of bridges. In prestressed concrete bridges, the primary problem encountered is the distress caused to the anchorages and unbonded tendons due to rusting under exposure to humid weather conditions.

The present day American, British and Draft revision of the Indian Standard Code[4] for prestressed concrete permits the use of type-3 or partially prestressed structural members in which tensile stresses of limited magnitude are permitted giving rise to local cracks in concrete of width of up to 0.2 mm under normal service loads. These cracks may provide access to passage of water to the cable ducts and tendons under severe exposure conditions resulting in rusting leading to explosive failure of the bridge structure due to the sudden breaking of the tendons as a result of their reduced cross-section. Bursting of the anchorage zone of prestressed concrete girders where large forces are concentrated over small areas may result in catastrophic failure of the entire bridge deck. Hence periodical surveillance and timely repairs form an integral part in the maintenance and rehabilitation of prestressed concrete bridges.

13.2 GENERAL FEATURES OF BRIDGE MAINTENANCE AND REHABILITATION

The main objective of bridge maintenance is to ensure the integrity of the structure during its life span so that it functions without any disruptions. To fulfill this objective, it is necessary to resort to periodical surveillance, repairs, rehabilitation and replacement depending upon the local conditions. The maintenance management system must also provide guidelines and methodologies to enable the local engineers to reach rational, cost effective decisions regarding maintenance and rehabilitation of distressed bridges.

According to Gokhale and Rohra[5], periodical inspections, repairs and rehabilitation, constitute the primary aspect of good and effective maintenance. This aspect is of paramount importance due to the rapid increase in the number of prestressed concrete bridges built during the last few decades of the 20th century as a result of easy availability of good quality cement, steel and epoxy compounds and various other building materials along with innovative methods of construction of various types of prestressed concrerte bridges.

Prestressed concrete bridges are likely to deteriorate especially in coastal regions mainly due to severe exposure conditions of temperature and humidity. It is a well established fact that the total number of prestressed concrete bridges built is much more in comparison with the new bridges under construction, but the amount of energy and the resources spent on preventive maintenance and or rehabilitation of distressed bridges is negligible in comparison with that spent on new bridges. Hence increasing number of bridges are becoming unserviceable and if let unserviced, they have to be pulled down. The cost of removing the damaged bridge may in some cases be several times that required for periodical maintenance and or rehabilitation. Hence it is more prudent to inspect the prestressed concrete bridges at regular intervals for detection of any signs of deterioration and or distress and initiate rehabilitation measures, thus restoring the structure to a state of full serviceability.

Many of the prestressed concrete bridges built on the West Coast National Highway (NH-17) have shown signs of distress due to severe environmental conditions typically prevailing in a coastal region. These bridges built during 1950's are more than 50 years old and they have not been maintained due to the absence of any systematic periodical surveillance and credible maintenance organization. Rehabilitation of bridge structures may become essential due to several reasons. Some common causes are design and or constructional deficiency, environmental effects, overloading of the structure either due to unanticipated loading or due to accidents and user made changes in the structure during the service life of the structure.

Every single bridge repair and rehabilitation is unique for the particular structure. Hence the use of common techniques for rehabilitation of various bridges is limited. As far as maintenance is concerned, several new cementatious materials and epoxy resins and compounds have been developed during the last decades which are highly effective in protecting the basic structure from the destructive effects of severe exposure conditions in the environment. It is important to note that engineers dealing with the problem of effective maintenance and rehabilitation of bridges must study the basic designs, history of construction, changes in loading patterns on the structure, and environmental changes, etc. before embarking on repairs. After a detailed analysis of all these parameters, the engineer will be in a better position to work out a strategy for long lasting rehabilitation measures for the distressed bridge.

13.3 MAINTENANCE METHODOLOGY

Maintaining Highway bridges and keeping them structurally sound and in a fit condition, so as to provide safe and uninterrupted traffic flow, is the primary function of a bridge maintenance engineer. Proper planning and investment towards effective periodical maintenance forms only a fraction of the cost to be incurred due to major repair and rehabilitation of the bridge structure. Hence it is always advisable to establish a programmed preventive maintenance system to detect any signs of distress in the initial stages itself through proper inspection procedures followed by appropriate repairs.

Structural and Highway engineers have recognized that maintenance is an important prerequisite to ensure safety and serviceability of the bridge structure during its intended life time. Recent developments in the domain of instrumentation have resulted in various types of instruments which could monitor the *in situ* strength of concrete, cracking in concrete, rusting in reinforcements. Also methods have been codified[6] to evaluate the *in situ* strength of slabs to sustain the designed loads by actual load testing of the slab panel and monitoring the deflections developed at the soffit of the slab. Structural slabs and beams exhibiting local distress can be repaired by external bonding of steel plates to the soffit by using epoxy adhesives. Honey combing, cracks and cavities in concrete can be repaired by the process of guniting, pressure grouting and shotcreting procedures.

13.4 INSPECTION OF BRIDGES

Periodical inspection of bridges is of prime importance in any maintenance system since any signs of distress detected in the early stages can be repaired with minimum costs. All types of remedial and preventive maintenance including minor repairs and replacement of bridge components should be planned at periodical intervals without disrupting the traffic on the bridge. The data collected from inspection reports should be scientifically evaluated from time to time to asses the need for remedial measures. According to Raina[7] the following categories of inspections are generally adopted for prestressed concrete bridges:

1. **Routine Inspection:** In this type, general inspections are undertaken frequently by Highway maintenance engineers having practical knowledge of Highway structures, though not necessarily experts in design, detailing and constructional aspects or experts in special problems of bridge inspection. This type of inspection is required to identify minor deficiencies which could lead to accidents or maintenance problems or major repairs at a future date. Such routine inspections are normally undertaken at intervals of one or two months.

2. **Detailed Inspection:** This type of inspection is further divided into general and major categories depending upon the frequency and intensity of inspection respectively.

    (a) **General Inspection:** This type of inspection is planned annually and it should cover all the structural elements of the bridge. It is mainly a visual inspection assisted by standard instrumental aids, invariably followed by a detailed written report.

    (b) **Major Inspection:** This type of inspection is comprehensive involving detailed examination of all structural elements. The process may involve setting up of special access facilities (such as inspection platforms to examine the soffits of deck slab and beams, articulation locations and bearings at supports, etc.) wherever required. Depending upon the importance of the structure, this type of inspection is conducted at intervals of 2 to 3 years or may be at smaller intervals for important bridges specially exposed to aggressive environments (e.g. Bridges located in coastal areas, marine locations and abnormal wind zones)

3. **Special Inspection:** Emergency or special inspections are planned under extraordinary situations such as earthquakes, high intensity/abnormal loadings, floods, etc. These inspections should be exhaustive including testing of structural elements (e.g. Non destructive testing using ultrasonic pulse technique to detect internal micro cracks and excessive deflections using dial gauges, etc.). The test results are examined in the light of structural analysis. For

this type of inspection, experienced bridge engineers should be involved in the investigating team.

Special inspections are timed such that the most critical evaluation of the performance of the structure is obtained. For example, structural elements such as foundations, bridge piers and protective works are inspected, before, during and after the floods. Bridge bearings and joints should be inspected during extremes of temperature while cracks at the soffits of prestressed concrete girders should be inspected under abnormal loading conditions. Exposure to aggressive environment may result in cracking and spalling of concrete in pretensioned girders with thin webs. The inspection team should be lead by a qualified and experienced bridge engineer who is familiar with the design and constructional aspects of the bridge structure to be inspected, so that the observations are properly and accurately assessed resulting in a meaningful technical report, containing details of distress/deficiency and recommendations for relevant repairs of the inspected bridge.

13.5 INSPECTION INSTRUMENTATION

Prestressed concrete bridge structures are likely to suffer different types of distress during their life time. The distress may be in the form of surface cracks, internal micro cracks rusting of reinforcements or spalling of concrete. Such structures should be subjected to detailed inspection with the help of instruments to assess the extent of damage to the structure. Modern testing equipments widely used by the specialized inspection team are listed below:

1. Snooper-crawler and adjustable ladders for providing access to the sides and soffit of the bridge deck.
2. Ultrasonic pulse velocity apparatus for detection of internal micro cracks
3. Rebound hammer (Schmidt hammer) for *in situ* evaluation of compressive strength or grade of concrete.
4. Mechanical extensometer or Demec Gauge with stainless steel targets for studying surface deformations in concrete under loads
5. Optical microscope with a light source to measure the width of cracks on the surface of concrete.
6. Electronic strain gauges for measurement of strains in steel and concrete due to loads and creep effects.
7. Hydraulic jacks, pressure transducers or load cells for measurement of forces, etc.
8. Magnetic detector for measuring thickness of concrete cover and for locating reinforcing bars.
9. Vibration measuring equipment.
10. Electrical resistance meter (for rust pockets).
11. Pachometer to locate and measure the size of steel bars embedded in concrete.

The instruments listed above should be judiciously used during inspection to assess the degree of distress and the test results should be critically examined before finalizing the reports of inspection.

The author[8] has used these instruments extensively for evaluating the strength of prestressed concrete lattice girders in Ram Kumar Mills at Rajajinagar, Bangalore. Rebound hammer was used by the author[9] for testing the compressive strength of concrete used in the slabs, columns and reinforced concrete beams at Darus Salam hostel complex located in Bangalore and also for testing the roof slab of explosion bunker[10] at M/S. Astra Indian Detonators Ltd, Bangalore. In the case of old bridges for which structural drawings are missing, the location and size of reinforcing bars embedded in concrete can be readily determined to an accuracy of 3 mm using Pachometer.

Revolutionary developments in the field of electronics have paved the way for a variety of sophisticated instruments for monitoring and inspection of distress in different types of bridge structures. Ultrasonic meters are used to determine the degree of deterioration, distress and/or material imperfections that are relevant to the structural integrity of the bridge. Bridges greater than 10.7 m in height and in those bridges which cannot be inspected from beneath due to rugged terrain or watery situations, a mechanical contraption widely known as Barin's snooper vehicle is ideally suited for inspection of the soffit of slabs and girders of the bridge deck. The snooper is mounted on a heavy duty truck with a swiveling platform and the underside of the bridge deck can be inspected by using the hydraulically operated platform. Figure 13.1 shows the typical schematic diagram of Barin's snooper system. The inspection of the under side of the bridge using the snooper platform is shown in Fig. 13.2.

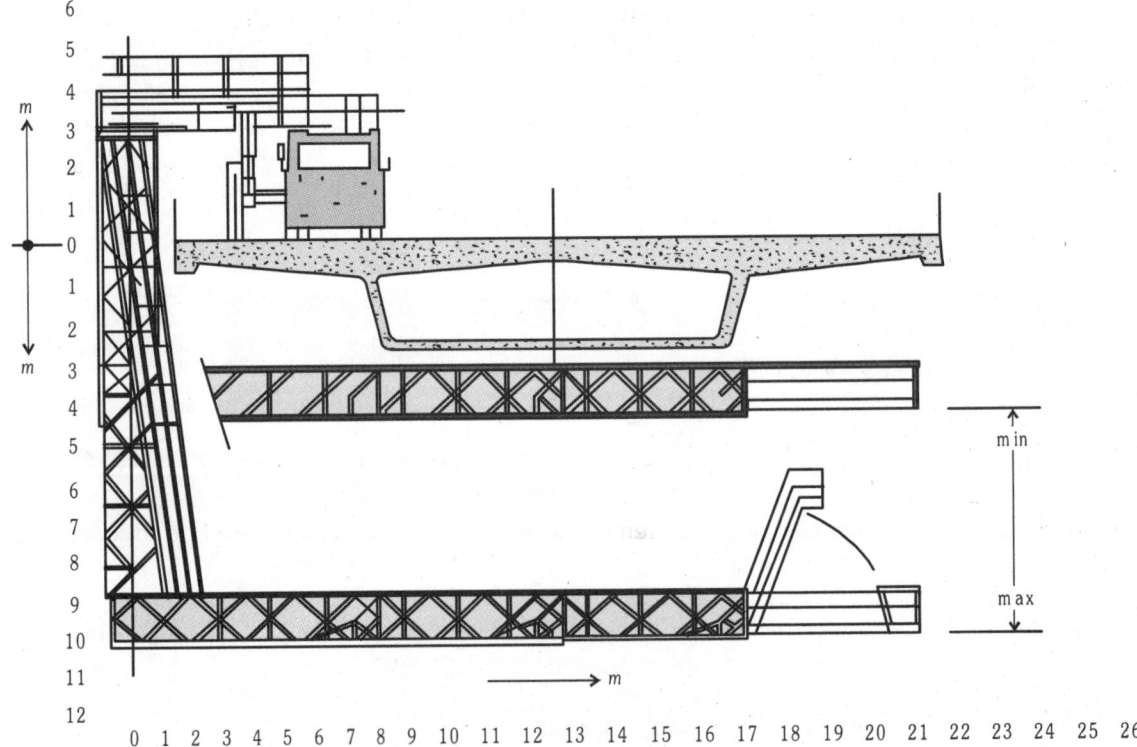

Fig. 13.1: Barin's Snooper System for Inspection of Bridges

13.6 CRACKS IN PRESTRESSED CONCRETE BRIDGES

Several types of cracks may develop in the various structural elements of prestressed concrete bridges. Plastic shrinkage and settlement cracks may develop in plastic concrete when it is not completely set. Concrete during its hardening phase (during the first 3 to 4 weeks after the final set) is likely to develop cracks due to the constraint to early thermal movement and /or drying shrinkage or due to differential settlement of supports. After hardening, concrete can crack due to overloads, faulty construction, inadequate and improper detailing of reinforcements, sulphate attack on cement concrete, alkali-aggregate reaction and long term drying shrinkage.

Fig. 13.2: Snooper Platform to Inspect the Soffit of Bridge Deck

The various types of cracks likely to develop in a bridge structure are shown along with their location and type in Fig. 13.3. These cracks are categorized and listed in Table. 13.1 as reported by Raina[7]. In the case of prestressed concrete I-girders with thin webs, cracks may develop due to shrinkage, temperature and loads in the web of the girders. This type of local distress can be prevented by suitable detailing of web reinforcements to resist the shear forces. Horizontal reinforcements well distributed over the depth of the web prevent the shrinkage cracks while vertical reinforcements resist the development of diagonal tension cracks. Figure 13.4 shows the typical remedial measures used in thin webs of I-girders to resist web shear, shrinkage and temperature cracks.

Typical cracks likely to develop in precast pretensioned double tee and I-section girders are compiled in Fig. 13.5 reported by Raina[11]. These various types of cracks may develop due to several causes like shrinkage of concrete, improper bond between steel and concrete, diagonal tension due to imposed loads and damage during transportation of the precast units.

In post tensioned prestressed concrete bridge girders, anchorages located at the end zones of girders transmit huge compressive forces on concrete spread over a small area. Consequently in the anchorage zone, splitting tension develops resulting in cracks as shown in Figs. 13.6 (a) and (b). Horizontally curved cables induce splitting tension cracks as shown in Fig. 13.6 (c). The reader may refer to a separate monograph[12] by the author for information regarding anchorage zone stresses and design of end blocks.

Table 13.1: Types of Cracks Concrete Structures, Causes, Location and Remedy (Ref.-7 and Fig. 13.3)

Types of Intrinsic Cracks (not Caused by Loading)	Letter Legend (see Fig. 13.3)	Sub Division	Most Common Location of Occurrence	Primary Cause (excluding restraint)	Secondary Causes/Factors	Remedy Assuming Basic Redesign is impossible	Time of Appearance
Plastic Settlement Cracks	A	Over the reinforcement	Deep sections	Excess Bleeding	Rapid early drying conditions	Reduce bleeding do air entrainment or revibrate mildly	Ten munutes to three hours
	B	Arching	Top of columns				
	C	Change of depth	trough and waffle slabs				
Plastic Shrinkage Cracks	D	Diagonal (may be normal to wind direction)	Roads and slabs	Rapid early drying	Low rate of bleeding & fast surface	Improve early curing and trowel	Thirty minutes to to six hours
	E	Random	Reinforced concrete slabs				
	F	Over Reinforcement (even mesh type)	Reinforced concrete slabs	Ditto plus steel near surface			
Early Thermal Contraction Crack	G	External restraint	Thick walls	Excess heat generation	Rapid cooling curing by relatively cold water	Reduce heat and/or insulate	One day to two or three weeks
	H	Internal restraint	Thick slabs	Excess temperature gradients			
Long-term Drying Shrinkage Cracks	I		Thin slabs and walls	Absence of movements joints, or inefficient joints	Excess shrinkage and inefficient curing	Reduce water content and improve curing	Several weeks or months

Types of Intrinsic Cracks (not Caused by Loading)	Letter Legend (see Fig. 13.3)	Sub Division	Most common location of Occurrence	Primary Cause (excluding restraint)	Secondary Causes/ Factors	Remedy Assuming Basic Redesign is impossible	Time of Appearance
Crazing-Cracks (occur only on surface)	J K	Against form work Floated concrete	Fair faced concrete slabs	Impermeable form work over trowelling	Rich mixes poor curing	Improve curing & finishing	One to seven days sometimes much later
Cracks due to Corrosion of Reinforcement (expansive reaction can lead to spalling of concrete)	L (and rust stains)	Natural and slow, or fast if excessive calcium chloride present	Columns and beams Precast concrete	Lack of cover and dampness Excess calcium chloride and dampness	Poor quality concrete	–	More than about two years
Cracks due to Alkali-aggregate Reaction (expansive reaction)	NC may show gel type or dried resin type deposit in crack)		(Damp locations)	Reactive silicates and carbonates in aggregates acting on alkali in cement	–	–	More than five years

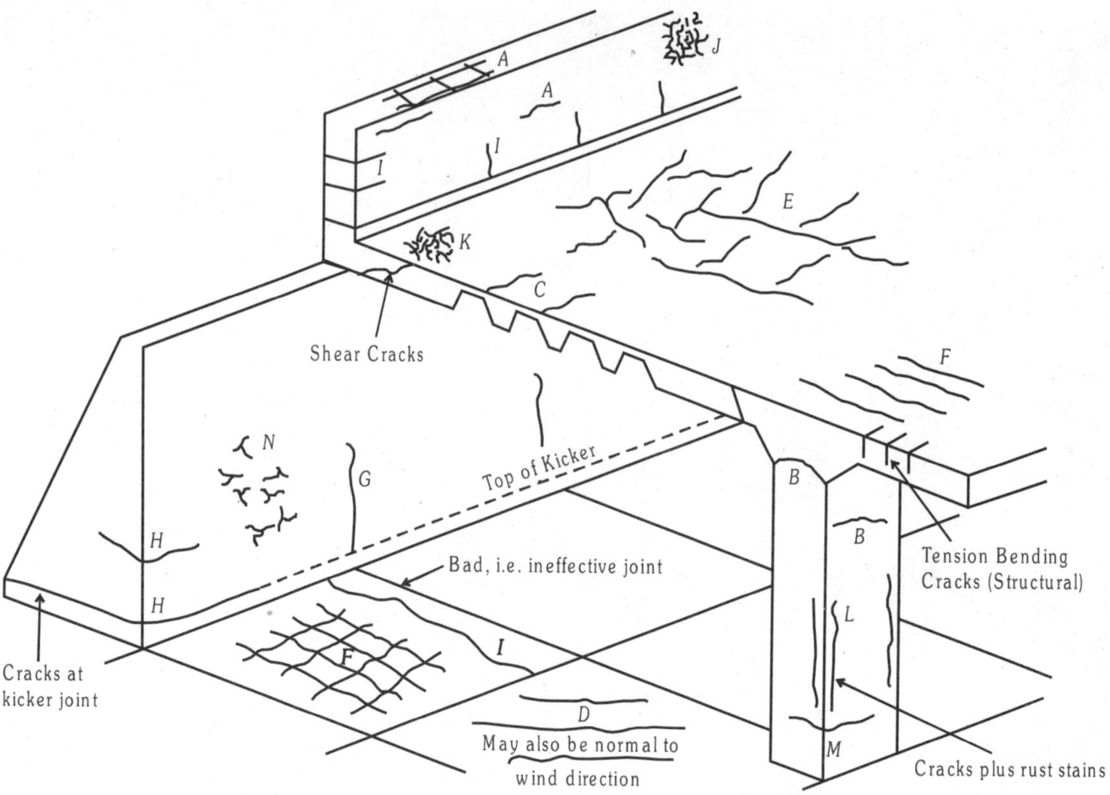

Fig. 13.3: Typical Examples of Intrinsic Cracks in a Hypothetical Concrete Structure (Refer Table 13.1 for Details of *A*, *B*, *C*, etc.)

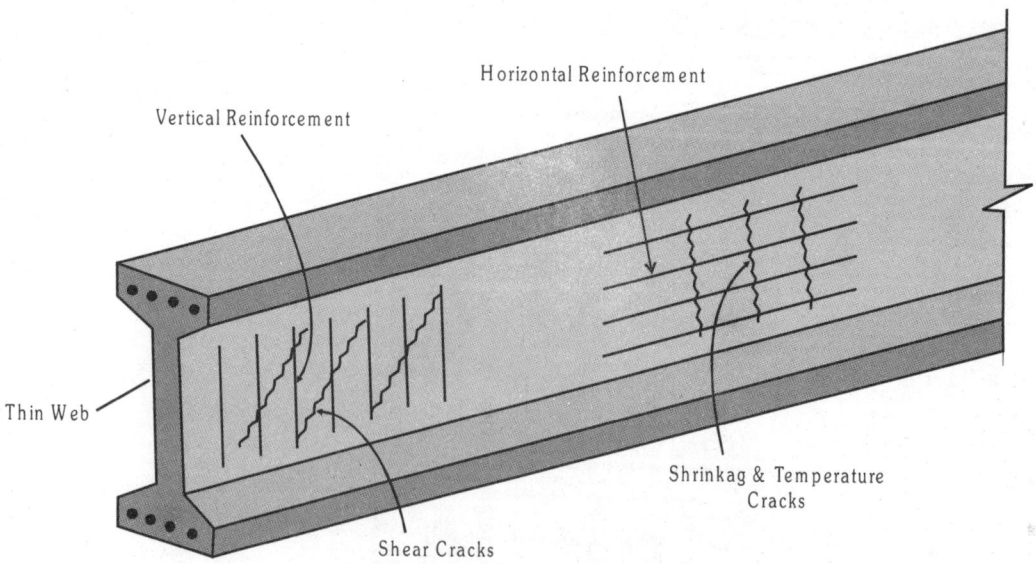

Fig. 13.4: Detailing of Reinforcements for Shrinkage and Shear Cracks in Pretensioned T-Beams with Thin Webs

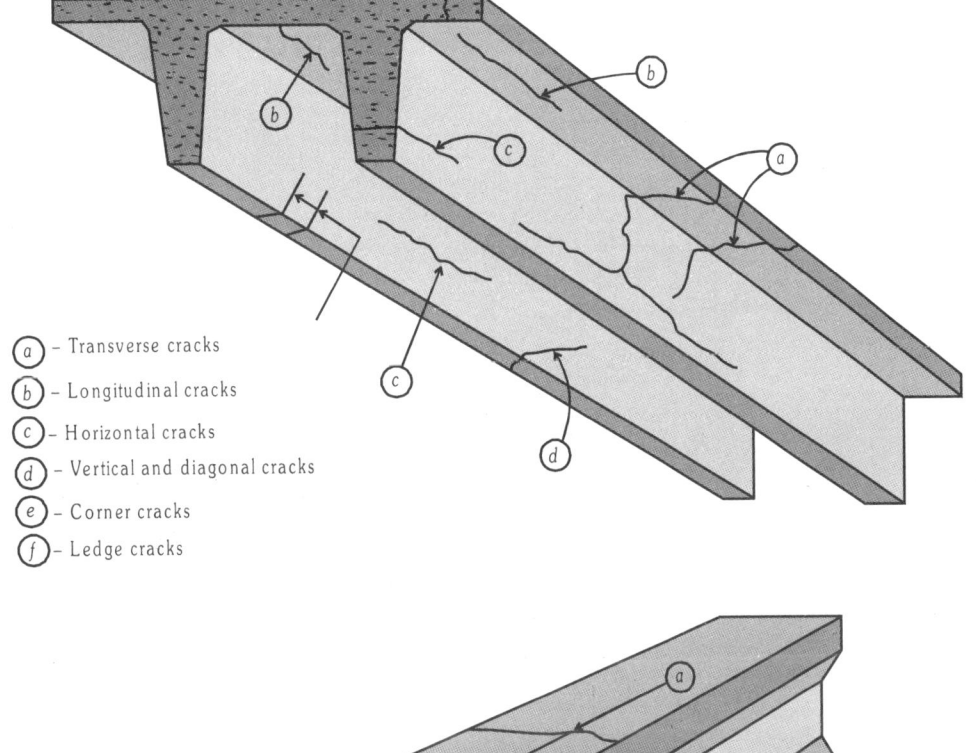

- ⓐ – Transverse cracks
- ⓑ – Longitudinal cracks
- ⓒ – Horizontal cracks
- ⓓ – Vertical and diagonal cracks
- ⓔ – Corner cracks
- ⓕ – Ledge cracks

Fig. 13.5: Typical Cracks in Pretensioned Double Tee and I-Section Girders

Multiple dormant random cracks are generally encountered in the soffit of deck slabs and the side faces of longitudinal and cross girders. In such cases, epoxy resin injection or grouting and sealing techniques are successfully adopted. Figure 13.7 (a) shows the epoxy resin sealing of cracks in the soffit of deck slab and webs of longitudinal girders while Fig. 13.7 (b) shows the epoxy sealing of cracks in the soffit of slab and cross girder of a bridge deck.

When the width of cracks is more than 1 mm, it is more economical to use the grouting and sealing technique. In this method, the crack is enlarged along its exposed face (recessing) and then

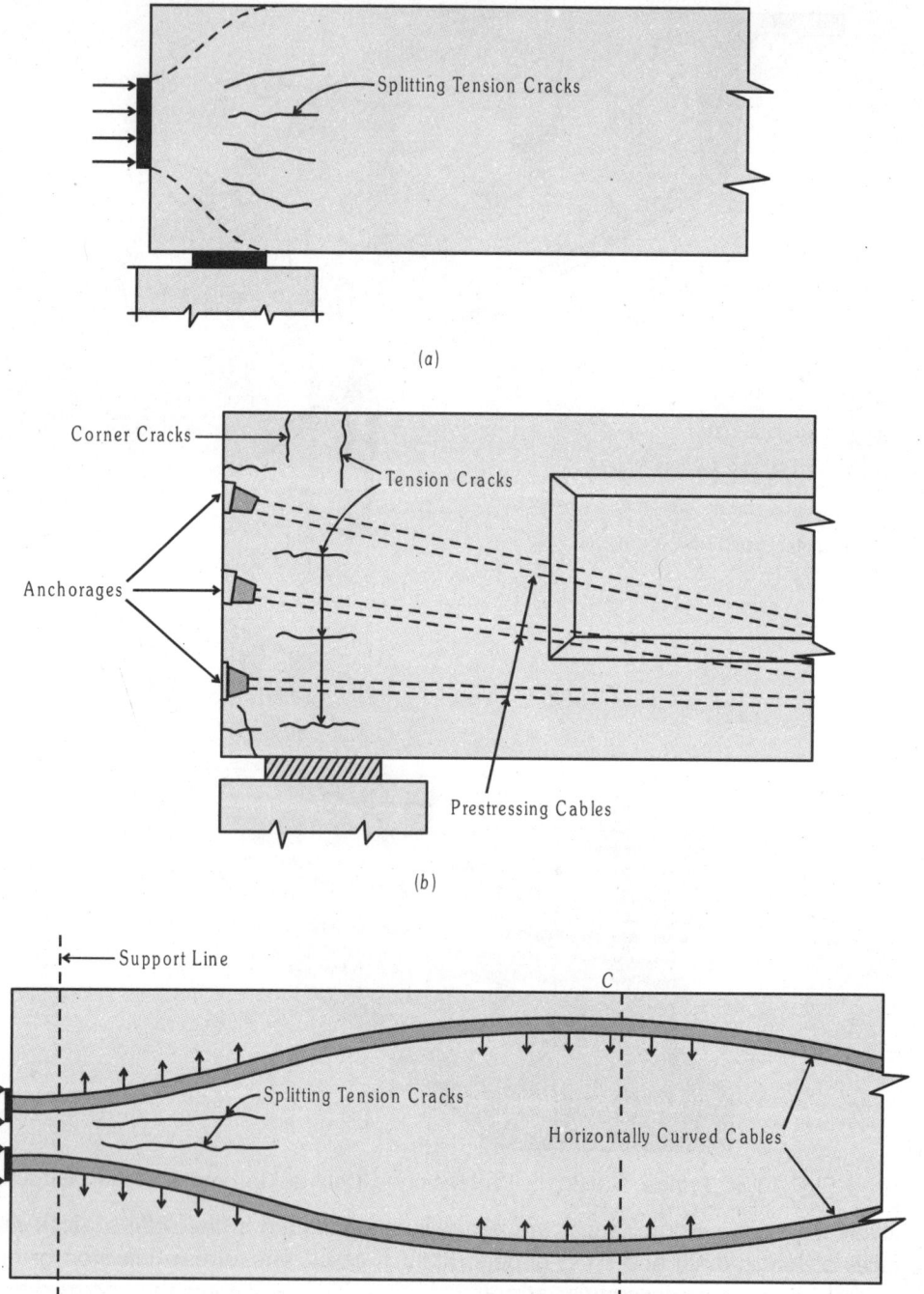

Fig. 13.6: Typical Cracks in the Anchorage Zone of Post Tensioned Girders

Maintenance and Rehabilitation of Prestressed Concrete Bridges

(a)

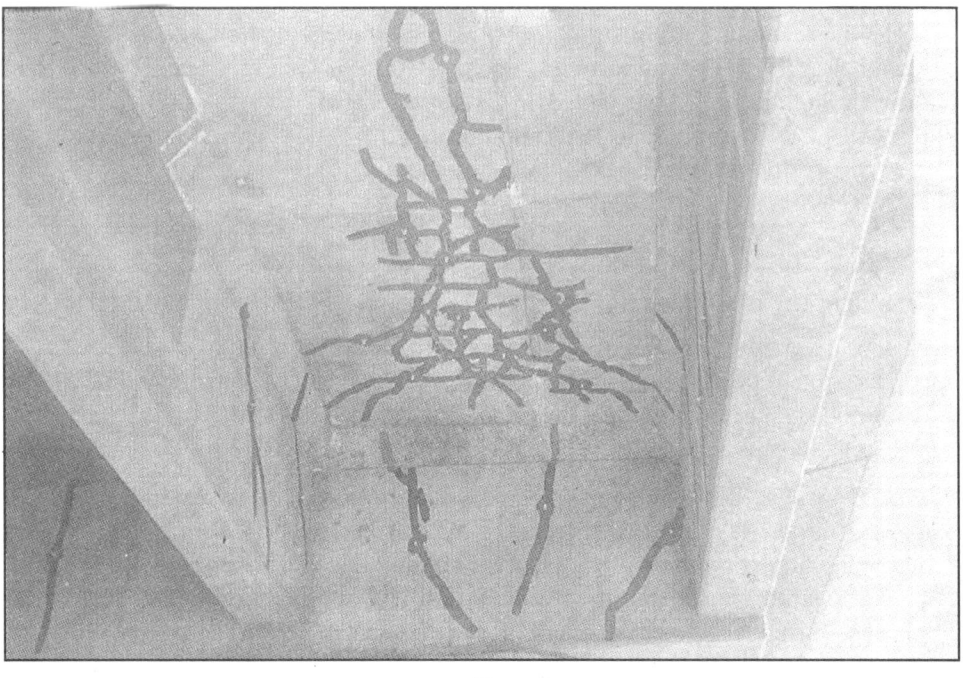

(b)

Fig. 13.7: Epoxy Sealing of Cracks in Deck Slab Main Girder and Cross Girder in Bridge Deck

the crack is cleaned of dust particles and grouted and the surface sealed using a suitable joint sealant. Various other techniques widely used for repair of dormant or dead cracks include dry packing, polymer impregnation,[13, 14] surface treatments and autogenous healing. The selection of the type of sealant is based on the amount of movement and the limitations imposed by the size of recess which can be cut together with the type of crack, *i.e.*, either vertical or horizontal. The following three types of sealants are generally used depending upon their suitability in a given situation:

1. Mastics,   2. Thermoplastics,   3. Elastomers

1. **Mastics:** These are asphaltic materials with a low melting point with added fillers or fibres. They are recommended where the movement does not exceed 15 per cent of the width of the groove. The groove should be cut so that it has a depth to width ratio of 2 : 1.

2. **Thermoplastics:** These comprise of asphalts, pitches and coal tar which become liquid or semi viscous when heated. The pouring temperature is usually above 100 Degrees Centigrade. The groove depth to width ratio should be 1 : 1 and the total design movement is of the order 25 per cent of the groove width. These materials soften less than mastics but they may extrude under high ambient temperatures and they may be degraded by exposure to ultra violet light, loosing elasticity after a few years of exposure to direct sun light.

3. **Elastomers:** These include a wide range of materials such as poly sulphides, epoxy poly sulphides, polyurethane, silicones and acrylics. These materials are advantageous since they can be used without heating. They have excellent adhesion to concrete and are not susceptible to softening within the normal range of ambient temperatures. Normally elastomers exhibit a higher degree of elongation of as much as 100 per cent extension but in practice, this should be limited to 50 per cent (i.e. ± 25 per cent). The groove depth to width ratio should be 1 : 2. The material should be prevented from adhering to the bottom so that the crack remains free as a live crack.

13.7 REPAIRS AND REHABILITATION OF BRIDGES

13.7.1 Classification of Damage

The nature and magnitude of damage in a prestressed concrete bridge should be first analysed based on inspection reports before deciding upon the type of repair and rehabilitation which depends upon the degree of damage suffered by the bridge structure. The following three major groups are identified based on the degree of damage:

1. **Minor Damage:** Shrinkage cracks and minor spalling of concrete on the surface of deck slabs and girders fall into this category. This requires superficial patching by using epoxy grout or guniting using shotcrete applied by a pneumatic gun. The damaged and delaminated concrete is first removed by hand tools and the surface is thoroughly cleaned before the application of epoxy grout. All cracks should be sealed by the epoxy grout applied under pressure injection from the soffit and rising to the top, along the cracks against gravity.

2. **Moderate Damage:** Moderate damage involves extensive spalling and cracking of concrete due to multifarious reasons. Epoxy grout or micro concrete is generally applied as in minor repairs. However it is recommended that welded wire fabric be attached to drilled dowels

placed at intervals of 500 mm or to the existing reinforcement in the damaged area. If the prestressing strands or reinforcements are exposed, sufficient care must be taken so as not to damage the steel during the cleaning operation. The exposed strands should be coated with epoxy resin bonding compound or cement slurry grout before patching.

3. **Severe Damage:** In cases where the bridge suffers severe damage, a detailed structural analysis and a design check based on the conditions of the damage should be conducted to establish the safety and serviceability of the structure. A comprehensive review of the calculations and detailed examination of the damage will help in selecting a cost effective and appropriate restoration technique for the damaged structure. If the loss of prestress is excessive resulting in tensile cracks, preloading method should be seriously considered in making concrete repairs in order to restore the equivalent of prestress effect as per original designs. The damaged girder may require restoration by post tensioning. The repair procedure may also include epoxy resin pressure injection, shotcreting and additional welded fabric with drilled anchors and guniting. Some typical restoration procedures are examined in the following sections.

13.7.2 Repair and Rehabilitation of Damaged Concrete

Concrete damaged in bridge decks due to spalling is repaired by removing the loose and unsound material by providing temporary supports to the girder to relieve dead load stresses. It is also advisable to conduct special stress check up in the structural elements before embarking on the repair works. Figure 13.8 shows the typical details of repairs to the spalled concrete in the web of a bridge girder. Expansion bolts or grout rebars are drilled into the sound concrete from the soffit and wire mesh is placed to the sides and welded to the existing bars. The web is built up to its original size by guniting or shotcrete applied as shown in Fig. 13.8.

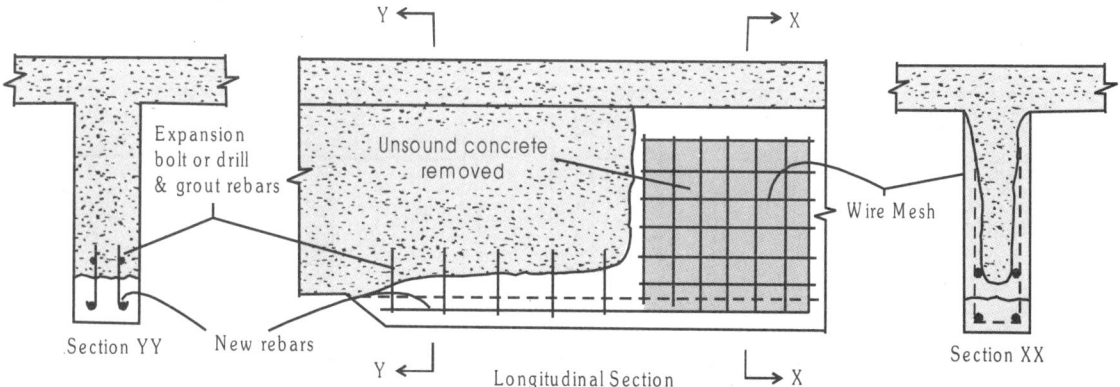

Fig. 13.8: Repair-Rehabilitation of Damaged/Spalled Concrete

In prestressed concrete tee beams with thin webs, diagonal tension cracks may develop in the web under heavy wheel loads on the deck if shear design and detailing or compaction of concrete in the support zone is not properly done. In such cases holes are drilled diagonally and rebars are placed and grouted to arrest the shear cracks as shown in Fig. 13.9. Alternatively, repairs against shear distress can also be done by jacking stirrups as shown in Fig. 13.10. In this method, the deck slab concrete is carefully removed to permit positive anchoring of additional vertical stirrups placed around the existing beam. After priming the exposed surfaces, epoxy mortaring or shotcreting or guniting is done providing a new concrete jacket over the old girder.

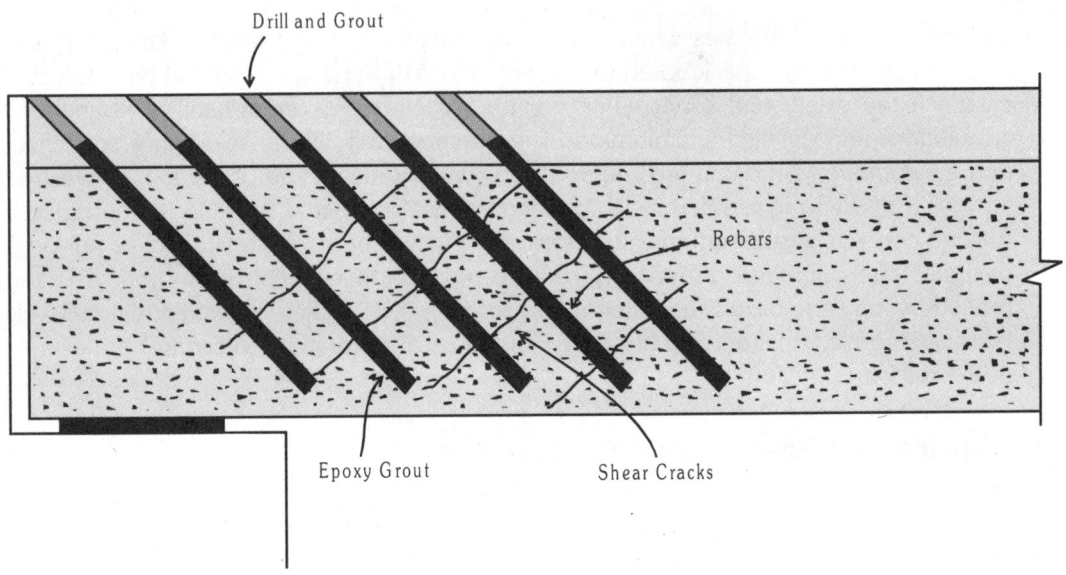

Fig. 13.9: Repairs of Shear Cracks by Stitching Rods and Epoxy Grouting

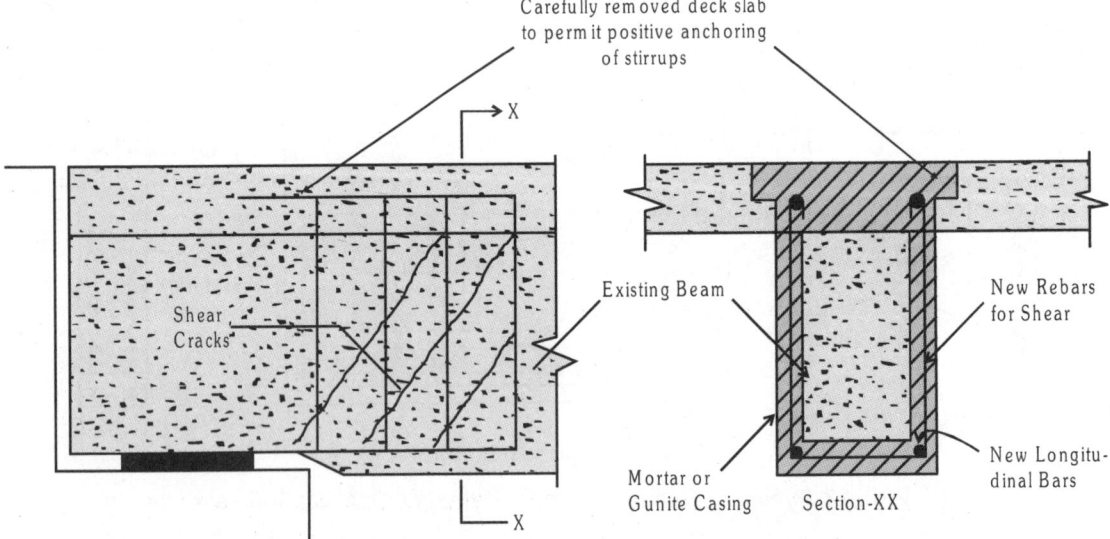

Fig. 13.10: Repairs against Shear Distress by Jacketing Stirrups

Prestressed concrete bridges located in coastal regions of very severe exposure conditions are likely to suffer extensive distress like spalling of concrete exposing the steel reinforcements. In such cases the unsound and loose concrete around the girder is removed and repairing against the loss of concrete section is done by jacketing the girder using a metal sleeve jacket as shown in Fig. 13.11. The metal jacket is fixed by bolts to the web of the girder and the gap between the steel box and web is filled by epoxy concrete grout, thus rehabilitating the distressed girder.

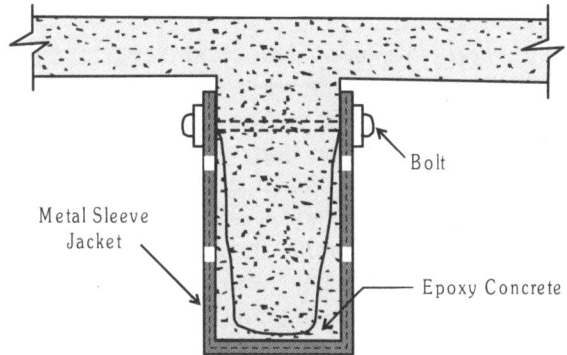

Fig. 13.11: Repair of Girder by Metal Sleeve Jacket

The propagation of cracks in tee girders of a bridge deck can be arrested by using the principle of post tensioning. Figure 13.12 shows this technique in which tensile cracks in the beams are arrested by inducing compression using tension ties fixed to the guide angles which in turn are fixed to the web of the tee beam at locations of tensile cracks. The rods or wires are tensioned by tightening the end nuts or by turning of turn buckles in the rods against the anchoring devices. It is necessary to check the stresses to guard against any possible adverse effects.

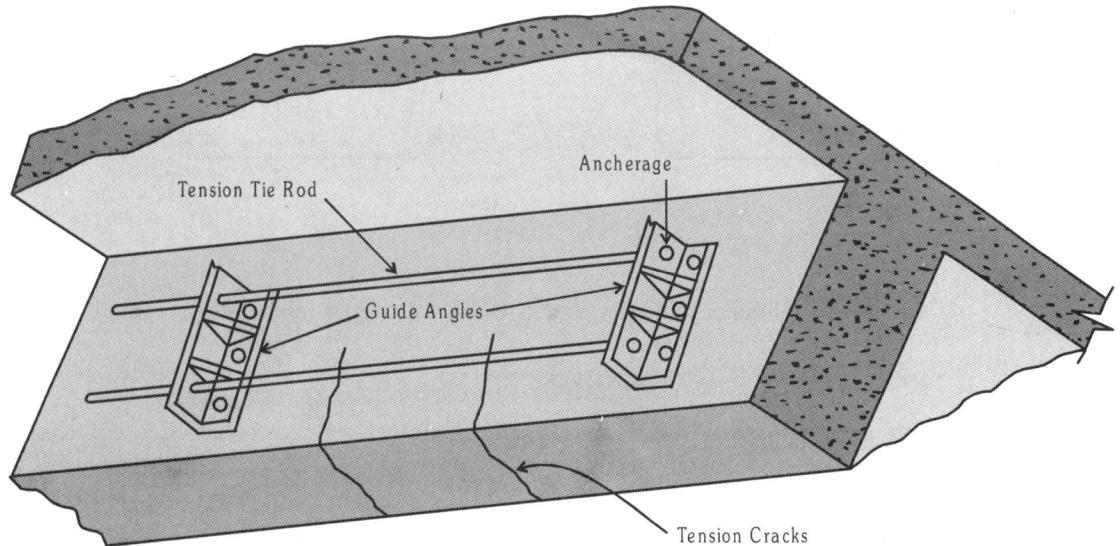

Fig. 13.12: Crack Arrest by Post Tensioning Principle

13.8 REPAIRS OF GIRDERS DAMAGED BY COLLISION

Many prestressed concrete bridges of short and medium spans are built by using precast pretensioned girders to avoid disruption of traffic. These pretensioned girders are likely to be damaged during transportation. In such cases the damaged portion can be repaired and rehabilitated provided the damage is only local and not extensive so as to rule out the use of the structural element.

Figure 13.13 shows the method of repairing the bottom flange of a pretensioned girder partially damaged at the sides so that the high tensile wires or strands are not exposed. The damaged portion is cleaned of dust and spalled concrete and dowel bars are introduced into the drilled holes to a depth of not less than 75 mm and then covered using mortar or non shrink grout as shown in Fig. 13.13 (a). If the damage extends over a larger depth covering the sloping sides, a steel wire mesh is embedded and repaired by applying mortar or non shrink grout as shown in Fig. 13.13 (b). If the damage is more severe and deeper so that the high tensile wires and strands are exposed, the damaged portion is repaired by using links and dowels along with wire mesh tied to the reinforcements. The entire damaged portion is repaired using mortar or non shrink grout as shown in Figs. 13.14 (a) and (b).

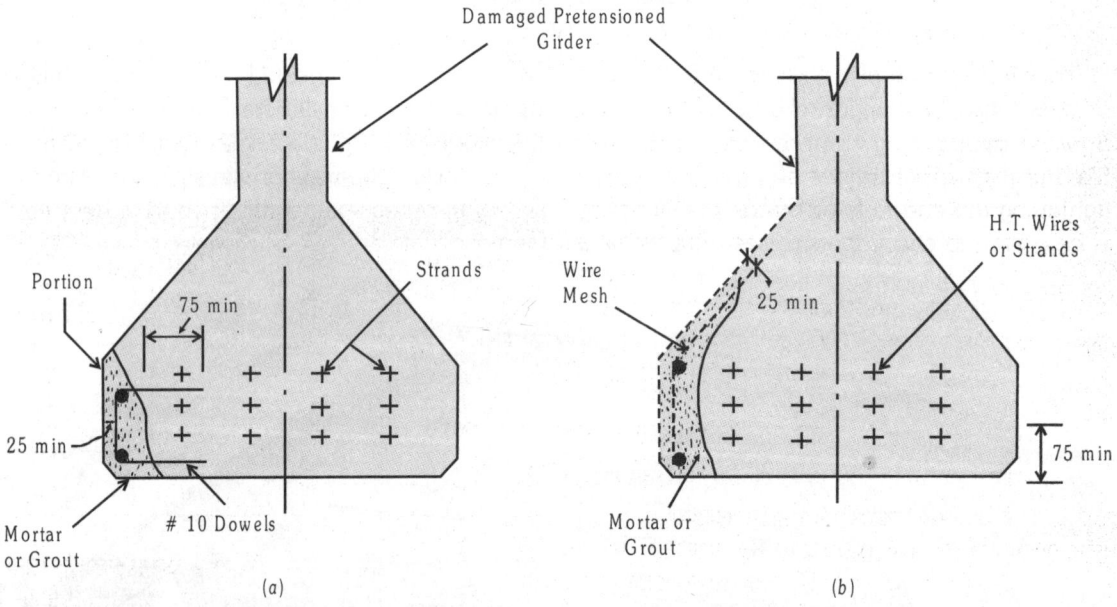

Fig. 13.13: Repairs to Pretensioned Girders Dmaged by Collision

13.9 RESTORATION OF DAMAGED PRESTRESSED CONCRETE BEAMS

Prestressed concrete girders located in coastal areas are likely to suffer extensive damage due to corrosion of reinforcements especially when unbonded tendons are used. The high tensile wires or strands may get damaged resulting in loss of prestress in concrete. In prestressed concrete girders, the tendons are highly stressed and with loss of cross-sectional area due to corrosion, the girder may fail suddenly due to fracture of steel in tension. The damaged tendons can be effectively spliced or strengthened by external forces[11]. The following methods have been found to be very useful in restoring the strength of damaged prestressed concrete girders:

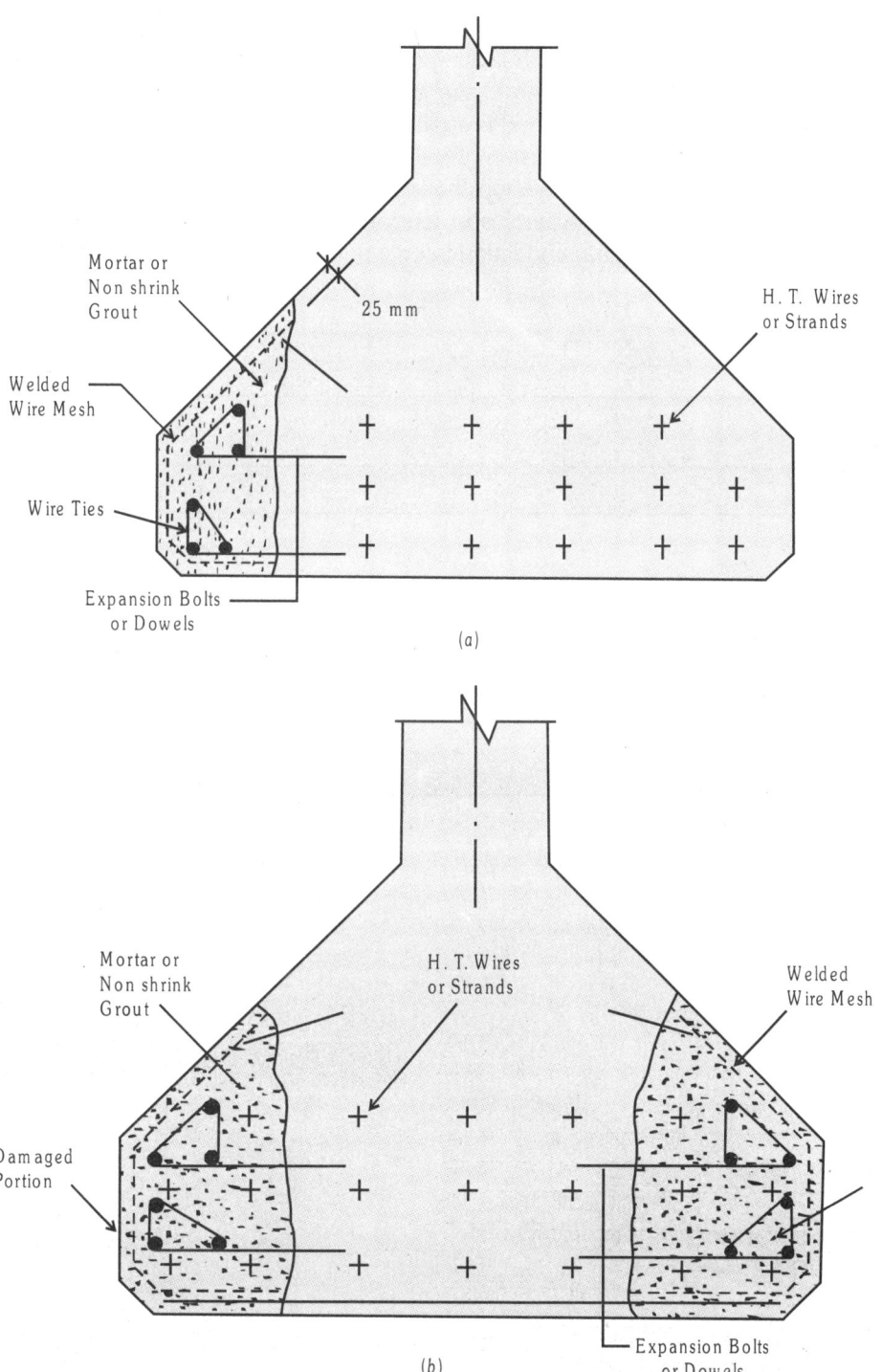

Fig. 13.14: Repairs to Damaged Pretensioned Girders

Method 1: In the case of damaged prestressed concrete I-girders, post tensioning rods (one on either side of the web) in conjunction with jacking (concrete) corbels located outside the damaged areas can be used to restore the lost area of the tendons due to corrosion and other destructive effects. To start with, the calculated pre load is applied and the damaged concrete is repaired. After the new concrete has gained the desired strength, the preload is removed. Figure 13.15 shows the method of locating the high strength rods on the roughened sloping bottom flange of the girder. After constructing the jacking corbels, final post tensioning of the rods is done as per the design strength requirements. Suitable spiral and link reinforcements should be used in the jacking corbels to strengthen the concrete.

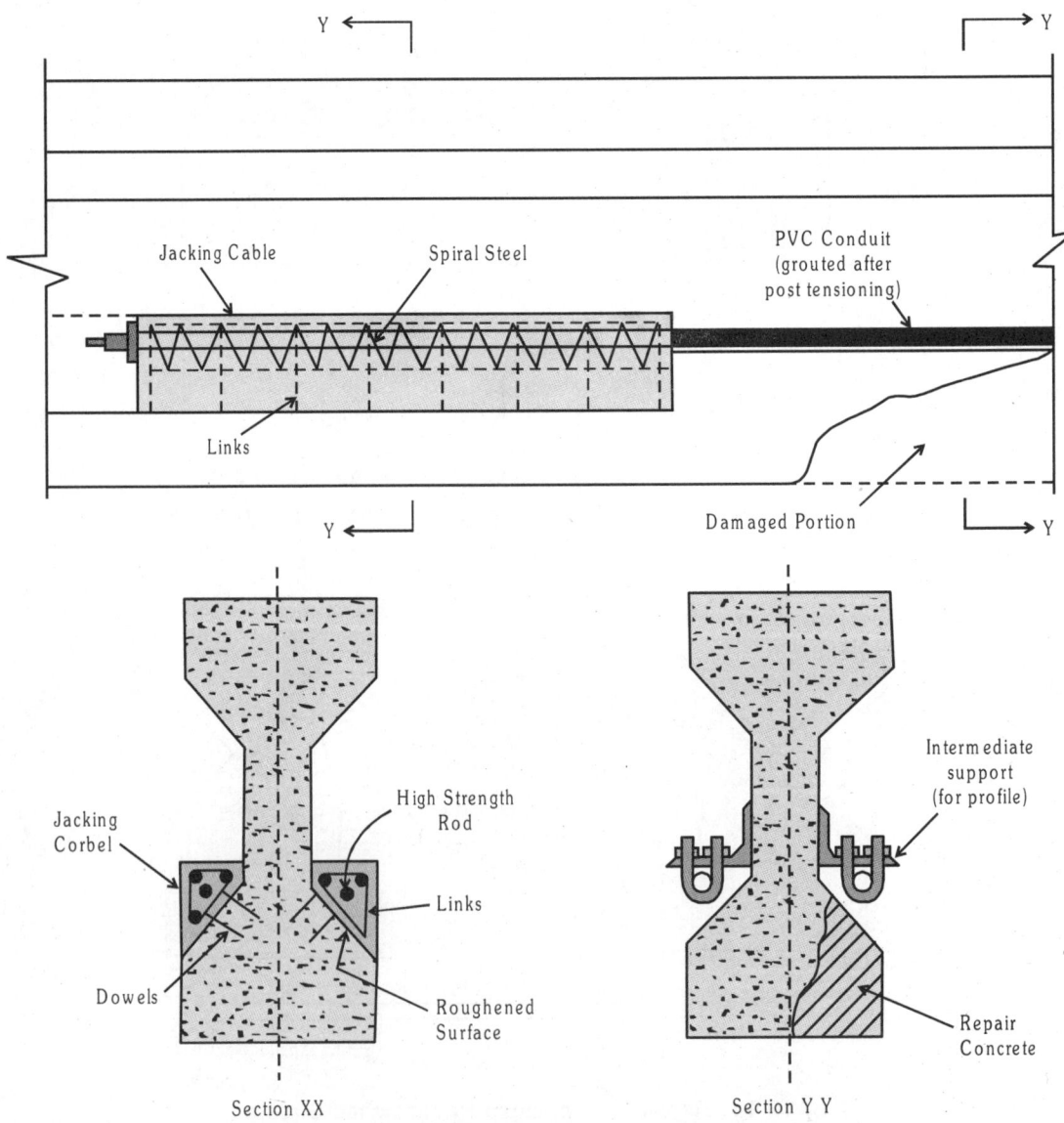

Fig. 13.15: Restoration of Girder by Post Tensioning

Method 2: In this method the damaged concrete is first repaired by applying the required preload and the concrete corbels are constructed with the required conventional steel reinforcement. Figure 13.16 shows the method of adding external reinforced concrete to restore the strength of the damaged girder. 16 mm diameter steel dowels are used to anchor the corbels to the bottom flange. In this method, repair and restoration is done by adding external reinforced concrete.

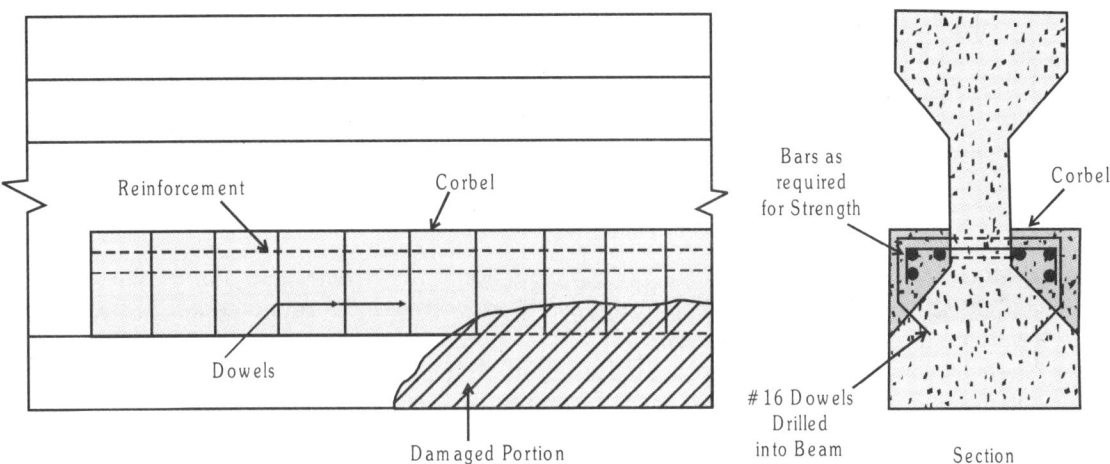

Fig. 13.16: Restoration by adding External Reinforced Concrete

Method 3: In this method, preload is applied prior to the repair of the damaged concrete and removed after the completion of repairs. A metal sleeve jacket is installed (as shown in Fig. 13.17) around and beyond the damaged area (up to a minimum of 1 m). The gap between the metal sleeve and girder is filled with epoxy grout by pressure injection.

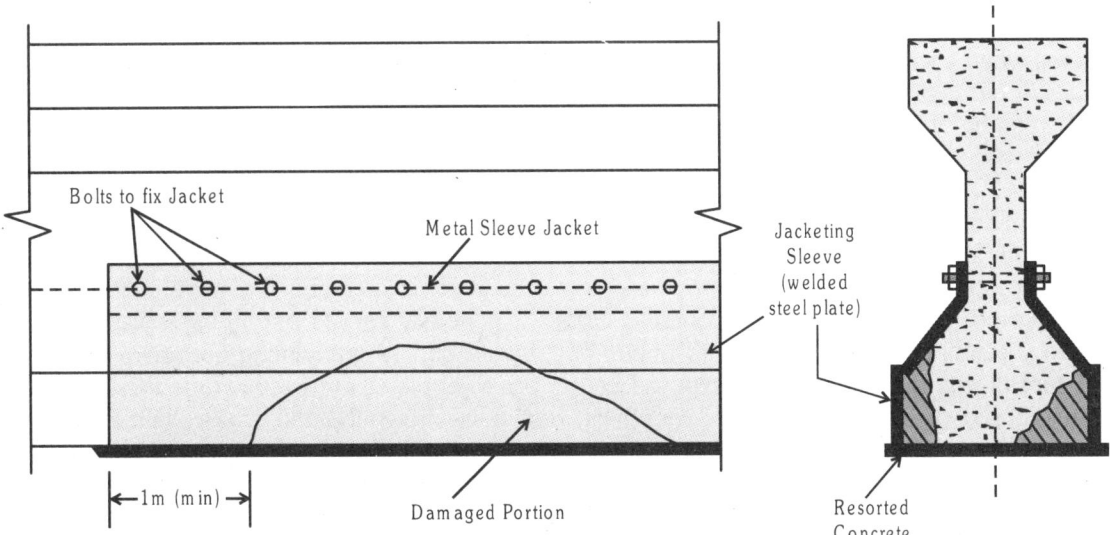

Fig. 13.17: Restoration by addition of Metal Sleeve Jacket

13.10 STRENGTHENING OF BEAMS BY EXTERNALLY BONDED STEEL PLATES

Due to increased intensity of Highway loading on existing bridge decks, it becomes necessary in many cases to increase the load bearing capacity of the existing reinforced or prestressed concrete girders. According to Raina[11], two fundamental methods of strengthening/repair are possible with epoxy resin adhesives which are listed below:

1. The depth of the structural element is increased by adding a new layer of concrete on top of an existing cross-section and bonding the old and new elements with modern epoxy resin adhesives.
2. The total reinforcement in the cross-section is increased by epoxy bonding of thin plates on the tension face of the beam to increase the flexural and shear strength.

The scope of application of this strengthening technique is ideally suited for the following situations:

(a) Restoration of structures by rectifying constructional deficiencies that impair the safety of the structure as a result of faulty dimensioning, corrosion of reinforcements, insufficient reinforcements, overloading, etc.
(b) Strengthening of an existing structural element by increasing its load bearing capacity
(c) Altering the load supporting structural system by changing spans by shifting or removing of supports, conversion of continuous beam to single span beam and *vice versa*, etc.

Research investigations to strengthen concrete flexural elements by externally bonded steel plates were first attempted in France around 1960. An inclined bridge on the French A6 motorway was strengthened by bonding additional tensile shear reinforcement plates as reported by Bresson[15]. Practical applications of this strengthening technique was followed in France, South Africa, Japan and Russia. In Switzerland[16], this method has been extensively used in both buildings and bridges for the last three decades. Experiments conducted at Transport and Road Research Laboratory in U.K. by Irwin[17] and MacDonald[18] have conclusively proved the efficacy of strengthening concrete beams by externally bonded steel plates.

Experiments conducted by Krishna Raju and Nadgir[19] have shown that reinforced concrete beams, when epoxy bonded with steel plates on the tension face exhibited significant increases of up to 30 per cent in the ultimate flexural strength in comparison with non plated beams. During the beginning of 1970s, the highway authorities of Tokyo undertook a project involving the repair and rehabilitation of elevated motorways in the entire metropolitan area. Nearly 10 km length of a total of 100 km of the motorway has been strengthened by bonded reinforcements.

Reinforcing plates of any suitable grade of structural steel can be used for strengthening of existing girders. Plate gauges below 3 mm are not suitable, because sand blasting can deform them. Steel plates between 6 to 16 mm thick were used in some strengthening works in Switzerland and England. Pretreatment of the concrete surface is generally carried out by sand blasting, shot blasting, grinding or roughening with pneumatic needle gun or granulating hammer. The grain structure of the concrete must be exposed before the steel plates are fixed. The adhesive epoxy resin joint is generally between 1 to 3 mm thick. Tests have shown that the tensile shear strength of the adhesive is initially proportional to the square root of the thickness. However the tensile shear strength reaches a maximum and then starts decreasing as the adhesive thickness is further increased. Hence thinner layers prove stronger and have greater resistance than thick ones. Figure 13.18 shows the soffit of a bridge deck strengthened by bonded steel plates to increase the load carrying capacity of the bridge without disruption of traffic.

Maintenance and Rehabilitation of Prestressed Concrete Bridges

Fig. 13.18: Rehabilitation of Bridge Deck by Epoxy Bonding of Steel Plates

Figure 13.19 shows the strengthening of beams to resist increased shear forces. The beam portion near the supports is bonded with steel plates to increase the ultimate shear strength of the beams. The method of improving the flexural and shear strength of concrete structural elements by epoxy bonding of steel plates has been recognized as a well established technique for rehabilitation of bridge girders

Fig. 13.19: Shear Strengthening of Beams by Epoxy Bonding of Steel Plates

13.11 CASE STUDIES OF REPAIRS AND REHABILITATION OF BRIDGES

Innumerable number of bridges have been built in various countries using prestressed concrete during the last six decades. Many of these structures built during the early period are showing signs of distress after decades of service. Rapid developments in the field of cement and polymer technology and chemical adhesives have paved the way for effective repair and rehabilitation of various structural elements of bridges. A brief survey of the typical bridge structures repaired, strengthened and rehabilitated is compiled below:

1. **Quinton Bridges in U.K:** Sommerard[20] has reported the rehabilitation work undertaken to strengthen the four bridges on the M-5 motorway at the Quinton interchange west of Birmingham. The super structure of the bridge with spans of 16.5, 27 and 16.5 m comprises a deck made up of voided slabs 900 to 1050 mm thick. Routine inspection indicated cracks in the soffit of end and central span sections. Review of design calculations indicated deficient tensile reinforcements at certain locations of the deck slab. The following two rehabilitation methods were examined:

   (a) Installation of prestressing elements
   (b) External reinforcement with bonded on steel plates

   A comparative analysis indicated the bonded steel plate strengthening to be more effective in spite of the fact that the technology was new in the year 1975. Accordingly the end sections were strengthened with steel plates 6 mm thick. A double layer of 6 mm thick plates were employed at the middle of central span. In highly distressed locations of the central span, three layers of 12 mm thick steel plates measuring about 3 m long and 250 mm wide were fastened to the soffits of the slab with screw plugs spaced at intervals of 450 to 900 mm. The soffit of the deck slab was pretreated to remove the unevenness and shoulders before the steel plates were bonded using epoxy resin. Subsequent tests indicated that the strengthened bridge slabs were flexurally stiffer indicating lesser deflections after the rehabilitation work.

2. **Obra Singrauli Bridge No. 93[5]:** The super structure of Obra Sigrauli Bridge, located on Eastern Railway in India comprises of 4 spans of 18.3 m and one span of 24.4 m. The bridge deck is made up of two prestressed concrete girders stressed with Freyssinet system of post tensioning. After 15 years of service, the prestressed girders developed large number of cracks at the junctions of girder and decks slab on both internal and external faces. Also longitudinal cracks were observed in the bottom flange of the girders and vertical cracks at the junction of diaphragm and main girders. Some of the cracks were as large as 3 mm.

   Investigations revealed that vertical stirrup reinforcements and the shear connectors at the junction of the top flange of the girder and deck slab were insufficient. The bridge being on a railway line was subjected to vibrations. The repeated vibrations in vertical direction might have contributed to the corrosion of steel. The bridge deck was strengthened by pumping low viscosity epoxy resin to seal the various cracks developed in the deck. After sealing of the cracks, the longitudinal girders were prestressed vertically at 1.2 m intervals using 12.5 mm strands with anchorages provided at the top of the deck in conjunction with I-section girder and a steel saddle at the soffit as shown in Fig. 13.20.

Maintenance and Rehabilitation of Prestressed Concrete Bridges

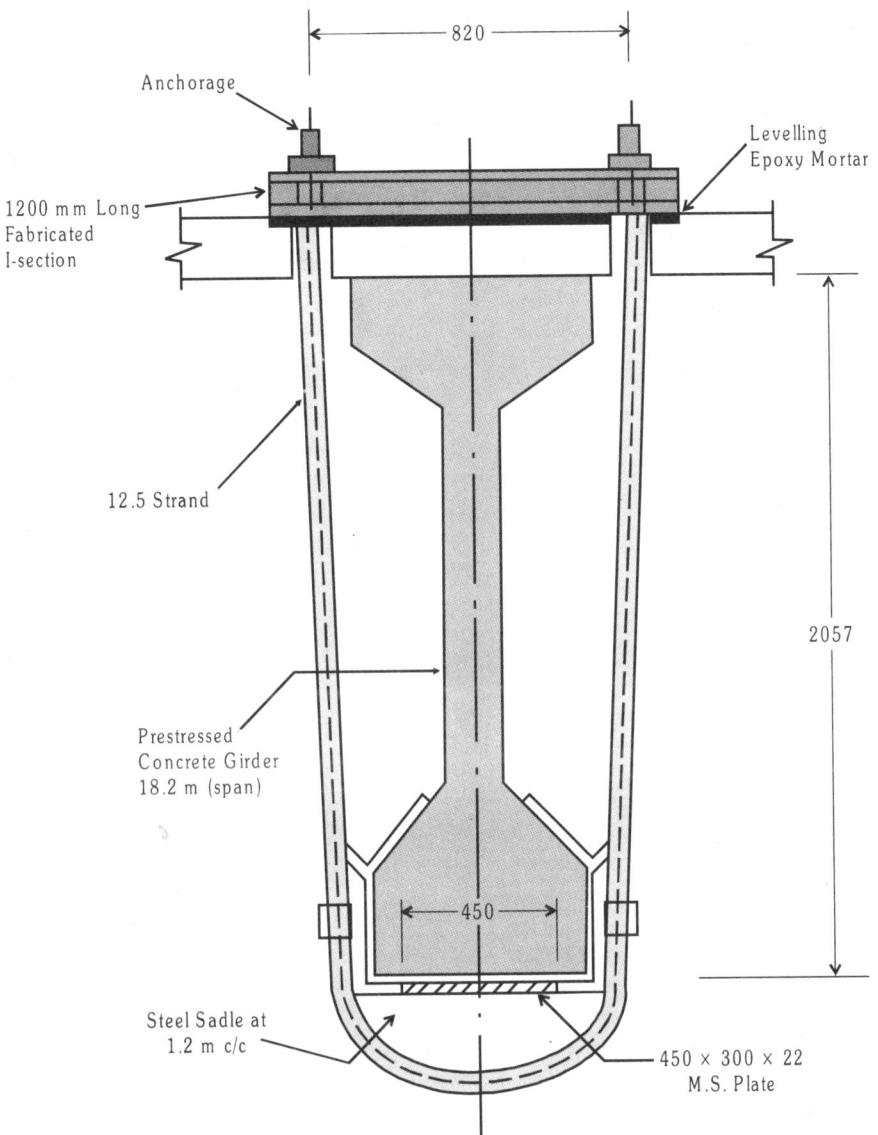

Fig. 13.20: Restoration of PSC Girder (Obra Singrauli Bridge No. 93)

3. **Swanley Bridges in U.K:** These bridges form part of the M-20 and M-25 motorway intersection. The super structure is made up of a continuous slab supported on inclined piers. Shortly after the bridge was opened for traffic, cracks were observed on the soffit of deck slab at the end sections. A design review indicated that the reinforcements at the cracked locations were inadequate. Hence the missing reinforcements were introduced in the form of bonded steel plates 6 mm thick, 250 mm wide and 3 to 6 m long plates bonded in three layers in each strip. Each strip of reinforcement was 12 m long and 15 strips were distributed over the entire width of the bridge. On top of the slab above the pier as many as

four steel plates were bonded together over a length of 12 m. All together 449 plates were epoxy bonded within a period of 20 days including pretreatment of concrete and plates. Somerard[20] has reported in detail the rehabilitation and testing of the repaired structure under dynamic loading.

4. **Gizenen Bridge, Muotta Valley, Switzerland[11]:** The bridge deck comprising of tee beam and slab built during 1911 and located in Switzerland had to be strengthened to withstand planned future loading. The damaged part of the bridge deck slab was repaired using epoxy resin mortar. A new cross beam was introduced at the centre of span. 15 mm thick steel plates were bonded to the soffit of the main girders and 10 mm thick plates to the cross girders. The plates were 200 and 150 mm wide respectively. The efficacy of the repair and rehabilitation was confirmed by load tests conducted by the Federal Material Testing Institute.

5. **Chambal Bridge[5]:** Chambal Bridge is a balanced cantilever type of bridge located on the state highway connecting Uttar Pradesh and Madhya Pradesh. The hybrid type bridge comprises both reinforced and prestressed concrete girders. The reinforced concrete box girders are used for cantilever spans and the prestressed concrete girders are used for the suspended spans. The bridge is located across the river Chambal near Etwah in Uttar Pradesh. The bridge is 592 m long with the deck comprising of single cell reinforced concrete box girders spanning 11.1 m and projecting on either side of the pier. The suspended span comprises of two prestressed concrete girders with reinforced concrete deck slab of span 40.6 m. Cast steel rocker and roller bearings have been used at articulations for supporting the suspended span as shown in Fig. 13.21.

Fig. 13.21: Prestressed Concrete Girder Supported on Rocker–Roller Bearings (Chambal Bridge)

After the bridge was completed and opened to traffic in 1975, it developed distress due to improper positioning of roller bearings. It was observed that the suspended span between tow intermediate piers shifted towards down stream side at roller end by about 20 mm. Subsequently heavy loads were transported over the bridge by the Department of Atomic Energy. The deviations towards the down stream gradually increased with the passage of time to about 110 mm as shown by the displacements of the bearings as shown in Fig. 13.22. Investigations revealed that the bearings were not at right angles to the axis of the bridge and the level of downstream side bearing was lower by 35 mm as compared to the elevation of upstream bearing. Hence due to transverse inclination of the bearings towards down stream, the span had a tendency to move in the transverse direction.

Fig. 13.22: Distress at Rocker-Roller Bearings (Chambal Bridge)

The basic rehabilitation scheme comprised of providing an access from the decking to the roller end articulation to inspect and replace the bearings. A steel inspection cum working platform was suspended from the bridge deck near the roller end articulation. Lifting and rotating the suspended span was done by placing a heavy steel truss over hammer head and roller end articulation. The trusses were tied to the hammer head at one end and to the prestressed concrete girder at the other end using steel suspenders. Freyssi flat jacks and PTFE/stainless steel sliding arrangements were introduced under the trusses near articulation on hammer head. The span was lifted by operating the flat jacks. The old roller bearings were replaced with new ones which were properly aligned and leveled with epoxy mortar. The traffic over the bridge was diverted only for 3 days during the course of the rehabilitation work which was carried out by Uttar Pradesh Bridge Corporation.

6. **Katepura Bridge, Maharashtra[5]:** The super structure of Katepura Bridge in Maharashtra comprises of simply supported prestressed concrete girders with reinforced concrete deck

slab with 4 spans of 37.8 m. the girders were cast in place over temporary staging and side shifted to position after necessary post tensioning of the girders. Each girder has 16 cables and the cables were stressed in two stages. During the construction period, when the stage prestressing was being carried out for a girder, it cracked with a loud sound. The cracks appeared near the end block as shown in Fig. 13.23. Detailed investigations revealed that the concrete in the end block portion of the girder was not homogeneous.

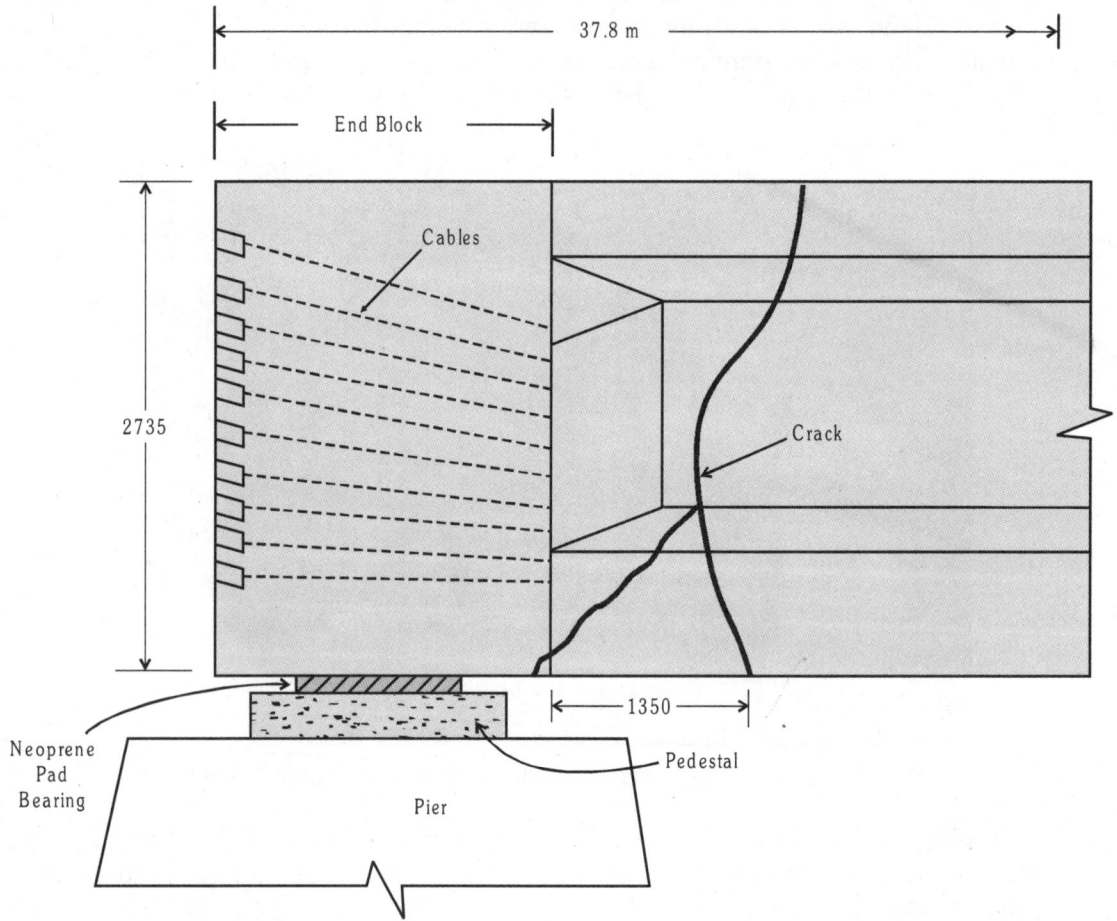

Fig. 13.23: Restoration of End Block of Prestressed Concrete Girder (Katepura Bridge (Maharastra))

Rehabilitation of the end block of the girder was done by completely dismantling the concrete in the end block after distressing the cables. New reinforcing bars were welded to the existing reinforcements of the girder. New concrete with vertical joint was provided with extra care. After the concrete attained the desired strength, prestressing operations were carried out in stages. After restoration work, the girder was load tested and found to be satisfactory at serviceability limit states.

REFERENCES

1. ROWE, R. E., CRANSTON, W. B., and BEST, B. C., New Concepts in the Design of Structural Concrete. *Structural Engineer*, Vol. 43, 1965, pp. 399-403.
2. FIB-CEB Joint Committee Practical Recommendations for the Design and Construction of Prestressed Concrete Structures, June 1966.
3. KRISHNA RAJU, N., Limit State Design for Structural Concrete. Proceedings of the Institution of Engineers (India), Vol. 51, Jan. 1971, pp. 138-143.
4. IS:1343 (Draft Second Revision—To be Published), Indian Standard Code of Practice for Prestressed Concrete, BIS, New Delhi.
5. GOKHALE, P. S. and ROHRA, M. R., Restoration of Distressed Concrete Structures, Case Studies, Concrete Structures of India, Compiled by Maharashtra India Chapter of American Concrete Institute and Presented at A.C.I's Fall Convention at New York, 1984, First Report, Jan. 1985, pp. 149-167.
6. IS: 456-2000, Indian Standard Code of Practice for Plain and Reinforced Concrete (Fourth Revision), BIS, New Delhi, 2000, pp. 30-31.
7. RAINA, V. K., *Concrete Bridge Practice, Construction, Maintenance and Rehabilitation*. Tata McGraw Hill-Publishing Co, New Delhi, 1988, pp. 287-326.
8. KRISHNA RAJU, N., Report of Rebound Hammer and Ultrasonic Pulse Velocity Tests on Prestressed Concrete Lattice Girders in Ram Kumar Mills, Rajajinagar, Bangalore, Civil Engineering Department, M. S. Ramaiah Institute of Technology, Bangalore, Oct. 1992, pp. 1-13.
9. KRISHNA RAJU, N., Technical Report on Rebound (Schimdt) Hammer Tests on R.C.C. Beams, Columns and Slabs in Darus Salam Hostel and Office Complex at Site No. 332, Queens Road, Bangalore, Civil Engineering Department, M. S. Ramaiah Institute of Technology, Bangalore, 1986, pp. 1-20.
10. KRISHNA RAJU, N., Technical Report on Ultrasonic Pulse Velocity Tests on Roof Slab of Explosion Bunker Complex at M/S. Astra Indian Detonators Limited, Bangalore, 1986, pp.1-15.
11. RAINA, V. K., *Concrete Bridges, Inspection, Repair, Strengthening, Testing, Load Capacity Evaluation*. Tata McGraw-Hill Publishing Co, New Delhi, 1994, pp. 10-63.
12. KRISHNA RAJU, N., *Prestressed Concrete* (Fourth Edition). Tata McGraw-Hill Publishers, New Delhi, 2007, pp. 265-293.
13. Concrete Polymer Materials—First Topical Report—B.N.L Report 53134 (T-509) Brookhaven National Laboratory, Upton, New York, Also U.S.B.R. General Report No. 31, Denver, December 1968.
14. KRISHNA RAJU, N., *Design of Concrete Mixes* (Fourth Edition). C.B.S. Publishers, New Delhi, 2002, pp. 210-215.
15. BRESSON, J., Reinforcement Par Collage d' Armatures du Passage Inferieur du CD 126 sous l'Autoroute du SUD' Annales de l'Institut Technique du Batiment et des Travaux Publics, Supplement No. 297, Sept. 1972, Concrete and Reinforcement Series No. 122.
16. HUGENSCHMIDT, F., Strengthening of Existing Concrete Structures with Bonded Reinforcements, EMPA, Switzerland, Sept. 1981.
17. IRWIN, C. A. K., The Strengthening of Concrete Beams by Bonded Steel Plates. Transport and Research Laboratory, Department of Environment, Supplementary Report 160 UC, Crowthorne, Berkshire, U.K, 1975.
18. MACDONALD, M. D., The Flexural Behaviour of Concrete Beams with Bonded External Reinforcement. Transport and Research Laboratory, Department of Environment, Supplementary Report-415, Crowthorne, Berkshire, U.K, 1978.
19. KRISHNA RAJU, N. and NADGIR, N. S., Limit State Behaviour of Concrete Beams Strengthened by Epoxy Bonded Steel Plates. *Indian Concrete Journal*, Vol. 65, No. 3 March 1991, pp. 124-129.
20. SOMMERARD, T., Swanley's Steel Plate Patch-Up, New Civil Engineer, London, 1977, No. 247, 16 June, 1977, pp. 18-19.

14

WORLD'S PROMINENT PRESTRESSED CONCRETE BRIDGES

14.1 GENERAL ASPECTS

Eugene Freyssinet, the originator of prestressed concrete toiled for seven long years to sell the history's most exciting building material but with no buyers. Although Freyssinet took patents for the revolutionary material '**Prestressed Concrete**' as early as in 1928, he had to wait for more than a decade to witness the wide spread application of his discovery in building bridge structures. The big boom in prestressed concrete was witnessed only in the post war years when the European countries led by France and Belgium pioneered the development and use of prestressed concrete in bridge construction. More than 300 prestressed concrete bridges were built in post war Germany during the years 1949 to 1953.

The application of prestressed concrete started early in India around 1948 starting with the construction of three railway bridges having spans ranging from 12.8 m to 19.2 m located on the Assam rail link in North Eastern India. The first prestressed concrete highway bridge built in India was the Palar Bridge near Chinglepet, built in 1954 with 23 spans of 27 m each located in the Madras Province. Prestressed concrete was first introduced in U.S.A. in 1949 for Magnel's Walnut Bridge. By 1960 the spans of beam bridges had reached 160 m and Morandi's bridge across the lake of Maricabo was constructed with spans of 235 m in Japan. Hamana Bridge in Japan is an excellent example of a growing family of long span prestressed concrete bridges erected by cantilever construction technique developed by Ulrich Finisterwalder[1] who is considered as one of the greatest bridge builders of our times.

The design, construction and maintenance of long span bridges poses a challenge to the ingenuity and creativity of the designer. The recent innovations in concrete technology has resulted in Rheodynamic concrete[2] generally referred to as self compacting concrete which can flow under its own weight and completely fill the form work containing dense reinforcements and attain high early strength and durability. This type of concrete is ideally suited for precasting of slabs, beams and cellular box girders widely used in the construction of pretensioned and post tensioned bridges decks. The developments in the field of cement, admixtures and polymers have paved the way for production high performance and high strength concrete[3] of grades M-80 to M-100 with minimum compacting effort.

Large span prestressed concrete girders necessarily require very high prestressing forces and suitable anchorages to transmit the compressive forces to concrete. Developments in high strength steel wires, strands along with prestressing anchorage systems like Freyssinet International multiwire and multi strand anchorages[4] can impart cable forces of the order of 10,000 kN. In the B.B.R.V. system, large capacity tendons can carry forces exceeding 10,000 kN. In the Lossinger-VSL system developed in Switzerland, 15 and 18 mm diameter multi strand cables can impart very large forces required for large span girders effectively reducing the number of cables.

14.2 WORLD'S LONG SPAN PRESTRESSED CONCRETE BRIDGES

During the last six decades, main span limits of prestressed concrete bridges have consistently increased and with the development of cable stays, the span limit of 300 m has been surpassed with the construction of Sunshine Sky Bridge with a main span of 365 m at Tampa bay, Florida[5]. Table 14.1 lists the location and main span lengths of prominent long span prestressed concrete girder and cable stayed bridges spread in different countries of the world based on the documentation of Victor[6], Raina[4], and the Japan Association of Prestressed Concrete Industry[7]. The combination of prestressing and cable stays techniques will facilitate construction of prestreesed concrete bridges with spans exceeding 500 m in the near future.

Table 14.1: Prominent Long Span Prestressed Concrete Bridges

Sl.No.	Name of Bridge	Location	Year	Main Span (m)
1.	Bassien Creek	Bombay, India	1982	115
2.	Mahatma Gandhi Sethu	Patna, India	1982	121
3.	Yamatogawa	Osaka, Japan	1970	120
4.	Zuari	Goa, India	1983	122
5.	Barak	Silchar, India	1982	122
6.	Lubha	Assam, India	1961	130
7.	Amakusa	Kumamato, Japan	1966	160
8.	Nagoya Ohashi	Saga, Japan	1967	176
9.	Urato	Kochi, Japan	1972	230
10.	Hamana	Imagiri Guchi, Japan	1976	240
11.	S. Joao	Porto, Portugal	1991	250
12.	Skye	Sky Island, U.K.	1995	250
13.	Schottwein	Semmering, Austria	1989	260
14.	Gateway	Brisbane, Australia	1986	260
15.	Varrod	Kristiansand, Norway	1994	260
16.	Humen	Pearl River, China	1998	279
17.	Raftundet	Lefeton, Norway	1998	298
18.	Stolmasundet	Austevoll, Norway	1998	301
19.	Brotonne	Caudebec, France	1977	320
20.	Sunshine Skyway	Tampa Bay, Florida	1988	365

14.3 NOTABLE EXAMPLES OF PRESTRESSED CONCRETE BRIDGES

Prominent examples of different types of prestressed concrete bridges built in India and various other countries are compiled in this section along with technical data comprising span length, type of bridge deck, total length of bridge and the technical organization executing the bridge work.

1. **Mahatma Gandhi Sethu:** Mahatma Gandhi Sethu also referred to as the Ganga Bridge is the longest river bridge in Asia extending over 5.575 km built across the mighty river Ganga at Patna. Figure 14.1 shows the Ganga Bridge built by using precast single cell segmental box girders of variable depth prestressed to form continuous spans. The bridge is made up of 46 spans of 121 m with a two lane deck. This is the first bridge in India built by using large scale precast segmental construction. The foundation for the piers comprises of single circular caisson sunk to a depth of approximately 50 m below the bed level. The bridge construction work was executed by Gammon India Ltd.

2. **Lubha Bridge:** Lubha Bridge, built across a 30 m deep gorge of the Lubha River in Assam is considered as one of the India's longest variable depth box girder bridge extending over a total length of 172 m. Figure 14.2 shows the prestressed box girder bridge having a central span of 132 m with side spans of 20 m having a deck width of 8 m. The Lubha bridge built in 1961 is one of the earliest prestressed concrete bridges built using the cantilever construction technique by M/S. Gammon India Ltd.

3. **Zuari Bridge:** Zuari Bridge located on the National Highway No. 17 in the west coast state of Goa comprises of prestressed concrete cantilever box girders of variable cross-section with four main spans of 122 m, two end spans of 69.5 m and a viaduct with 5 spans of 36 m. The precast segments 3 m long, varying in depth from 2.07 m to 8.05 m were assembled and prestressed using Freyssinet system. Figure 14.3 shows the Zuari Bridge built in 1963 by M/S. Gammon India Ltd.

4. **Yamuna Bridge:** Yamuna Bridge located at Kalpi in Uttar Pradesh is 767 m long with a two lane carriage way of 7.5 m and 1.5 m wide foot paths on either side. The bridge is made up of 8 intermediate spans of 85 m and two end spans of 43.5 m. The super structure comprises of a single cell box girder deck prestressed by Freyssinet system. The deck is monolithic with piers and built *in situ* by free cantilever construction method. Figure 14.4 shows the Yamuna Bridge deck supported by diamond shaped cellular piers resting on circular caissons some of which were sunk up to depth of 50 m below bed level. The bridge was built by the combined efforts of P.W.D. Uttar Pradesh, STUP Consultants Ltd. and the National Building Construction Corporation.

5. **Sone Bridge:** Figure 14.5 shows the Sone Bridge located in Uttar Pradesh with a total length of 1006 m having 22 spans of 45.73 m. The deck is made of prestressed concrete tee beams and slab with constant depth girders. Each span of the deck weighing 85 tons is supported on colcrete piers and wells sunk 21.34 m below the bed level. The bridge work was executed by the Public Works Department of Uttar Pradesh in association with the contracting firm Gammon India Ltd.

6. **Buxar Bridge:** Buxar Bridge built in the state of Bihar is a post tensioned prestressed concrete box girder bridge with a two lane traffic width of 7.62 m. The bridge extending over a total length of 1122 m has 10 intermediate spans of 101 m and two end spans of 55 m. The bridge is of continuous type with variable depth box girder. Figure 14.6 shows the prestressed concrete bridge deck supported on reinforced concrete piers built on well foundations. The bridge was constructed by M/S. Gammon India Ltd.

World's Prominent Prestressed Concrete Bridges

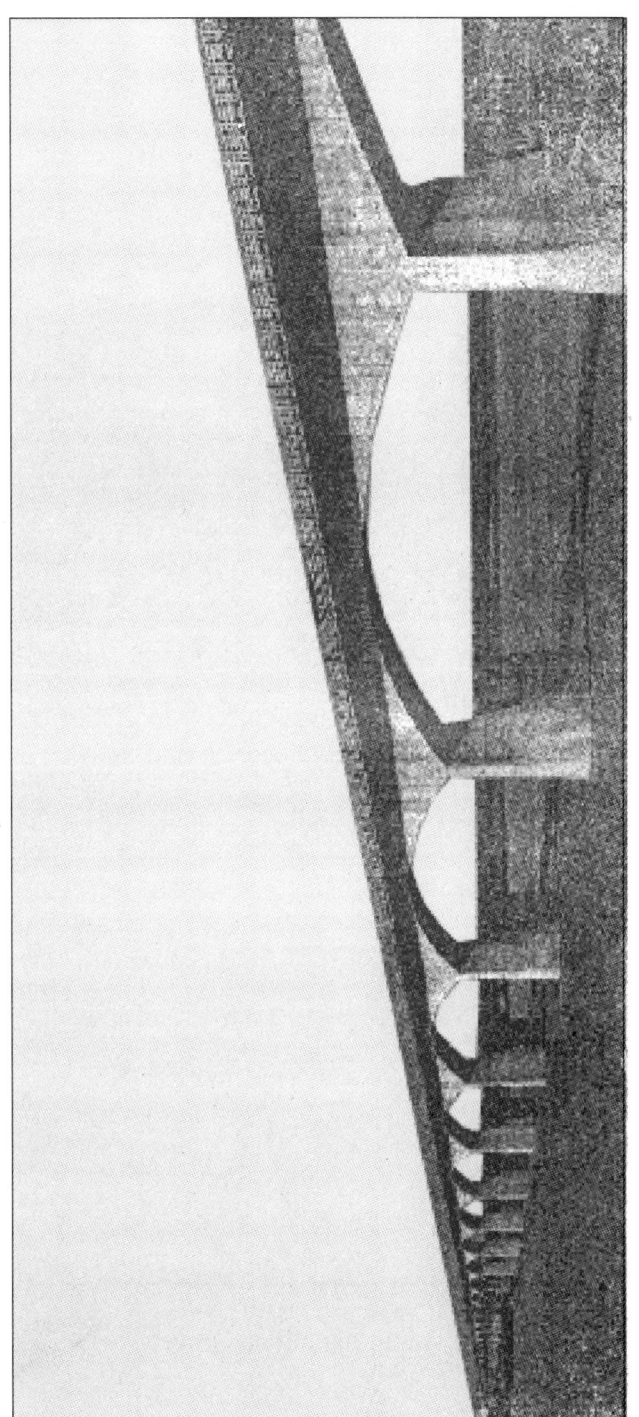

Fig. 14.1: Mahatma Gandhi Sethu at Patna

Prestressed Concrete Bridges

Fig. 14.2: Lubha Bridge, Assam

Fig. 14.3: Zuari Bridge, Goa

World's Prominent Prestressed Concrete Bridges

Fig. 14.4: Yamuna Bridge at Kalpi, U.P.

Fig. 14.5: Sone Bridge, Uttar Pradesh

Fig. 14.6: Buxar Bridge, Bihar

7. **Gambhirkad Bridge:** Gambhirkad Bridge located in the state of Punjab is of hybrid type built with a central suspended span of precast prestressed concrete over a length of 40 m and two side spans of reinforced concrete. The tee girders are prestressed by the Freyssinet system The 107 m long bridge with 7 m wide road way has a central span of 55 m with reinforced concrete deck slab. Figure 14.7 shows the bridge deck supported on V type reinforced concrete piers. The bridge was built for the public works department of Punjab by M/S. Gammon India Ltd.

8. **Bhaghirathi Bridge:** Figure 14.8 shows the Bhaghirathi Bridge located in the state of West Bengal. The length of the bridge is 280 m with three intermediate spans of 78 m each and two end spans of 23 m. The bridge deck comprises of balanced cantilever units supporting suspended spans. The balanced cantilever bridge deck has the advantage of continuity but without the disadvantages of a statically indeterminate structure. Balanced cantilever bridges are ideally suited for situations where minor settlements are likely due to movement of foundations. The central suspended spans consist of prestressed concrete beams launched in position using a traveling scaffolding tower. The bridge deck is supported on 15 m high concrete piers founded on brick caissons sunk to a depth of 18 m below bed level. The bridge deck was constructed by M/S. Gammon India Ltd. for the Public Works Department of West Bengal.

9. **Barak Bridge:** Barak Bridge built at Silchar in the north eastern state of Assam is a long span prestressed concrete bridge which ensures navigation under the main span. The bridge has a main central span of 122 m and two end spans of 56 m each. The 282 m long main bridge and the 187 m long viaduct at the approaches have a clear road way of 7 m. Figure 14.9 shows the Barak Bridge with its main spans built by the cantilever method of construction. The bridge work was executed by M/S. Gammon India Ltd. for the Public Works Department of Assam.

Fig. 14.7: Gambhirkad Bridge, Punjab

Fig. 14.8: Bhagirathi Bridge, West Bengal

Fig. 14.9: Barak Bridge, Silchar, Assam

10. **Bhima Bridge:** Figure 14.10 shows the Bhima Bridge located in the state of Maharashtra built in the year 1956. The bridge having a total length of 328 m comprises 7 spans of 47m with a road width of 7 m. The super structure consists of prestressed concrete simply supported girders with deck slab. The bridge deck is supported on concrete piers over well foundations. The bridge was constructed by M/S. Gammon India Ltd. for the Public Works Department of Maharashtra.

Fig. 14.10: Bhima Bridge, Maharashtra

11. **Sunshine Sky Way Bridge:** Figure 14.11 shows the Sunshine Sky Way Bridge built over Tampa Bay at Florida. At present, this bridge is considered to have the longest main span among the various prestressed concrete bridges of the world. The main span portion of 365 m

Fig. 14.11: Sunshine Sky Bridge, Tampa Bay, Florida

is cable stayed with a prestressed concrete box girder deck supported by two pylons. The bridge deck is supported by single plane mixed type cable system coinciding with the centre line of the road way dividing the traffic lanes. The towering pylons with the sleek prestressed concrete deck presents an imposing view over the Tampa Bay.

12. **Chaco-Corrientes Bridge:** Chaco-Corrientes Bridge shown in Fig. 14.12 and located in Argentina is considered as one of the longest precast prestressed concrete cable stayed box girder bridge in South America. The bridge deck is supported by two lateral layers of high tensile cables passing over saddles located on top of A-type towers. This innovative structure was preferred to the conventional steel suspension bridge mainly due to the reduction in bending moments in the stiffening deck leading to economy in overall costs together with the added advantage of aesthetics and aerodynamic stability.

Fig. 14.12: Chaco-Corrientes Bridge, Argentina

13. **King Fahd Causeway:** Figure 14.13 shows the aerial view of King Fahd Causeway located in Bahrain connecting the main land with an island. The high level bridge deck comprises of precast box girder segmental units assembled by traveling gantry and post tensioned to form a continuous bridge deck. The assembly of modular cellular box units using a traveling gantry on the top of the pier is shown in the left corner of the figure. Free cantilever method was used for the construction of the bridge. The bridge deck is supported on concrete piers resting on caisson foundations.

14. **Coatzacoalcos Bridge:** Coatzacoalcos Bridge located in Mexico is an excellent example of long span prestressed concrete cable stayed bridge, an aerial view of which is shown in Fig. 14.14. The bridge with a total length of 513 m has a main central span of 288 m with two side spans of 112.50 m. The deck is made up of a trapezoidal shaped box girder of 3.4 m depth with a road way width of 18.10 m. The bridge was designed by the French Consultants SOGELERG for the Mexican ICA contractors. The Freyssinet stay cables comprise of 37 and 61 HC 15 type provided in a single plane coinciding with the central median and

arranged in a mixed type when passing over the top of the pylons raising over 100 m above the bed level.

Fig. 14.13: King Fahd Causeway, Bahrain

15. **Yonegami Bashi Bridge:** Figure 14.15 shows the soffit portion of the Yonegami Bashi Bridge located in the Kanagawa Perfecture in Japan. The 125 m long prestressed concrete box girder bridge with 4 spans of 31.2 m each built in 1960 is credited to be the first curved prestressed concrete built ever built in Japan. The main girder itself is a circular arc with a

radius of 120 m and the high tensile cables are also curved in plane along with the main girder. The bridge deck provides a road way 8.5 m wide and the box girders are prestressed using the Freyssinet system.

Fig. 14.14: Coatzacoalcos Bridge, Mexico

16. **Arakawa Railway Bridge:** Arakawa Railway Bridge located in Tokyo comprises of through type prestressed girders built for the National Railways of Japan. Figure 14.16 shows the bridge deck with a provision for double track railway lines with a width of 11.6 m. The 157.7 m long bridge is made up of 4 spans of 39.4 m. The cross-section of the bridge deck is made up of L-shaped girders with inclined webs prestressed by the Freyssinet system. The bridge located on the Tohoko Main Line is designed to withstand Japanese Railway Loading Standards.

17. **Nagoya Ohashi Bridge:** Nagoya Ohashi Highway bridge located in a place of scenic beauty in the Genka National Park, was considered as the longest span prestressed concrete bridge in Japan when it was completed in the year 1967. Figure 14.17 shows the panaromic view of the bridge having a total length of 258 m with a main central span of 176 m and side spans of 41 m. The three span continuous bridge built by using the *cast-in-situ* cantilever construction technique has adopted box girder cross-section for the bridge. Dywidag method of prestressing was adopted for constructing the bridge.

World's Prominent Prestressed Concrete Bridges

Fig. 14.15: Yonegami-Bashi Bridge, Japan

Fig. 14.16: Arakawa Railway Bridge, Japan

Fig. 14.17: Nagoya Ohashi Bridge, Japan

18. **Kawaoto-gawa Bridge:** Figure 14.18 shows the Kawaoto-gawa Bridge located on the Tokyo-Nagoya Expressway in Japan. The continuous prestressed concrete bridge extending over

a length of 600 m has 2 main spans of 90.3 m, 4 spans of 90 m and two approach spans of 29.7 m. The bridge deck is of 2 span rigid frame type with box girders of depth 5 m and total width of road way being 22 m. The special feature of the bridge is that it does not use many expansion joints. The bridge has equal spans and constant girder height with a beautiful appearance to match the surrounding natural landscape. The bridge completed in the year 1969 by Japan Highway Corporation adopted Dywidag method for prestressing the girders.

Fig. 14.18: Kawaoto-gawa Bridge, Japan

19. **Yamato-gawa Bridge:** Yamato-gawa Bridge located on the Osaka-Kobe Expressway is long span prestressed concrete bridge covering a length of 592.19 m. The 6 span continuous girder bridge with box section and central hinges is owned by Hanshin Expressway Corporation and completed in the year 1970. Figure 14.19 shows bridge under construction using the *cast-in-situ* cantilever technique method. The maximum and minimum span lengths are 120 and 75.69 m respectively. The total road way width at the top is 17 m and the box girders are prestressed using the Dywidag method. The two box girders are supported by single piers built on caisson foundations.

20. **Amakusa Bridge:** Figure 14.20 shows the Amakusa Bridge located in the Kumamoto Perfecture in Japan. The three span continuous prestressed concrete bridge with box girder section was built using the cantilever construction method with *cast-in-situ* technique. The continuous prestressed concrete girder bridge completed in the year 1966 extends over a total length of 361 m. The main central span is 160 m with two side spans of 100 m. The road bridge owned by the Japanese Highway Corporation is situated in the prominent international tourist route running across the island of Kyushu. The bridge deck of 6.5 m width was built using the Dywidag method.

14.4 GENERAL REMARKS

The notable examples of prominent prestressed concrete bridges compiled above are by no means exhaustive and only major bridges built in India in particular and in various other countries like

World's Prominent Prestressed Concrete Bridges

Fig. 14.19: Yamato-gawa Bridge, Japan

Fig. 14.20: Amakusa Bridge, Japan

Argentina, Mexico, United States and Japan in general are presented. During the post war years, Germany pioneered the development and massive construction of bridges destroyed during the war using the revolutionary material 'Prestressed Concrete' developed by Finisterwalder. Thereafter the European countries, U.S.A., India, U.S.S.R. and Japan used this new material in preference to the traditional material steel for medium and long span bridges. Innovations like cable stayed technique in conjunction with prestressed concrete has pushed further the upper span limits of bridges.

In present day scenario cable stayed prestressed bridges are invariably preferred to steel bridges mainly due to the superior resistance characteristics of prestressed concrete to aggressive environmental exposure conditions. The reader may refer to the excellent publications by Shirley-Smith[8], Virola[9], Plowden[10], Subba Rao[11], Leonhardt[12], Gerwick[13] Muller[14], Lin et al.[15] and others for detailed information regarding analysis, design and construction of prestressed concrete bridges.

Presently, India has embarked on a gigantic highway development project with the planning of the Golden Quadrilateral connecting the capital cities of various states spread between the north-south and east-west corridors. This massive highway project involves the crossing of number of rivers and valley's using suitable bridges. In this context, the use of prestressed concrete for the construction of innumerable number of bridges ensures both economy and durability.

REFERENCES

1. FINISTERWALDER, U., Free Cantilever Construction of Prestressed Concrete and Mushroom Shaped Bridges, First International Symposium on Concrete Bridge Design, ACI Special Publication SP-23 American Concrete Institute, Detroit, 1969, pp. 467-494.
2. KRISHNA RAJU, N., *Prestressed Concrete* (Fourth Edition). Tata McGraw-Hill Publishing Co, New Delhi, 2007, pp. 696-714.
3. KRISHNA RAJU, N., *Design of Concrete Mixes* (Fourth Edition). C.B.S. Publishers New Delhi, 2002, pp. 210-215.
4. RAINA, V. K., *Concrete Bridge Practice, Analysis, Design and Economics*. Tata McGraw-Hill Publishing Co, New Delhi, 1991, pp. 41-59.
5. NAWY EDWARD, G., *Prestressed Concrete, A Fundamental Approach*. Prentice Hall, Englewood Cliffs, New Jersey, 1989, pp. 3-700.
6. JOHNSON VICTOR, D., *Essentials of Bridge Engineering* (Fifth Edition). Oxford and IBH Publishing Co., New Delhi, 2001, pp. 414-419.
7. Prestressed Concrete in Japan, Japan Association of Prestressed Concrete Industry, Tokyo, Japan, May, 1970, pp. 1-56.
8. SHIRLEY-SMITH, H., *World's Great Bridges*. English Language Book Society, London, 1964, pp. 250.
9. VIROLA, J., *The World's Greatest Bridges*. American Society of Civil Engineers, Vol. 38, No. 10, Oct. 1968, pp. 52-55.
10. PLOWDEN, D., *Bridges—The Spans of North America*. The Viking Press, New York, 1974, pp. 328.
11. SUBBA RAO, T. N., Long Span Prestressed Concrete Bridges in India, Seminar on Problems of Prestressing, Indian National Group of the I.A.B.S.E., Preliminary Publication, Madras, Jan/Feb 1970, pp. I-113 to I-130.
12. LEONHARDT, F., New Trends in Design and Construction of Long Span Bridges and Viaducts (Skew, Flat Slabs, Torsion Box), Preliminary Publication of Eighth Congress of I.A.B.S.E., New York, 1968.
13. GERWICK, B. C. (Jr), Precast Segmental Construction for Long Span Bridges, Civil Engineering, New York, January, 1964.
14. MULLER, J., Long Span Precast Prestressed Concrete Bridges Built in Cantilever, First International Symposium on Concrete Bridge Design, ACI Publication SP-23, American Concrete Institute, Detroit, 1969, pp. 705-740.
15. LIN, T. Y. and GERWICK, B. C., Design of Long Span Concrete Bridges with Special Reference to Prestressing, Precasting, Erection, Structural Behaviour and Economics, First International Symposium on Concrete Bridge Design, ACI Publication SP-23, American Concrete Institute, Detroit, 1969, pp. 693-704.

APPENDICES

APPENDICES

APPENDIX 1

PROPERTIES OF PRESTRESSING STEELS

Table A.1.1: Prestressing Wires (IS: 1785-Part-1-1983)

Nominal Diameter (mm)	Area (A_p) (mm^2)	Weight (kg/m)	Ultimate Tensile Strength f_p (N/mm^2)	$0.8 f_p A_p$ (kN)	$f_p A_p$ (kN)
2.50	4.9	0.037	2010	7.87	9.84
3.00	7.0	0.053	1865	10.44	13.05
4.00	12.5	0.095	1715	17.15	21.43
5.00	20.0	0.150	1570	25.12	31.40
7.00	38.5	0.292	1470	45.27	56.59
8.00	50.0	0.390	1375	55.00	68.75

Table A.1.2: Prestressing Bars (IS: 2090-1983)

Nominal Diameter (mm)	Area (A_p) (mm^2)	Weight (kg/m)	Ultimate Tensile Strength f_p (N/mm^2)	$0.8 f_p A_p$ (kN)	$f_p A_p$ (kN)
10	79	0.62	980	61.93	77.42
12	113	0.89	980	88.59	110.74
16	201	1.58	980	157.58	196.98
20	314	2.47	980	246.17	307.72
22	380	2.98	980	297.92	372.40
25	491	3.85	980	384.94	481.18
28	616	4.83	980	482.94	603.68
32	804	6.31	980	630.33	787.92

Table A.1.3: Prestressing Strands (IS: 6006-1983) (Seven wire Strand)

Nominal Diameter (mm)	Area (A_p) (mm^2)	Weight (kg/m)	Ultimate Tensile Strength f_p (N/mm^2)	$0.8 f_p A_p$ (kN)	$f_p A_p$ (kN)
6.3	23.2	0.182	1723	31.97	39.97
7.9	37.4	0.294	1723	51.55	64.40
9.5	51.6	0.45	1723	71.12	88.90
11.1	69.7	0.548	1723	96.07	120.09
12.7	92.9	0.730	1723	128.05	160.06
15.2	139.4	1.094	1723	192.14	240.18
9.5	54.8	0.432	1862	81.63	102.03
11.1	74.2	0.582	1862	110.52	138.16
12.7	98.7	0.775	1862	147.02	183.77
15.2	140.0	1.102	1862	208.54	260.68

APPENDIX 2

CONSTANTS FOR BEAM SECTIONS

Table A.2.1: Constants for T-Sections

Section	$\dfrac{b_w}{h}$	$\dfrac{D_f}{h}$	A	Y_b	Y_t	I	r^2	k_t	k_b
1	0.1	0.1	0.19 bh	0.714 h	0.286 h	0.0179 bh^3	0.0945 h^2	0.132 h	0.333 h
2	0.1	0.2	0.28	0.756	0.244	0.0192	0.0688	0.0910	0.282
3	0.1	0.3	0.37	0.755	0.245	0.0193	0.0520	0.0689	0.212
4	0.1	0.4	0.46	0.735	0.265	0.0202	0.0439	0.0597	0.165
5	0.2	0.1	0.28	0.629	0.371	0.0283	0.1010	0.1610	0.272
6	0.2	0.2	0.36	0.678	0.322	0.0315	0.0875	0.1290	0.272
7	0.2	0.3	0.44	0.691	0.309	0.0319	0.0725	0.1050	0.234
8	0.2	0.4	0.52	0.684	0.316	0.0316	0.0616	0.0900	0.195
9	0.3	0.1	0.37	0.585	0.415	0.0365	0.0985	0.1690	0.237
10	0.3	0.2	0.44	0.626	0.374	0.0408	0.0928	0.1480	0.248
11	0.3	0.3	0.51	0.645	0.355	0.0417	0.0819	0.1270	0.231
12	0.3	0.4	0.58	0.645	0.355	0.0417	0.0720	0.1120	0.203
13	0.4	0.1	0.46	0.559	0.441	0.0440	0.0954	0.1710	0.216
14	0.4	0.2	0.52	0.592	0.408	0.0486	0.0935	0.1580	0.229
15	0.4	0.3	0.58	0.609	0.391	0.0499	0.0860	0.1410	0.220
16	0.4	0.4	0.64	0.612	0.388	0.0502	0.0785	0.1280	0.205
17	1.0	1.0	1.00	0.500	0.500	0.0833	0.0833	0.1670	0.167

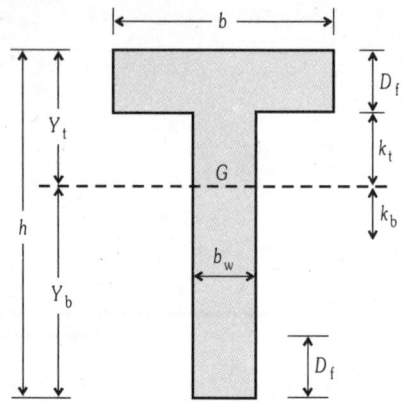

Table A.2.2: Constants for Unsymmetrical I-Sections (Ratio of bottom to top flange width = 0.3)

Section	$\dfrac{b_w}{b}$	$\dfrac{D_f}{h}$	A	Y_b	Y_t	I	r^2	k_t	k_b
1	0.1	0.1	0.21 bh	0.650 h	0.350 h	0.0260 bh^3	0.1236 h^2	0.190 h	0.354 h
2	0.1	0.2	0.32	0.675	0.325	0.0345	0.1080	0.160	0.332
3	0.1	0.3	0.43	0.672	0.328	0.0387	0.0900	0.184	0.274
4	0.2	0.1	0.29	0.610	0.390	0.0316	0.1090	0.179	0.280
5	0.2	0.2	0.38	0.647	0.353	0.0378	0.0994	0.153	0.282
6	0.2	0.3	0.47	0.655	0.345	0.0402	0.0856	0.131	0.248

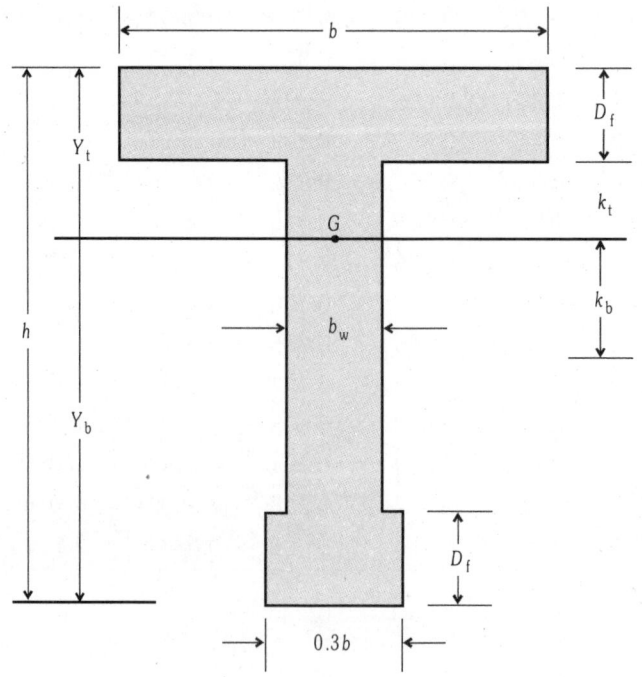

Appendix 2

Table A.2.3: Constands for Unsymmetrical I-Sections (Ratio of bottom to top flange width = 0.5)

Section	$\dfrac{b_w}{b}$	$\dfrac{D_f}{h}$	A	Y_b	Y_t	I	r^2	k_t	k_b
1	0.1	0.1	0.23 bh	0.597 h	0.403 h	0.0326 bh^3	0.1420 h^2	0.238 h	0.352 h
2	0.1	0.2	0.36	0.611	0.389	0.0464	0.1288	0.210	0.331
3	0.1	0.3	0.49	0.606	0.394	0.0535	0.1090	0.180	0.274
4	0.2	0.1	0.31	0.572	0.428	0.0373	0.1204	0.210	0.282
5	0.2	0.2	0.42	0.595	0.405	0.0488	0.1160	0.195	0.286
6	0.2	0.3	0.53	0.599	0.401	0.0540	0.1020	0.170	0.254
7	0.3	0.1	0.39	0.557	0.430	0.0443	0.1103	0.198	0.250
8	0.3	0.2	0.48	0.582	0.418	0.0510	0.1065	0.183	0.255
9	0.3	0.3	0.57	0.592	0.408	0.0553	0.0970	0.164	0.238

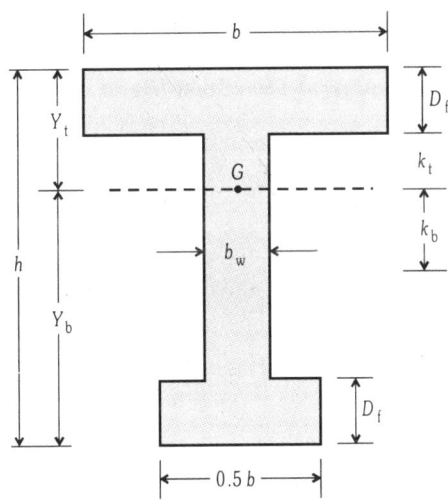

Table A.2.4: Constants for Unsymmetrical I-Sections (Ratio of bottom to top flange width = 0.7)

Section	$\dfrac{b_w}{b}$	$\dfrac{D_f}{h}$	A	Y_b	Y_t	I	r^2	k_t	k_b
1	0.1	0.1	0.25 bh	0.554 h	0.446 h	0.0381 bh^3	0.1525 h^2	0.276 h	0.342 h
2	0.1	0.2	0.40	0.560	0.440	0.0560	0.1391	0.248	0.316
3	0.1	0.3	0.55	0.557	0.443	0.0651	0.1182	0.212	0.267
4	0.2	0.1	0.33	0.540	0.460	0.0425	0.1290	0.239	0.280
5	0.2	0.2	0.46	0.552	0.448	0.0578	0.1250	0.228	0.281
6	0.2	0.3	0.59	0.553	0.447	0.0657	0.1113	0.202	0.249
7	0.3	0.1	0.41	0.534	0.466	0.0467	0.1140	0.214	0.244
8	0.3	0.2	0.52	0.546	0.454	0.0598	0.1150	0.210	0.254
9	0.3	0.3	0.63	0.550	0.450	0.0663	0.1051	0.191	0.234

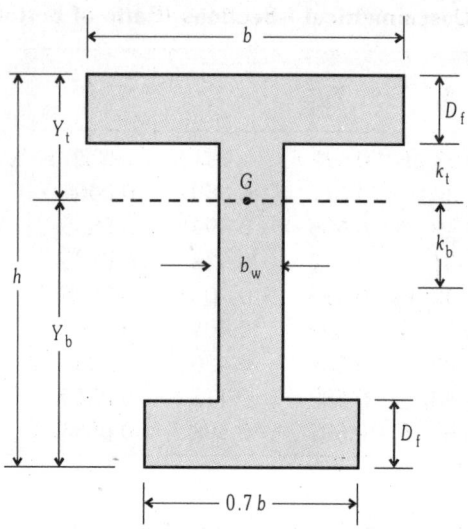

Table A.2.5: Constants for Unsymmetrical I-Sections (Ratio of top to bottom flange width = 0.3)

Section	$\frac{b_w}{b}$	$\frac{D_f}{h}$	A	Y_b	Y_t	I	r^2	k_t	k_b
1	0.1	0.1	0.21 bh	0.350 h	0.650 h	0.0260 bh^3	0.1236 h^2	0.354 h	0.190
2	0.1	0.2	0.32	0.325	0.675	0.0345	0.1080	0.332	0.160
3	0.1	0.3	0.43	0.328	0.672	0.0387	0.0900	0.274	0.134
4	0.2	0.1	0.29	0.390	0.610	0.0316	0.1090	0.280	0.179
5	0.2	0.2	0.38	0.353	0.647	0.0378	0.0994	0.282	0.153
6	0.2	0.3	0.47	0.345	0.655	0.0402	0.0856	0.248	0.131

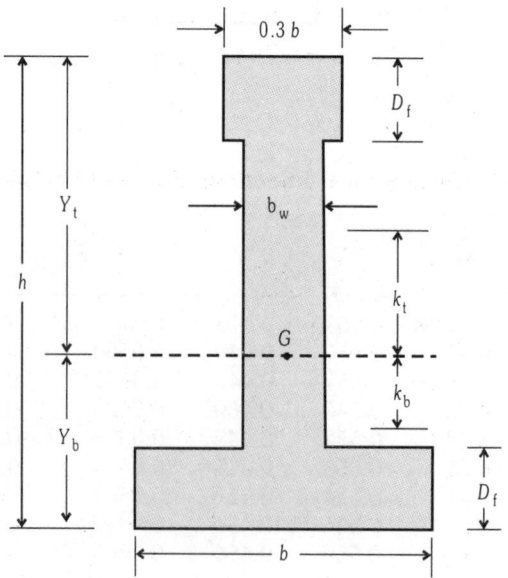

Table A.2.6: Constants for Symmetrical I and Box Sections

Section	$\dfrac{b_w}{b}$	$\dfrac{D_f}{h}$	A	Y_b	Y_t	I	r^2	k_t	k_b
1	0.1	0.1	0.28 bh	0.500 h	0.500 h	0.0449 bh^3	0.160 h^2	0.320 h	0.320 h
2	0.1	0.2	0.46	0.500	0.500	0.0671	0.146	0.292	0.292
3	0.1	0.3	0.64	0.500	0.500	0.0785	0.123	0.246	0.246
4	0.2	0.1	0.36	0.500	0.500	0.0492	0.137	0.274	0.274
5	0.2	0.2	0.52	0.500	0.500	0.0689	0.132	0.264	0.264
6	0.2	0.3	0.68	0.500	0.500	0.0791	0.117	0.234	0.234
7	0.3	0.1	0.44	0.500	0.500	0.0535	0.121	0.243	0.243
8	0.3	0.2	0.58	0.500	0.500	0.0707	0.122	0.244	0.244
9	0.3	0.3	0.72	0.500	0.500	0.0796	0.111	0.222	0.222
10	0.4	0.2	0.64	0.500	0.500	0.0577	0.111	0.222	0.222
11	0.4	0.2	0.64	0.500	0.500	0.0725	0.113	0.226	0.226
12	0.4	0.3	0.76	0.500	0.500	0.0801	0.105	0.211	0.211

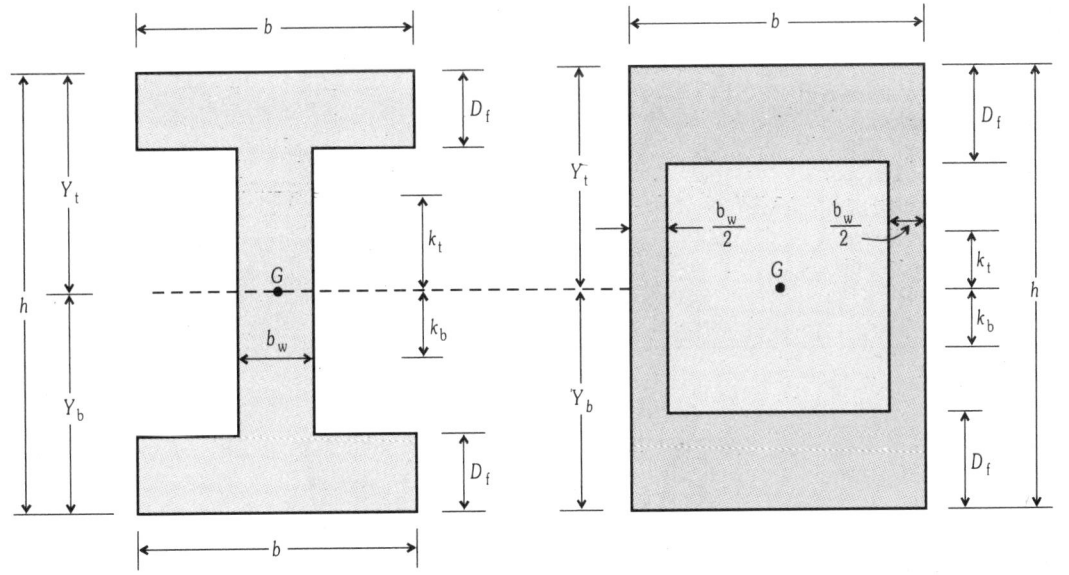

APPENDIX 3

POST-TENSIONING SYSTEMS

Table A.3.1: Freyssi Anchorage Cones (Freyssinet Prestressed Concrete Co. Ltd., Bombay)

Tendon Reference	$12\phi5$	$12\phi5$ $E \times T$	$12\phi7$	$12\phi7$ $E \times T$	$12\phi8$	6T13
Number of Wires	12	12	12	12	12	6(12.7 mm) 7 ply strands
Dimensions of male and female cones (mm)						
(a) Outer diameter. ϕ_c	96	120	120	150	150	150
(b) Height of female cone, h	100	100	120	125	125	125
(c) Nominal outside diameter of connection sleeve, ϕ_e	32	32	38	38	51	51
Diameter of grout hole, ϕ_g	8	—	11	—	11	10
Height of male cone, b	74	—	95	—	108	110

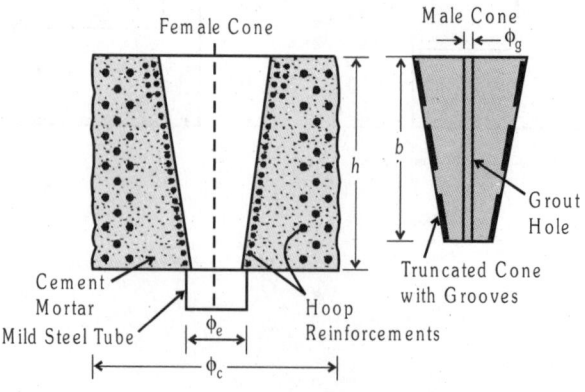

346

Appendix 3

Table A.3.2: Standard Freyssinet Prestressing Cables

Tendon System	Units	12 φ 5	12 φ 7	12 φ 8	24 φ 7	24 φ 8	6T13	12T13
Nominal diameter of wire/strand	mm	5	7	8	7	8	13	13
Nominal U.T.S. of cable	N/mm^2	1600	1500	1400	1500	1400	1800	1800
Number wires/strand per cable	no	12	12	12	24	24	6	12
Nominal steel area of cable	mm^2	235	462	603	922	1206	557	1115
Nominal ultimate breaking force of cable	kN	376	691	844	1382	1688	1002	2004
Maximum allowable prestressing Force	kN	300	553	675	1105	1350	801	1603
Maximum allowable initial stress	N/mm^2	1280	1200	1120	1200	1120	1440	1440
Approximate weight per unit length of cable	kg/m	1.8	3.6	4.7	7.2	9.4	4.5	9.0

Table A.3.3: P.S.C. Freyssinet System (Anchorage Details of K-Range System)

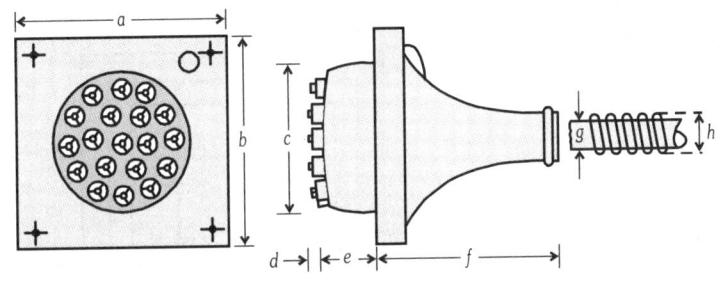

Tendon Reference		4 K 15	7 K 15	12 K 15	19 K 15	27 K 15	37 K 15
No. of 15 mm strands	n	4	7	12	19	27	37
Anchorage block and sheathing dimensions (mm)	a	130	170	200	250	300	420
	b	170	225	260	340	400	495
	c	115	140	160	220	260	320
	d	10	10	10	10	10	10
	e	50	60	60	65	80	95
	f	100	155	175	412	450	480
	g	55	65	75	95	110	130
	h	61	71	81	101	118	140

Characteristic strength of each strand = 265 kN

Table A.3.4: C.C.L System (Wire Spiral Anchorage Details)

Tendon	Ultimate Prestressing Force (kN)	Type	Anchorage Flange Size	Length	Dia	A	B	C (mm)	D	E	F	G	H	J	K
12 φ 7	715	S 4	124 × 121	165	42	70	65	65	65	130	100	150	100	150	175
8 φ 7	475	S 3	140 × 76	178	39	65	60	65	65	130	90	100	120	165	75
4 φ 7 / 2 φ 7	237 / 118	S 2	70 × 70	83	25	50	45	60	60	115	85	95	85	95	75

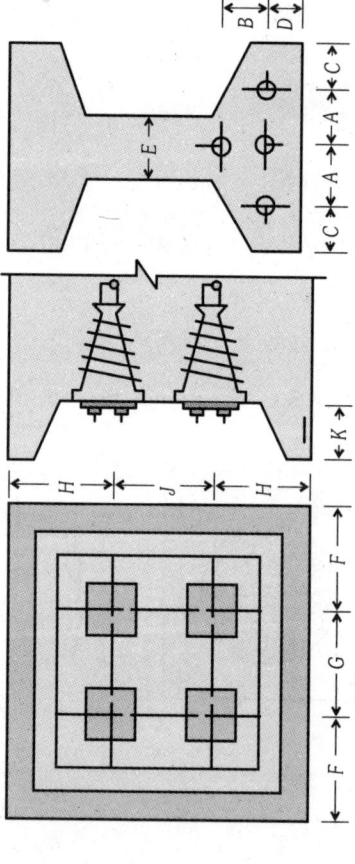

Appendix 3

Table A.3.5: B.B.R.V. System (A-Type Stressing Anchorage Details)

Tendon Reference		A24	A34	A42	A55	A61	A73	A85	A97	A109
No. of 7 mm wires	n	24	34	42	55	61	73	85	97	109
Characteristic strength	kN	1542	2185	2699	3535	3920	4692	5463	6234	7005
Jacking force 70%	kN	1080	1530	1890	2474	2744	3284	3824	4364	4904
Jacking force 75%	kN	1157	1639	2024	2651	2940	3519	4097	4676	5254
Jacking force 80%	kN	1234	1748	2159	2828	3136	3753	4370	4987	5604
Bearing plate:										
Side length	A	205	240	270	305	325	355	380	410	430
Thickness	B	25	30	35	40	45	55	60	65	75
Trumpet O/D	C	58	68	73	83	88	93	103	108	113
Anchor Head:										
Thread Diameter	D	110	124	139	158	158	172	185	202	202
Standard Length	E	44	50	56	64	69	74	81	81	
Chocks	F	117	131	147	168	168	182	196	214	214
Sheathing I/D	G	50	60	65	75	80	85	95	100	105
Sheathing O/D	H	58	68	73	83	88	93	103	108	113

All dimensions in mm

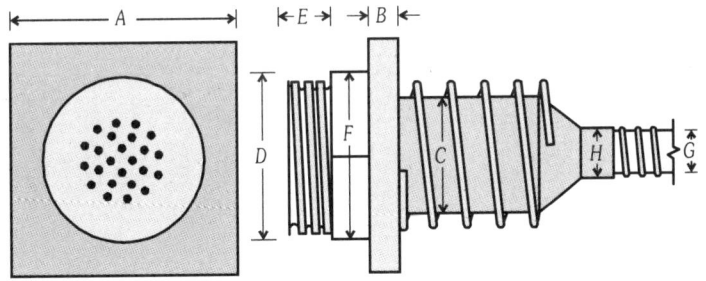

Table A.3.6: Dywidag Single Bar Anchorage Tendon Characteristics

Nominal diameter	mm	15.87	15.87	25.4	25.4	31.75	31.75	34.92
Actual diameter	mm	15.0	16.0	26.5	26.5	32.0	32.0	36.0
Ultimate strength	N/mm²	1082	1585	1034	1103	1034	1103	1034
Area	mm²	180.6	200.0	548.4	548.4	806.5	806	10019.4
Ultimate load (P_u)	kN	193.5	313.5	567.1	604.9	834.0	889.6	1054.1
$0.8 P_u$	kN	154.8	250.8	453.6	483.9	667.2	711.6	843.2
Anchorage details								
Bar diameter	mm	15.87	15.87 S	25.4	25.4	31.75		34.92
Solid anchor plate	mm	76 × 76 × 19	101 × 101 × 25	127 × 140 × 32	127 × 140 × 32	152 × 178 × 38	178 × 190 × 44	178 × 190 × 44
	mm	51 × 127 × 25	76 × 127 × 25	101 × 65 × 32		127 × 203 × 38	127 × 241 × 44	127 × 241 × 44
Nut extension	mm	25.4	41.3	47.6		63.5		69.8
Bar protrusion	mm	50.8	76.2	63.5		76.2		88.9
Pocket former								
Height	mm	111.1	111.1	177.8		203.2		220.3
Maximum O.D.	mm	79.3	79.3	130.1		165.1		165.1
Coupling details								
Length	mm	88.9	134.7	134.7		171.4		219.0
Diameter O.D.	mm	28.5	31.7	50.8		60.3		67.9
Sheathing								
Bar sheathing: O.D.	mm	25.4	25.4	38.1		44.4		50.8
I.D	mm	19.0	19.0	31.7		38.1		44.4
Coupling sheathing								
Outer cover	mm	44.4	44.4	69.8		82.5		95.2
Inner diameter	mm	34.9	34.9	60.3		73.0		85.7

Appendix 3

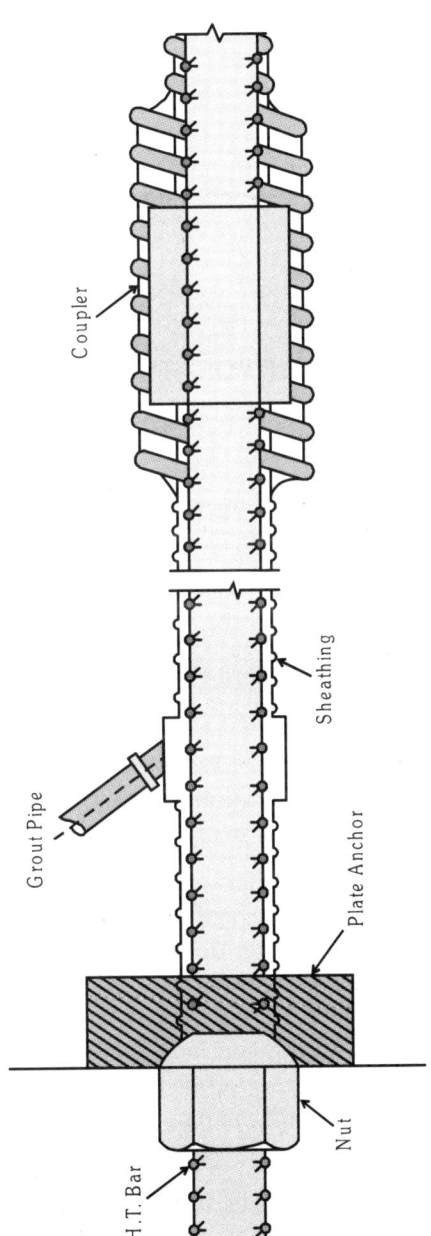

Fig. 6: Dywidag Single Bar Anchorage

APPENDIX 4

GROUTING OF POST-TENSIONED DUCTS

The main objective of grouting the ducts of post-tensioned concrete members are

(i) to prevent corrosion of tendons
(ii) to ensure efficient transfer of stress between the tendons and the concrete member
(iii) to improve the serviceability and strength characteristics of the concrete member.

The salient recommendations of the Indian Roads Congress (IRC: 18-2000) code are summarized below:

1. **Composition of Grout:** The grout is prepared by using Ordinary Portland cement and water. Fine sand passing 150 micron I.S sieve may be used if the internal diameter of the ducts exceed 150 mm. The weight of sand in the grout shall be not more than 10 percent of the weight of cement unless proper workability can be ensured by addition of suitable plasticizers. Acceptable admixtures conforming to IS: 9103 may be used if tests have shown that these improve the properties of the grout.

2. **Formation of Ducts:** The cable ducts are formed by sheathing ducts made of cold rolled cold annealed (CRCA) mild steel having thickness of not less than 0.3 mm, 0.4 mm and 0.5 mm for ducts having internal diameter up to 50 mm, 75 mm and 90 mm respectively. For bigger diameter of ducts, thickness of sheathing shall be based on recommendations of the supplier of the prestressing system. Alternatively corrugated High density polyethylene (HDPE) sheathing ducts conforming to the specifications mentioned in Clause.3.6.2 of IRC: 18-2000 may be used. Vents are provided at crests in the duct profile and at intervals not greater than 15 m. Anchorage vents are also provided and it should be possible to close all the vents.

3. **Batching and Mixing of Grout:** The materials of grout should be batched by mass. For a neat cement grout the optimum water/cement ratio should not normally exceed 0.45. Before grouting, the properties of the grout mix should be tested in a laboratory. The compressive strength of 100 mm cubes of the grout shall not be less than 17 N/mm^2 at 7 days. The grout should be mixed using a mixer capable of producing a homogeneous

colloidal grout and after mixing, keeping the grout in slow continuous agitation until it is ready to be pumped into the tendon ducts.

The mixing of the grout is done by adding water to the mixer first followed by cement. Mixing time depends on the type of mixer but will normally be between 2 and 3 minutes and not more than 4 minutes. Where admixtures are used, the manufacturer's recommendations should be followed. All the piping, pumping and mixing equipment should be thoroughly washed with clean water after each series of operations.

4. **Grouting Procedure:** The grouting operation should be carried out as early as possible, but not later than 2 weeks of stressing the tendons. The injection procedure should ensure that ducts are completely filled. Ducts should be grouted at a continuous and steady rate of 6 m/min to 12 m/min for horizontal ducts and 2 to 3 m/min for vertical ducts at a pressure not exceeding 1 N/mm^2.

Grouting should continue until the fluidity or density of the grout flowing from the free ends of the vent openings is the same as that of the injected grout. The vents should be closely successively as the filling of the ducts continue and after closing the last vent, the pressure should be held at 0.5 N/mm^2 for 5 minutes.

Vertical and inclined ducts should be grouted from the lowest point, the maximum length grouted in one operation being 50 m. Vents and all other openings should be sealed after grouting to prevent the ingress of moisture, deicing minerals and other corrosive agents. Grout not used within 30 minutes of mixing should be rejected. Also care should be taken to avoid leaks from one duct to another at joints of precast members.

The effectiveness of the grouting can be checked by using non destructive testing techniques like Gamma radiography.

AUTHOR INDEX

A

Abeles, P. W. 10, 16
Adams, H. C. 185
Alexander the Great 1

B

Baikov, V. 69
Bate, S. C. C. 53, 69
Baur-Leonhardt, 51
B. B. R. V. 51
Bennett, E. W. 53, 69
Bessemer, 1
Best, B. C. 289
Birkenmeaier, M. 289
Bresson, J. 317
Burns, N. 17

C

Campbell-Allen, D. 211
Chettoe, C. S. 185
Cheung, Y. K. 210
Chhauda, J. N. 96, 266
Chukh, E. 210
Collins, M. P. 69
Courbon, J. 139
Cranston, W. B. 69, 289
Cussens, A. R. 139, 210

D

Dean, E. E. 10, 16

Dischinger, F. 8
Dudnik 210
Dywidag 51

E

Edwards, A. D. 210
Erntroy, H. C. 44, 52
Evans, R. H. 69
Eugene Freyssinet 5, 16

F

Faulkes, K. A. 69
Finisterwalder, U. 8, 16, 163, 210, 266, 289
Franklin, R. E. 44, 52
Freyssinet, E. 51

G

Galambos, C. F. 42
George Washington 2
Gerwick, B. C. (Jr) 52, 336
Gifford-Udall 51
Gokhale, P. S. 317
Guyon. J. 102

H

Hambly, E. C. 210
Hansen, N. W. 69
Hendry, A. W. 139
Hognestad, E. 69
Hugen Schmidt, F. 317

I

Irwin, C. A. K. 317

J

Jacques Fauchart, 210
Jaegar, L. G. 139
Jessica Binns 210, 336
Johnson Victor 210, 336

K

Kapla. M. S. 96, 266
Krishna Raju, N. 17, 52, 139, 185, 254, 317
Krishna Reddy, Y. 52

L

Laksmanan, N. 139
Le Corbusier 2
Lee, D. J. 289
Lee McCall 51
Leonhardt, Fritz 8, 10, 210, 254, 266, 336
Lim, P. K. 210
Lin. T. Y. 17, 163, 336
Little, G. 139

M

Magnel, G. 6
Marshall, W. T. 185, 254
Massonet, C. 139
Mattock, A. H. 69
McDonald, M. D. 317
Mitchell, D. 69
Morice, P. B. 139
Muller, J. 289, 336
Murshev, V. 69

N

Nadgir, N. S. 317
Naidu, M. P. 210, 289
Nawy Edward, G. 336
Nelson, H. M. 185
Norman Foster, 16

O

Owen, D. R. J. 210

P

Pennells, E. 210
Phillips, D. V. 210
Plowden, D. 336
Podolony, W. 289
Pranesh, R. N. 69
Prasada Rao, A. S. 96, 266

R

Raghavendra, N. 163
Raiker, R. N. 210
Raina, V. K. 16, 39, 254
Rajagopalan, K. S. 42
Reynolds, C. E. 163
Richmond, B. 210
Roebling 1
Rohra, M. R. 317
Rowe, R. E. 42, 96, 289
Ruesch, H. 87, 96

S

Sarkar, S. 96, 139, 262, 266
Sankaralingam, C. 210, 289
Schwier, F. 254
Scordelis, A. C. 210
Seni, A. 42
Shacklock, B. W. 44, 52
Shirley Smith, H. 336
Sieman 1
Sigalov, E. 69
Sisodiya, R. G. 210–11
Smith, B. S. 254
Smulski, E. 185
Sommerard, T. 317
Steedman, J. 163
Steinman 2
Subba Rao, T. N. 210, 266, 336
Suryaprakasha Rao, D. 96

T

Tamhankar, M. G. 96
Tang, M. C. 254
Taylor, F. W. 185
Teychenne, D. C. 44, 52
Thomas, P. K. 42
Thompson, S. E. 185

Timoshenko, S. 96
Trikha, D. N. 210
Troitsky, M. S. 254

U

Ulrich Finisterwalder 8, 16, 289, 336

V

Victor, J. D. 76, 96, 139
Virola, J. 336
Visvesvaraya, H. C. 163

W

Warner, R. F. 69
Wedgewood, R. J. L. 211
Witecki, A. A. 210
Woinwsky-Krieger, S. 96

Z

Zellner, W. 16
Zienkiwicz, O. C. 210

Subject Index

A

AASHTO loading 25, 35
ACI Committee-211 52
ACI method 44
Advantages of prestressed concrete 8
Aerodynamic stability 237
Aggregate-cement ratio 45
Akkar Bridge 236
Amakusa Bridge 319, 334
Analysis of slab decks 72
Anchorage diameter 68
Anchorage force 68
Anchorage plates 162
Anchorages 51
Anchorage zone reinforcement 209
Approximate methods 223
Arakawa Railway Bridge 332
Assembly of prestressing steel 274

B

Balanced cantilever 3
Barak Bridge 11, 319, 325
Barin's snooper system 294
Bassien Creek Bridge 263, 319
Batching 269
Bearing plate 349
Bearing stress 50
Bending concordant profile 176
Bending moment 38
Bhaghirathi Bridge 188, 325
Bhima Bridge 329
Boussens Bridge 188, 276

Bow string 3
Box girder 189, 197
Box section 61
Bridge loading standards 18
British method 44
British standard loadings 23, 35
Broad gauge 40
Brotonne Bridge 319
Bursting tensile force 62
Buxar Bridge 320

C

Cable duct 209
Cable profile 177
Cable stayed bridge 212, 243
Cable stays 8, 217
Cable system 214
Cable zone 175
Caisson foundations 16
Cantilever construction 275, 283
Cantilever construction method 240, 286
Cantilever portion 100
Cap cable 143, 195
Case studies of repairs and rehabilitation 312
Castejon Bridge 188
Cast-in-situ concrete 275
Cellular box girder 196
Cellular box girder bridges 187
Cement and Concrete Association 11
Chaco Corrientes Bridge 330
Chambal Bridge 314
Choice of spans 278
Check for stresses 132

Coefficient of dynamic augument 41
Coatazacoalcos Bridge 330
Cold drawn stress relieved wires 46
Coleron Bridge 6
Column 166, 171
Compaction 270
Comparative analysis 113
Composite bridge construction 10
Composition of grout 352
Compressive strength 43
Concordant cables 156
Concrete technology 13
Construction joints 271
Construction methods 240
Continuous girder 3
Continuous prestressed structures 141
Continuous span bridge decks 141
Courbon's method 101, 114
Cracks in prestressed members 294
Cross girders 98, 137, 162
Creep 272
Crushed granite 268
Crushing of concrete 58
Curved cables 10

D

Dead loads 170, 182
Deck slab 99
Deck system 214
Deflections 141
Degree of redundancy 225, 235
Demec gauge 293
Design aids and tables 87
Design of end block 136, 161, 208
Design of high strength concrete 44
Design of stay cables 249
Design philosophy 54
Design of web girder 200
Detailed inspection 292
Determination of cable forces 230
Differential shrinkage 272
Diaphragms 98
Distribution coefficient 103
Dispersion of loads 75
Displacements 272
Durability 336
Dynamic behavior 237
Dynamic effects 40

E

Eccentricity 57, 64
Eccentricity of cables 247
Economical aspects 256
Economic evaluation 264
Economic studies 241
Elastic analysis 87
Elastic design coefficients 54
Elastic shortening 272
Elastomers 302
Electrical resistance meter 293
Electronic strain gauge 293
End blocks 50, 62
Epoxy bonding 284
Epoxy resins 284
Epoxy sealing 301
Expansion joints 271
Externally bonded steel plates 310
Extradosed bridges 16

F

Failure by crushing of concrete 135
Failure by yielding of steel 135
Finite element method 191
Finite strip method 191
Flanged section 61
Flexural parameter 103
Foot paths 98
Folded plate theory 191
Forces in cables 225
Forces in end blocks 62
Formation of ducts 352
Fran 225
French highway loadings 25, 35
Fully anchored system 237

G

Gateway Bridge 319
Gambhirkad Bridge 325
Ganga Bridge 8
General inspection 292
German highway loading 35
Girders damaged by collision 306
Giznen Bridge 314
Glenuium sky 271
Grade of concrete 43, 268
Grouting 352
Grouting of ducts 274

Subject Index

Grouting procedure 353
Guyon-Massonet method 101, 103, 116

H

Hamana Bridge 8, 319
Hand rails 98
H.A. type of loading 23
H.B. loading 18, 23
Harumi Railway Bridge 12
Hendry-Jaegar method 101, 110, 120
Hinged footing 181
High strength concrete 10, 43, 268
High tensile indented wires 46
High tensile steel 46
High tensile steel bars 47
High way loadings of Austria 28
Highway loadings of Belgium 29
Highway loadings of Germany 26
Highway loadings of Italy 29
Highway loadings of Japan 27, 35
Highway loadings of Netherlands 29
Highway loadings of New Zealand 27
Highway loadings of Norway 31
Highway loadings of Sweden 28
Humen Bridge 319
Hydraulic jacks 293

I

Impact allowance 18, 35, 37
Imposed dead load 58
Impact factors 34
Incremental launching 285
Indian Roads Congress 18
Indian Railway bridge loading standards 39
Indian standard loadings 34
Innovative construction techniques 284
Inspection of bridges 292
Inspection instrumentation 293
IRC Class A loading 20, 34
IRC Class B loading 20

J

Joao Bridge 319

K

Katepura Bridge 315
Kawaotogawa Bridge 333
Kerbs 98

King Fahd causeway 330
Krishna Bridge 265

L

Lane width 37
Lightweight aggregate 45
Limit state approach 54
Limit state design 54
Limiting zone 175
Live load combinations 20
Load distribution methods 101
Load factor 54
Longitudinal cable profiles 221
Long line system 11
Long span bridge construction 274
Long span bridge decks 11, 187
Loss factor 89
Loss of stress 48
Loss ratio 59
Lubha Bridge 11, 319, 320

M

Magnetic detector 293
Mahatma Gandhi Sethu 319
Maintenance methodology 291
Maintenance and rehabilitation 290
Major inspection 292
Mastics 302
Materials for prestressed concrete 43
Maximum shear stress in concrete 56
Maximum water-cement ratio 44
M-beams 11
Millau Bridge 213
Mechanical extensometer 293
Messina straights 213
Minimum cement content 44
Minimum prestressing force 57, 246
Minimum reinforcements in slabs 89
Minimum section modulus 56, 240
Minor damage 302
Moderate damage 302
Modified prestressing force 65
Modulus of elasticity of concrete 45
Modulus of elasticity of steel 48
Morice and Little version 101
Multicell box segments 11

N

National Highways 19
Nagayooshoshi Bridge 319, 322

Negative moments 88
Normandie Bridge 213

O

Obra Singrauli bridge 312
Open spandrel arch bridge 3
Optical microscope 293
Optimum inclination of cables 226
Otogawa railway bridge 12
Over reinforced section 58

P

Pachometer 293
Parabolic cable profile 156
Parapets 99
Partially anchored system 237
Partially prestressed concrete 10
Permanent dead load 58
Permissible shear stress in concrete 56
Permissible stress in concrete 49
Permissible stress in steel 50
Permissible tendon zone 132
Portal frames 167
Positive moments 88
Post tensioned cables 69
Prestressing continuous bridge decks 142
Prestressing cables 282
Prestressing force 92, 156, 203
Pretensioned beams 11
Priniipcal tensile stress 59, 194
Progressive placement method 286
Push out method 240, 285
Pylons 214

Q

Quinton bridges 312

R

Racking forces 40
Raftundet bridge 319
Reaction factor 102, 114, 151
Rebound hammer 293
Rectangular section 60
Rehabilitation of damaged concrete 303
Relaxation of stress in steel 48
Repairs and Rehabilitation of Structures 303

Repairs to damaged pretensioned girders 307
Restoration 306
Restoration of PSC girder 313
Rheodynamic concrete 271
Rheofit 271
Rigid frame bridges 165
Road research laboratory 52
Rigorous methods 225
Routine inspection 292

S

Schmidt hammer 293
Schottwein Bridge 319
Second Hooghly Bridge 8, 263
Second Vivekananda Bridge 176
Sections cracked in flexure 59
Sections uncracked in flexure 59
Segmental box girder 187, 284
Self anchored system 237
Severe damage 303
Shear reinforcement 66
Shear strength 58–59
Shear stress 61
Sheathing ducts 51
Skye Bridge 319
Slab bridge decks 72
Slab type rigid frames 166
Snooper crawler 293
Solid cantilever slab 74
Solid slabs spanning in one direction 73
Sone Bridge 320
Special inspection 292
Staging method 240, 284
Standard loading train 18
State highways 19
Statical indeterminacy 235
Stiffening girders 232, 244, 250
Stolmasundet Bridge 319
Strands 47
Strengthening of beams 310
Stress 225
Stromsund Bridge 8
Structural analysis 223
Structural anchorages 237
Structural components 98
Strudl 225
Sunshine Skyway Bridge 319, 329
Superimposed dead load 58
Supplementary reinforcement 48, 136
Swanley Bridges 313

Subject Index

T

Tee beam and slab 3, 98
Tendons 143
Tendon profile 177
Tensile stresses 49
Thin plate 87
Thermal expansion 272
Thermoplastics 302
Torsion 195
Torsional analysis of box girders 191
Torsion coefficient 117
Torsional parameter 103
Torsional resistance 60
Torsional shear stress 61
Towers 214
Transom 175
Transom tendon profile 177
Transverse bending 193
Types of aggregate 268
Types of high tensile steel 46

U

Ultimate flexural strength 135, 159, 180
Ultimate shear strength 135, 160, 181
Ultimate load 58
Ultimate moment of resistance 59
Ultrasonic pulse velocity 293
Under reinforced section 58
Urato Bridge 319

V

Varrod Bridge 319
Verrazano Narrows Bridge 2, 3
Vidyasagar Sethu 8, 213
Vivekananda Tollway Bridge 16, 213

W

Walnut Bridge 6
Water-cement ratio 44, 269
Wearing course 99
Weight of cables 233
Wind and earthquake forces 40
Workability 43, 44, 268

Y

Yamatogawa Bridge 319, 334
Yamuna Bridge 320
Y-beams 11
Yielding of steel 58
Yonegami Bashi Bridge 331

Z

Zero energy system 271
Zuari Bridge 319–20

Reader's Notes

Reader's Notes

Reader's Notes